Mathematics and Visualization

Series Editors

Gerald Farin
Hans-Christian Hege
David Hoffman
Christopher R. Johnson
Konrad Polthier

Springer

Berlin
Heidelberg
New York
Hong Kong
London
Milan
Paris
Tokyo

Hans-Christian Hege
Konrad Polthier (Editors)

Visualization and Mathematics III

With 240 Figures, 47 in Color

Springer

Hans-Christian Hege

Konrad-Zuse-Zentrum
für Informationstechnik Berlin (ZIB)
Takustraße 7
14195 Berlin, Germany
e-mail: hege@zib.de

Konrad Polthier

Institut für Mathematik, MA 8-3
Technische Universität Berlin
Straße des 17. Juni 136
10623 Berlin, Germany
e-mail: polthier@math.tu-berlin.de

Cover image: Stereographic projection of a compact minimal surface in S^3 by H.B. Lawson. By K. Polthier (TU-Berlin).

Cataloging-in-Publication Data applied for

A catalog record for this book is available from the Library of Congress.

Bibliographic information published by Die Deutsche Bibliothek
Die Deutsche Bibliothek lists this publication in the Deutsche Nationalbibliografie;
detailed bibliographic data is available in the Internet at http://dnb.ddb.de

Mathematics Subject Classification (2000): 68U05, 68U10, 68U20; 53-04, 65D18, 65D17, 65S05, 97-04, 97U70

ISBN 3-540-01295-8 Springer-Verlag Berlin Heidelberg New York

Springer-Verlag Berlin Heidelberg New York
a member of BertelsmannSpringer Science+Business Media GmbH

http://www.springer.de

© Springer-Verlag Berlin Heidelberg 2003
Printed in Germany

Typeset in TeX by the authors
Cover design: *design & production* GmbH, Heidelberg

Printed on acid-free paper   46/3142db - 5 4 3 2 1 0 –

Preface

Mathematical Visualization aims at an abstract framework for fundamental objects appearing in visualization and at the application of the manifold visualization techniques to problems in geometry, topology and numerical mathematics. The articles in this volume report on new research results in this field, on the development of software and educational material and on mathematical applications.

The book grew out of the third international workshop "Visualization and Mathematics", which was held from May 22-25, 2002 in Berlin (Germany). The workshop was funded by the DFG-Sonderforschungsbereich 288 "Differential Geometry and Quantum Physics" at Technische Universität Berlin and supported by the Zuse Institute Berlin (ZIB) and the DFG research center "Mathematics for Key Technologies" (FZT 86) in Berlin. Five keynote lectures, eight invited presentations and several contributed talks created a stimulating atmosphere with many scientific discussions.

The themes of this book cover important recent developments in the following fields:

– Geometry and Combinatorics of Meshes
– Discrete Vector Fields and Topology
– Geometric Modelling
– Image Based Visualization
– Software Environments and Applications
– Education and Communication

We hope that the research articles of this book will stimulate the readers' own work and will further strenghten the development of the field of Mathematical Visualization.

We appreciate the thorough work of the authors and reviewers on each of the individual articles, and we thank you all. Beside the editors, the reviewers and members of the program committee were:

Helmut Alt	Ulrich Kortenkamp
Tom M. Apostol	Jens-Peer Kuska
James Arvo	Carsten Lange
Chandrajit Bajaj	Gregory Leibon
Thomas Banchoff	Nelson L. Max
Philippe Bekaert	Heinrich Müller
Werner Benger	Gregory M. Nielson
Alexander Bobenko	Ronny Peikert
Alexander Bogomjakov	Ulrich Pinkall
Philip L. Bowers	Helmut Pottmann
Ken Brakke	Jürgen Richter-Gebert
Claude Bruter	Martin Rumpf
Matthieu Desbrun	Dietmar Saupe
Peter Deuflhard	Roberto Scopigno
Thomas Ertl	Hans-Peter Seidel
Gerald E. Farin	James Sethian
George Francis	Marc Stamminger
Hans Hagen	John Sullivan
Andrew J. Hanson	Nobuki Takayama
Joel Hass	Gabriel Taubin
David Hoffman	Daniel Weiskopf
Victoria Interrante	Rüdiger Westermann
Chris Johnson	Ross Whitaker
Michael Joswig	Luiz Velho
Alexander Keller	Jarke van Wijk
Leif Kobbelt	Günter M. Ziegler

Special thanks to Robert Staufenbiel for his help in compiling the manuscripts and creating the LaTeX source of this book.

Berlin, 2002
Hans-Christian Hege
Konrad Polthier

Table of Contents

Part II Discrete Vector Fields and Topology

Part III Geometric Modelling

Part IV Image Based Visualization

Part V Software Environments and Applications

Part VI Education and Communication

Films: A Communicating Tool for Mathematics   393
Michele Emmer

The Potentials of Math Visualization and their Impact on the Curriculum   407
Beau Janzen

Part I

Geometry and Combinatorics of Meshes

Planar Conformal Mappings of Piecewise Flat Surfaces

Philip L. Bowers and Monica K. Hurdal

Department of Mathematics, The Florida State University, Tallahassee, FL 32306, USA. {*bowers,mhurdal*}*@math.fsu.edu*

Introduction[†]

There is a rich literature in the theory of circle packings on geometric surfaces that from the beginning has exposed intimate connections to the approximation of conformal mappings. Indeed, one of the first publications in the subject, Rodin and Sullivan's 1987 paper [10], provides a proof of the convergence of a circle packing scheme proposed by Bill Thurston for approximating the Riemann mapping of an arbitrary proper simply-connected domain in $\mathbb{C}$ to the unit disk. Bowers and Stephenson's work in [4], which explains how to apply the Thurston scheme on nonplanar surfaces, may be viewed as a far reaching generalization of his scheme to the setting of arbitrary equilateral surfaces. Further, in [4] Bowers and Stephenson propose a method for uniformizing more general piecewise flat surfaces that necessitates a truly new ingredient, namely, that of inversive distance packings. This inversive distance scheme was introduced in a very preliminary way in [4] with some comments on the difficulty involved in proving that it produces convergence to a conformal map. Even with these difficulties, the scheme has been encoded in Stephenson's packing software `CirclePack` and, though all the theoretical ingredients for proving convergence are not in place, it seems to work well in practice. This paper may be viewed as a commentary on and expansion of the discussion of [4]. Our purposes are threefold. First, we carefully describe the inversive distance scheme, which is given only cursory explanation in [4]; second, we give a careful analysis of the theoretical difficulties that require resolution before conformal convergence can be proved; third, we give a gallery of examples illustrating the power of the scheme. We should note here that there are special cases (e.g., tangency or overlapping packings) where the convergence is verified, and our discussion will give a proof of convergence in those cases.

Each oriented piecewise flat surface has a natural conformal structure defined on its interior by a complex atlas with conformal charts of two types. First, each interior edge gives rise to an edge chart that isometrically maps the interior of the two Euclidean triangles meeting along that edge to the plane, preserving orientation. Overlap maps between the intersections of two such

[†] This work is supported by NSF grant DMS-0101329, NIH grant MH-57180 and FSU grant FYAP-2002.

charts are Euclidean isometries and therefore conformal. Second, each interior vertex gives rise to a vertex chart that uses a power map defined on a small open neighborhood of the vertex to rescale an angle sum Θ different from 2π to one equal to 2π. The vertex charts are chosen to have pairwise disjoint domains and the local form of the chart map is the power map $z \mapsto z^{2\pi/\Theta}$. The overlap mapping between any edge chart and vertex chart is conformal as the vertex, where the derivative is zero, is not in the overlap. In this way any orientable piecewise flat surface becomes a Riemann surface. Notice that though there are in general cone points at the vertices in a piecewise flat surface, these are singularities of the piecewise Euclidean metric only and not singularities of the conformal structure. Indeed, the total angle sum at each vertex given by the conformal structure is 2π and the Euclidean angle α between two arcs emanating from a vertex is measured as $2\pi\frac{\alpha}{\Theta}$ in the conformal structure. The conformal structure thus measures the "market share" of the angle α with respect to the total Euclidean angle Θ.

Though the inversive distance scheme for conformal mapping may be presented in the full generality of arbitrary piecewise flat surfaces, of arbitrary genus with an arbitrary number of boundary components, we have chosen to restrict our attention to the simply connected case so as to illuminate the essential features of the algorithm and so that we may discuss the details of the proof of convergence without the added difficulty of having to work with moduli spaces. In fact, we will consider piecewise flat quadrilaterals and ask for a method to conformally map them to rectangles.

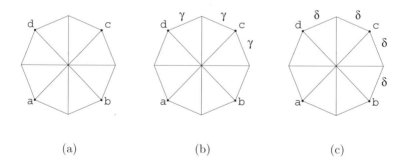

(a)   (b)   (c)

Fig. 1. Three piecewise flat conformal quadrilaterals.

Perhaps an example will help illustrate the problem the algorithm addresses. Consider the simple triangulation K of a topological quadrilateral with eight faces with a common central vertex and four distinguished boundary vertices a, b, c, and d as in Fig. 1. There are many ways to define a metric on K making each face a flat Euclidean triangle. Three examples are indicated in Fig. 1 where each side is given unit length, except for the sides labeled with γ and δ, which are given side lengths $\gamma = 0.3473$ and $\delta = 0.2611$. These examples are discussed in greater detail in Section 5, but for now realize that these labels encode a piecewise flat metric on K by identifying the faces with Euclidean triangles of side lengths given by the edge labels. This in turn produces three different conformal structures on K that each realizes K as a conformal quadrilateral. By standard theorems on conformal mapping, any conformal quadrilateral maps conformally to a Euclidean rectangle unique up to scaling. Approximations to this conformal mapping in each case are indicated in Fig. 2, where we see approximations to the image triangulations under the conformal mapping to a rectangle. Note that the first and third rectangles of Fig. 2 are both squares, but the conformally correct shapes of the faces in the two examples are different, and the conformal modulus of the second is approximately $\mu = 1.2031$.

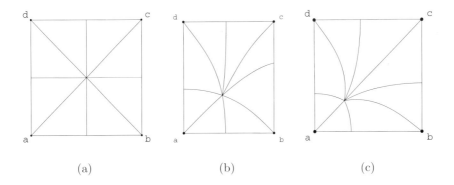

(a)   (b)   (c)

Fig. 2. Uniformizations of quadrilaterals from Fig. 1.

Complete proofs for convergence of the scheme in the more specialized setting of equilateral surfaces, where each edge has unit length and each face is identified with a unit equilateral triangle, are found in [4]. The original motivation for developing the scheme of [4] was to construct fundamental domains for the equilateral surfaces that arise in Grothendieck's theory of dessins d'enfants and to approximate their associated Belyĭ maps. Since then, Hurdal *et al* [7] have adopted this method to construct flat mappings of surfaces in $\mathbb{R}^3$ and have applied the method to obtain flat mappings of the human brain, which is of current interest in the neuroscience community. The desire to obtain better conformal integrity in these brain mappings has inspired us to investigate further this preliminary suggestion in [4] that a modification of their scheme using inversive distance packings could be used to build conformal mappings in this piecewise flat—as opposed to piecewise equilateral—setting. Stephenson's software `CirclePack` was used for the brain mappings of [7] as well as for calculating and rendering our examples. We note here that we do not present the circle packing algorithm used in `CirclePack` to calculate the packing for given inversive distance data as this has been discussed amply in [5].

The ingredients of this conformal mapping scheme are inversive distances of circles in the Riemann sphere, circle patterns in the Riemann sphere, and hexagonal refinement. The first three sections of the paper are centered around these three respective themes. We find that many mathematicians, even those who specialize in complex analysis and conformal geometry, are not familiar with the inversive distance between pairs of circles in the Riemann sphere. In Section 1, we present an inversive distance primer and prove some results about the conformal placement of circles in the Riemann sphere. In Section 2, we review the basics of piecewise flat structures on surfaces and introduce circle patterns with inversive distances encoded along edges. These patterns, generalizations of circle packings where edges encode tangencies, have been studied in the case where neighboring circles overlap with some angle between 0 and $\pi/2$. Bowers and Stephenson [4] introduced the notion of circle patterns where neighboring circles may not overlap, but where they do satisfy á priori inversive distance requirements. We emphasize again that the theoretical underpinnings of this topic are not entirely in place and are a matter of current research by Bowers, Stephenson, Hurdal, and others, but the good news is that the algorithm seems to work well in practice. This iterative algorithm for producing a sequence of patterns that are hoped to approximate more and more closely the desired conformal mapping is presented in Section 3, where hexagonal refinements are introduced. The difficulties in the proof of convergence to the desired conformal mapping are discussed in Section 4. We detail three main theoretical problems that must be addressed for a complete resolution of the question of convergence to a conformal map, and we prove convergence under the assumption that these problems have been resolved. Section 5 presents a gallery of examples that illustrate the algorithm by approximating conformal mappings to rectangles of conformal quadrilaterals that arise from piecewise flat metrics on topological disks, as in the examples of this introduction. This allows us to approximate the conformal moduli of quadrilaterals that arise from piecewise flat metrics, and to view the conformally correct shapes of the faces of the triangulation after mapping to the plane. We shall point out how well the algorithm works in practice, producing image triangulations with exactly the expected properties. Finally, Section 6 discusses practical implementation issues in applications, and computational and theoretical issues surrounding these.

1 An Inversive Distance Primer

The inversive distance between two oriented circles in the Riemann sphere $\widehat{\mathbb{C}}$ is a conformal invariant of the location of the circles in the sphere and their relative orientations; see [1]. Indeed, given oriented circle pairs C_1, C_2 and C_1', C_2' of $\widehat{\mathbb{C}}$, there exists a Möbius transformation T of the Riemann sphere with $T(C_i) = C_i'$ for $i = 1, 2$, respecting their relative orientations, if and only if the inversive distance between C_1 and C_2 equals that between C_1' and C_2'. An oriented circle C is the boundary of a unique open disk $\overline{C}$, called the interior of C, that lies to the left of C as C is traversed in the direction of its orientation. The precise definition of inversive distance may be stated elegantly with the aid of cross ratios and circle interiors.

Definition 1. *Let C_1 and C_2 be oriented circles in the Riemann sphere $\widehat{\mathbb{C}}$ bounding the respective disks $\overline{C}_1$ and $\overline{C}_2$, and let D be any oriented circle mutually orthogonal to C_1 and C_2. Denote the points of intersection of D with C_1 as z_1, z_2 ordered so that the oriented subarc of D from z_1 to z_2 lies in the disk $\overline{C}_1$. Similarly denote the ordered points of intersection of D with C_2 as w_1, w_2. The inversive distance between C_1 and C_2, denoted as $\mathrm{InvDist}(C_1, C_2)$, is defined in terms of the cross ratio*

$$[z_1, z_2; w_1, w_2] = \frac{(z_1 - w_1)(z_2 - w_2)}{(z_1 - z_2)(w_1 - w_2)}$$

by

$$\mathrm{InvDist}(C_1, C_2) = 2[z_1, z_2; w_1, w_2] - 1.$$

Recall that cross ratios of ordered 4-tuples of points in $\widehat{\mathbb{C}}$ are invariant under Möbius transformations. This implies that which circle orthogonal to both C_1 and C_2 is used in the definition is irrelevant as a Möbius transformation that setwise fixes C_1 and C_2 can be used to move any one orthogonal circle to another. Also, which one of the two orientations on the orthogonal circle D is used is irrelevant as the cross ratio satisfies $[z_1, z_2; w_1, w_2] = [z_2, z_1; w_2, w_1]$. This equation also shows that the inversive distance is preserved when the orientation of both circles is reversed so that it is only the relative orientation of the two circles that is important for the definition. When C_1 and C_2 overlap, the oriented angle of overlap may be defined unambiguously as the angle between the tangents to the circles at a point of overlap formed by one tangent pointing along the orientation of its parent circle and the other pointing against the orientation of its parent circle. We distinguish six different ways that two circles may overlap and describe the inversive distance in each case.

1.1 Six Cases

The inversive distance is always a real number since the cross ratio of four points that lie on a circle is always real. The way to dissect the inversive distance is through the auxiliary function

$$T(z) = 2[z_1, z_2; z, w_2] - 1 = 2\frac{(z_1 - z)(z_2 - w_2)}{(z_1 - z_2)(z - w_2)} - 1,$$

which is a Möbius transformation that takes the triple z_1, z_2, w_2 to the triple $-1, 1, \infty$. The function T takes the orthogonal circle D to the real line, the circle C_1 to the unit circle centered at the origin, and the circle C_2 to the vertical line orthogonal to the real axis at the point $T(w_1)$; see Fig. 3. Notice that $\mathrm{InvDist}(C_1, C_2) = T(w_1)$ may take on any real value and we distinguish the six cases according to the two relative orientations for each of the three possibilities for intersection of C_1 with C_2. Fig. 3 provides a snapshot of all the possibilities labeled according to whether the orientations are aligned or opposite, and whether the intersection consists of none, one, or two points.

Figs. 3(a) and 3(b) illustrate the possibilities for disjoint circles. If the orientations are opposite, the inversive distance is in the range from $-\infty$ to -1 exclusive, and if aligned, in the range from $+1$ to $+\infty$ exclusive. Figs. 3(c) and 3(d) illustrate those for tangent circles where the inversive distance is ± 1 depending on relative orientation. Figs. 3(e) and 3(f) illustrate those for intersecting circles where the inversive distance is between -1 and 0 for intersection angles between π and $\pi/2$ and between 0 and $+1$ for angles between $\pi/2$ and 0. Referring to the angle labels in Fig. 3, we may read off the inversive distances as

$$\mathrm{InvDist}(C_1, C_2) = \sec \alpha$$

for disjoint circles, where α is the indicated angle, and

$$\mathrm{InvDist}(C_1, C_2) = \cos \alpha$$

for intersecting circles, where α is the oriented angle of intersection of C_1 with C_2. Notice for intersecting circles, since the overlap angle α may be determined without regard to the normalizing transformation T, the inversive distance has an immediate, easily understood meaning. One can look at two overlapping circle pairs and estimate whether they are Möbius equivalent, a task of great difficulty for disjoint circle pairs. For those with a finely developed intuition for hyperbolic space and the Poincaré extensions of Möbius transformations, there is a more geometric understanding of inversive distance available.

1.2 An Alternate Description in Terms of Hyperbolic Geometry

Notice that if the orientation of only one member of a circle pair is reversed, the inversive distance merely changes sign. This follows from the immediate relation $[z_1, z_2; w_2, w_1] = 1 - [z_1, z_2; w_1, w_2]$. We therefore define

Definition 2. *The absolute inversive distance between any pair of unoriented circles is the absolute value of the inversive distance between the two circles when given either relative orientation. We use the same notation,*

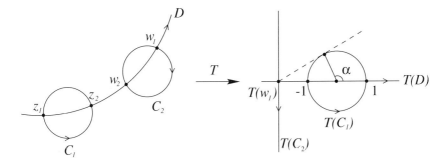

(a) Disjoint circles, opposite orientations.

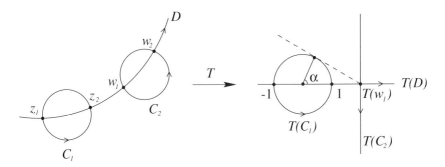

(b) Disjoint circles, aligned orientations.

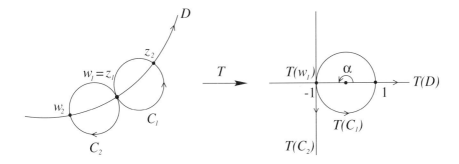

(c) Tangent circles, opposite orientations.

Fig. 3. Three of six ways that two circles overlap. Here, $T(z_1) = -1, T(z_2) = 1$.

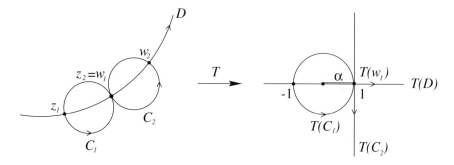

(d) Tangent circles, aligned orientations.

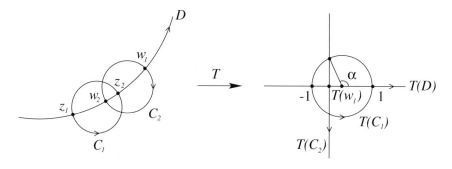

(e) Intersecting circles, opposite orientations.

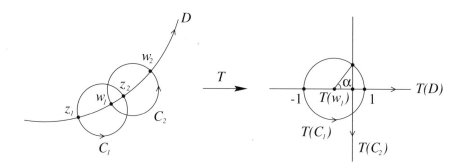

(f) Intersecting circles, aligned orientations.

Fig. 3. Remaining three ways that two circles overlap. Here, $T(z_1) = -1, T(z_2) = 1$.

InvDist(C_1, C_2), *for the absolute inversive distance between unoriented circles C_1 and C_2.*

It is clear then that there is a Möbius transformation taking an unoriented circle pair C_1, C_2 to another unoriented pair C_1', C_2' if and only if their absolute inversive distances agree. When C_1 and C_2 overlap with acute angle α the absolute inversive distance is $\cos \alpha$ and when they are tangent it takes the value 1. In this subsection our aim is to expose a geometric understanding of the absolute inversive distance between two disjoint circles in terms of hyperbolic geometry. This is a great intuitive aid for understanding inversive distances between disjoint circles. Toward this end assume C_1 and C_2 are disjoint and by appropriate choices of orientation map via T so that $T(C_1)$ is the unit circle and $T(C_2)$ is the vertical line through the point $\Delta = $ InvDist$(C_1, C_2) > 1$. Consider the extended complex plane as the sphere at infinity for the hyperbolic 3-space realized as the upper half-space model with metric $ds = |dx|/x_3$ on $\mathbb{H}^3 = \{x = (x_1, x_2, x_3) \colon x_3 > 0\}$. The Poincaré extension of T, denoted $\widetilde{T}$, is an isometry of $\mathbb{H}^3$. The circle C_1 bounds a hyperbolic plane P_1 in $\mathbb{H}^3$ that is realized as the upper hemisphere of the sphere in $\mathbb{R}^3$ with the same Euclidean center and radius as C_1, and similarly C_2 bounds the hyperbolic plane P_2. We calculate the hyperbolic distance δ between the planes P_1 and P_2.

First, since $\widetilde{T}$ is an isometry, we work with $\widetilde{T}(P_1)$, which is the upper hemisphere of the unit sphere in $\mathbb{R}^3$, and with $\widetilde{T}(P_2)$, which is the vertical half plane $\{x \in \mathbb{H}^3 \colon x_1 = \Delta\}$. There is a unique geodesic segment Σ in $\mathbb{H}^3$ meeting both $\widetilde{T}(P_1)$ and $\widetilde{T}(P_2)$ orthogonally at the respective points A and B. This geodesic segment lies on the circle in the vertical $x_1 x_3$-plane that is mutually orthogonal to $\widetilde{T}(P_1)$, $\widetilde{T}(P_2)$, and to the x_1-axis; see Fig. 4. Elementary geometry shows this circle to be centered at the point $(\Delta, 0, 0)$ and of Euclidean radius $\sqrt{\Delta^2 - 1}$, and the points A and B to be given by $A = (\cos \sec^{-1} \Delta, 0, \sin \sec^{-1} \Delta)$ and $B = (\Delta, 0, \sqrt{\Delta^2 - 1})$. A calculation of the hyperbolic length of Σ by integrating the line element $ds = |dx|/x_3$ along Σ from A to B gives the value of δ as

$$\delta = \ln \left| \Delta + \sqrt{\Delta^2 - 1} \right| = \cosh^{-1} \Delta.$$

This proves that the absolute inversive distance between C_1 and C_2 is precisely the hyperbolic cosine of the hyperbolic distance between the planes P_1 and P_2 bounded by C_1 and C_2. Experience with this understanding of inversive distance for disjoint circles coupled with the fact that Poincaré extensions of Möbius transformations are isometries of $\mathbb{H}^3$ has proved invaluable in our research, particularly for gaining intuition in working with disjoint circle patterns.

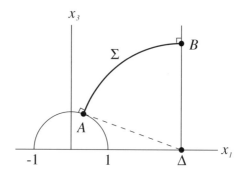

Fig. 4. The hyperbolic length of Σ is $\cosh^{-1}\Delta$.

1.3 A Euclidean Formula

The simplest formula for the absolute inversive distance between two circles in the complex plane is the one that the algorithm for conformal flattening uses. Though simple, it is not at all transparent that it should yield a Möbius invariant for the placement of two circles in the plane. We leave it as an exercise to verify that if C_i is the circle in the complex plane $\mathbb{C}$ centered at a_i of radius R_i, for $i = 1, 2$, then the absolute inversive distance is given by

$$\text{InvDist}(C_1, C_2) = \left| \frac{R_1^2 + R_2^2 - |a_1 - a_2|^2}{2R_1 R_2} \right|. \tag{1}$$

2 Piecewise Flat Surfaces and Circle Packings

A *combinatorial quadrilateral* is an abstract oriented simplicial 2-complex K that triangulates a closed topological disk with four distinguished boundary vertices $\{a, b, c, d\}$ ordered respecting the boundary orientation. The sets of vertices, edges, and faces of K are denoted respectively as $\mathbf{V}$, $\mathbf{E}$, and $\mathbf{F}$. A *piecewise flat structure* for K is determined by an *edge length function* $|-|: \mathbf{E} \rightarrow (0, \infty)$ that satisfies the triangle inequality condition, namely, that for every three edges e_1, e_2, e_3 that bound a face of K, the inequality

$$|e_1| \leq |e_2| + |e_3|$$

holds. An edge length function $|-|$ for K determines a piecewise Euclidean metric by assigning the length $|e|$ to each edge e of K and identifying each face $\langle v_1, v_2, v_3 \rangle$ with a flat Euclidean triangle of edge lengths $|e_i|$, where the edge $e_i = \langle v_j, v_k \rangle$ and $\{i, j, k\} = \{1, 2, 3\}$. The resulting piecewise Euclidean metric space is denoted as $|K|$ and, as explained in the introduction, carries the structure of a Riemann surface. Each interior vertex of K is a cone point singularity for the piecewise Euclidean metric but is not a singularity of the conformal structure. Our aim is to describe a scheme for approximating the conformal mapping of $|K|$ to a rectangle that maps the four distinguished boundary vertices to the four corners of the rectangle. Of course we do not have a candidate for the target rectangle since we do not know the modulus of the conformal quadrilateral $|K|$; however, the algorithm ideally will produce a sequence of target rectangles that converges to the correct one as well as a curvilinear triangulation of the target rectangle with the combinatorics of K that shows the correct conformal shapes of the faces.

A plentiful supply of piecewise flat surfaces is available from triangular grids in $\mathbb{R}^3$ where we read off side lengths of edges of actual Euclidean triangles. This, though, gives but a limited supply of the piecewise flat surfaces available, as many such surfaces admit no isometric embedding in $\mathbb{R}^3$, and many admit no embedding in any Euclidean space that isometrically embeds each edge as a straight Euclidean line segment.

The iterative algorithm for conformally mapping $|K|$ to a rectangle uses as seed certain inversive distance data calculated from the edge length function $|-|$. This gives rise to a piecewise flat surface with inversive distance information encoded along edges by a function $\Phi \colon \mathbf{E} \to [0, \infty)$. We abstract this by not assuming an á priori piecewise flat structure from which the edge function Φ arises.

Definition 3. *Let* $\Phi \colon \mathbf{E} \to [0, \infty)$ *be a function on the edge set of the complex K. A circle packing for (K, Φ) is a collection*

$$\mathcal{C} = \{C_v \colon v \in \mathbf{V}\}$$

of circles in the plane $\mathbb{C}$, each oriented counterclockwise, such that the inversive distance of neighboring circles is given by Φ, i.e., $\mathrm{InvDist}(C_u, C_v) = \Phi(\langle u, v \rangle)$ for each edge $\langle u, v \rangle$ in $\mathbf{E}$.

Perhaps a more descriptive term would be circle 'pattern', as opposed to 'packing', whenever the circles are disjoint. Nonetheless, we shall use the term 'packing' to describe a collection of circles, disjoint or not, that has a combinatorial pattern encoded in a complex K governing the placement of the circles in the plane. Tangency packings, which use only the combinatorial information encoded in K and not any varying inversive distance data, are used in [4] to uniformize piecewise equilateral surfaces. When the surface is piecewise flat where faces are generally not equilateral, more than the combinatorics of K must be used to build approximate conformal maps. We now describe how the metric information of $|K|$ may be used to embellish the combinatorics of K with inversive distance data, which turns out to be sufficient for generating candidates for approximate conformal maps.

Let $|-|$ be an edge length function for K and let $R \colon \mathbf{V} \to (0, \infty)$ be a positive function on the vertices that, for each edge $\langle u, v \rangle$, satisfies the condition

$$R(u)^2 + R(v)^2 \leq |\langle u, v \rangle|^2. \tag{2}$$

This inequality guarantees that if the edge $e = \langle u, v \rangle$ is drawn in the plane as a segment of length $|e|$, and circles C_u and C_v both oriented counterclockwise of respective radii $R(u)$ and $R(v)$ are centered at the vertices of e, then the oriented overlap, if the circles intersect nontrivially, is at most $\pi/2$, and the interiors, if the circles are disjoint, are also disjoint. The resulting *radius function* $R\colon \mathbf{V} \to (0, \infty)$ determines an inversive distance function Φ_R on the edge set by Equation 1:

$$\Phi_R(e) = \mathrm{InvDist}(C_u, C_v) = \frac{|\langle u, v \rangle|^2 - R(u)^2 - R(v)^2}{2R(u)R(v)}. \tag{3}$$

A circle packing $\mathcal{C}$ for (K, Φ_R), if it exists, gives rise to a *discrete conformal mapping* of $|K|$ to the plane by mapping the vertices of K to the centers of their corresponding circles and extending affinely on the metric faces. The image of such a discrete conformal mapping is the *carrier* of the circle packing $\mathcal{C}$, and is the union of the triangles corresponding to the faces of K formed by connecting centers of three mutually neighboring circles by straight line segments. The circle packing $\mathcal{C}$ is said to be *oriented* if the orientations of all of these nondegenerate triangles inherited from the orientation on K are compatible. Equivalently, $\mathcal{C}$ is oriented if the discrete conformal mapping f is an orientation preserving map from $|K|$ to the plane. When $\mathcal{C}$ is oriented and Φ takes values in the unit interval, this discrete conformal mapping is quasi-conformal, but it may fail to be so for general Φ values. Moreover, when $\mathcal{C}$ is oriented, it maps the triangulation of $|K|$ to a triangulation of the image of this map, though there may be degeneracies. We describe an algorithm in the next section that produces a sequence of these discrete conformal mappings, which serve as the candidates for approximating the conformal mapping of $|K|$ to a rectangle. To force convergence we need to normalize the boundary circles in some way, and we do so by making a further demand on our circle packings that will force a rectangular shape upon the image. In general there are many different circle packings for the same data (K, Φ). For example, in the case of tangency packings, each specification of boundary radii for the circles that correspond to boundary vertices determines a unique oriented packing with the combinatorics of K. Alternately, each specification of boundary angle sums at boundary vertices also uniquely determines an oriented packing for K. We shall call a circle packing $\mathcal{C}$ for (K, Φ) a *rectangular packing* if it is oriented and the angle sum of the faces at a boundary vertex in the image triangulation are all π, except at the four distinguished boundary vertices, where the angle sums are $\pi/2$. The carrier of a rectangular packing is a rectangle. Fig. 5 shows two packings and their carriers for the same piecewise flat surface $|K|$ and inversive distance data Φ; the packing on the right is rectangular.

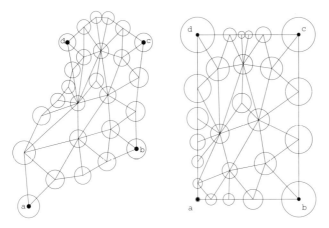

Fig. 5. Two packings for the same data (K, Φ).

3 Hexagonal Refinement

We now fix a combinatorial quadrilateral K with an edge length function $|-|$ that produces the piecewise flat conformal quadrilateral $|K|$. Let R be a constant radius function that satisfies Inequality 2 at each edge and $\mathcal{C}$ a rectangular packing for (K, Φ_R), where Φ_R satisfies Equation 3. It is important for proving convergence that R be a constant function. Let f be the discrete conformal mapping determined by $\mathcal{C}$. The seed data for our conformal mapping algorithm is the 4-tuple

$$(K_0, |-|_0, R_0, \Phi_0) = (K, |-|, R, \Phi_R),$$

from which we produce the mapping data

$$(\mathcal{C}_0, f_0) = (\mathcal{C}, f).$$

We think of f as the zeroeth approximation to the conformal mapping that maps $|K|$ to a planar rectangle. The constant radius function may be chosen to have any positive value between 0 and $\lambda/\sqrt{2}$, where λ is the minimum of $|e|$ as e ranges over all the edges of K.

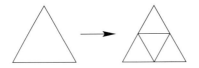

Fig. 6. Hexagonal refinement, $K \to K'$.

For better approximations we employ *hexagonal refinement*, or hex-refinement for short, which subdivides a triangle into four subtriangles as in Fig. 6; see [4]. The complex thus obtained from K by subdividing each face as in Fig. 6 is denoted as K'. There is a natural edge length function $|-|'$ on K' obtained by reading off the lengths of edges obtained by placing a vertex at the midpoint of each metric edge in $|K|$ to hex-subdivide each metric face of $|K|$ into four similar copies of itself, scaled by $1/2$. Then $|K|$ and $|K'|'$ are isometric and thus indistinguishable as metric spaces. If the constant radius $R' = \frac{1}{2}R$ is used for K', the induced inversive distance function $\Phi_{R'}$ replicates on the edges of a face of K' the three inversive distances of its parent face in K. Starting then with the seed data $(K_0, |-|_0, R_0, \Phi_0)$, we generate an infinite sequence recursively by

$$(K_{n+1}, |-|_{n+1}, R_{n+1}, \Phi_{n+1}) = (K'_n, |-|'_n, R'_n, \Phi'_n),$$

for which $|K| = |K_n|_n$ for all n. This produces an infinite sequence of mapping data $(\mathcal{C}_n, f_n)$, where $f_n \colon |K| = |K_n|_n \to \mathbb{C}$ is the discrete conformal mapping of the piecewise flat surface $|K|$ to the plane determined by the rectangular packing $\mathcal{C}_n$ for (K_n, Φ_n). Recall that there are four distinguished boundary vertices $\{a, b, c, d\}$ of K ordered respecting the orientation of the boundary. We assume one more normalization condition, easily accomplished by Euclidean similarities, by requiring the first two distinguished vertices a and b of K to map to the respective points 0 and 1 under each f_n, and the two others to map to the upper half plane. The image of each discrete conformal mapping f_n is then a rectangle in the upper half plane one of whose sides lies along the unit interval $[0, 1]$.

The main convergence result of [4] may be used to prove, in the special case of tangency packings where $|-|$ gives a unit length to each edge, R is identically $1/2$, and Φ_R is identically 1, that the sequence of mappings $f_n \colon |K| \to \mathbb{C}$ exists and converges uniformly to the unique conformal mapping F of $|K|$ to a rectangle in the plane with $F(a) = 0$, $F(b) = 1$, and $F(c)$ and $F(d)$ in the upper half plane. Moreover, the pointwise quasi-conformal dilatations of the maps f_n are bounded above and converge uniformly to unity on compact subsets of the complement of the vertices of $|K|$. Our analysis of the proof will show in the next section that this holds in the piecewise flat case when R can be chosen so that Φ_R has values in the unit interval, and our goal is to understand precisely what is lacking in extending the proof to the general case.

When the sequence f_n does converge to the expected conformal map F, the conformal modulus of the conformal quadrilateral $|K|$ is thus determined to be $\mu = |F(d)|$, the height of the image rectangle $F(|K|)$. One might expect then that the maximum quasi-conformal dilatations of the sequence f_n converge to unity, but this is not the case. In fact, the maximum quasi-conformal dilatations of the sequence are in general bounded away from unity since, at any vertex v of K whose angle sum Θ determined by $|-|$ is different from

2π, there is always high distortion at the vertices of K_n neighboring v; see [4]. Nonetheless, this high local distortion is relegated to smaller and smaller neighborhoods of the original vertices of K as we progress along the sequence f_n. The result is that the limit mapping F has local dilatation 1, i.e., is conformal, at every point of $|K|$ other than those of the original vertex set $\mathbf{V}$. Removability of isolated singularities then comes into play to guarantee that the dilatations at the original vertices are 1 and, therefore, the limit mapping F is conformal.

4 Proving Convergence and Conformality

There are three main problems associated with the inversive distance scheme for approximating the conformal mapping of $|K|$ to a rectangle. The first is that of the existence of a rectangular packing $\mathcal{C}_n$ for (K_n, Φ_n), the second is that of quasi-conformality of the mappings f_n with globally bounded dilatations, and the third is that of the rigidity of infinite hexagonal packings of the plane with prescribed periodic inversive distance data. The first problem concerns the existence of the approximating sequence f_n, the second concerns the convergence of the sequence f_n to a quasi-conformal mapping F, and the third concerns the conformality of the limit mapping F. We shall discuss each of these in turn after some general comments on inversive distance packings with Φ-values restricted to lie in $[0, 1]$, i.e., in which two neighboring circles intersect nontrivially. This has been the subject of a large body of theoretical research over the past decade and a half and there is an extensive literature on the subject of existence and uniqueness of packings, particularly in the tangency case where Φ is identically 1. The understanding of the existence and uniqueness of tangency circle packings with prescribed combinatorics, as well as rigidity of infinite packings, is crucial in the work of [10] and [4] where mapping algorithms are shown to converge to the correct conformal mappings. Using existence, uniqueness, and rigidity results now in place allows us to adapt the proofs of [4] to the nontangency but overlapping case where Φ may take values in the interval $[0, 1]$. This section will give just such a proof that also covers the general case of unrestricted Φ values if the three problems that we analyze in this section are found to have appropriate resolutions.

The problem of existence. The existence and uniqueness of tangency circle packings for a complex K was first proved in [2] for arbitrarily assigned boundary radii or angle sums. This is viewed in [2] as the discrete analogue of the classical Perron method of solving the Dirichlet Problem on planar domains. Existence and uniqueness results for overlapping packings with prescribed angles of overlap, where Φ has values in the unit interval, are proved in [12] and [6]. It follows from this work that a rectangular packing for the data (K_n, Φ_n) exists as long as the values of Φ_n lie in the unit interval and two technical conditions first described by Thurston in [12] are satisfied. These *Thurston conditions* are, for an inversive distance assignment Φ,

T1 If a simple loop in the complex K formed by the three edges e_1, e_2, e_3 separates the vertices of K, then $\sum_{i=1}^{3} \cos^{-1} \Phi(e_i) < \pi$;

T2 If $v_1, v_2, v_3, v_4 = v_0$ are distinct vertices of K forming edges $\langle v_{i-1}, v_i \rangle$ and $\Phi(\langle v_{i-1}, v_i \rangle) = 0$ for $i = 1, 2, 3, 4$, then either $\langle v_0, v_2 \rangle$ or $\langle v_1, v_3 \rangle$ is an edge of K.

The problem of existence persists when neighboring circles are allowed to be disjoint, where the Φ values may be greater than unity. In this case the general boundary value problem is not always solvable, i.e., there are examples of inversive distance assignments Φ where no circle packing in the plane with the combinatorics of K can realize the inversive distance data. Even when such packings do exist, they may not exist with predetermined boundary radii or angle sums. Thus, there are examples of data (K, Φ) for which there are no rectangular packings. These will be detailed in forthcoming publications, but for now their existence points to the fact that the moduli space of data for which there do exist general inversive distance packings is a much more complicated object than those for the special cases of tangency and overlapping packings.

For the present work, this lack of a complete understanding of the existence of a circle packing for (K, Φ) means that we cannot guarantee that the seed packing $\mathcal{C}_0$ for our algorithm exists. However, when it does exist, the algorithm produces a sequence of approximate conformal mappings. The examples of inversive distance packing data without rectangular packings require some gymnastics to construct and do not seem to arise naturally from, for example, polyhedral surfaces embedded in $\mathbb{R}^3$. We have never encountered a surface in practice where the lack of existence prevented us from building a seed packing for the algorithm. This problem does not seem to be a major practical impediment to the widespread application of the inversive distance scheme for conformally mapping piecewise flat surfaces to the plane.

The problem of quasi-conformality. Assume the rectangular packings $\mathcal{C}_n$, and therefore the discrete conformal mappings f_n, exist for all n. The argument of [4] for proving convergence of f_n to a limit mapping F in the tangency case uses the classical theory of normality of families of quasi-conformal mappings found, for instance, in [9]. The argument, which is given for conformal quadrilaterals in the proof of the theorem below, requires that f_n be a sequence of quasi-conformal mappings with *bounded dilatations*, meaning that there is a global bound κ on the maximal dilatations $\kappa(f_n)$ of all the maps in the sequence. In this case, it will be shown that the sequence f_n converges uniformly to a κ-quasi-conformal mapping of $|K|$.

Quasi-conformality of each map f_n as well as a global bound on their dilatations in the tangency case is guaranteed by the ring lemma of [10]. Forthcoming publications will show that the ring lemma generalizes to those inversive distance packings for which Φ never takes the value 0, i.e., the case of non-orthogonal overlaps, and for which the Thurston conditions hold. However, this generalized ring lemma provides quasi-conformality only in the overlapping case where the Φ values lie in the half-closed interval $(0, 1]$. The lemma does not provide quasi-conformality in the setting of disjoint circle neighbors where Φ may take values greater than unity. In fact, there are examples of inversive distance assignments given by Φ where $\mathcal{C}$ exists, so that the discrete conformal mapping f exists, for which f is not quasi-conformal. Thus there is no guarantee that even if the sequence f_n exists that each mapping is quasi-conformal, and even if each is, there is no guarantee that the sequence has bounded dilatations. Again these examples require some gymnastics to construct and seem not to appear among, for example, polyhedral surfaces in $\mathbb{R}^3$, and again this problem does not seem to be a major practical impediment to the widespread application of the inversive distance scheme.

The problem of rigidity of infinite hexagonal packings. Assume now that the first two problems have been resolved for $|K|$ and we have a sequence of discrete conformal mappings f_n, each quasi-conformal, with dilatations bounded by κ. In this case there will exist a κ-quasi-conformal limit mapping F. The final step for uniformizing $|K|$ is the verification of the conformality of F. This step is accomplished for tangency packings in both [10] and [4] by use of the hexagonal packing lemma of [10], which depends on a rigidity result about infinite circle packings of the complex plane. The analogous rigidity result for overlapping packings is proved in [6], and so the ingredients are in place to verify conformality of the limit mapping whenever Φ takes values in the unit interval. We believe that a very general rigidity result holds for arbitrary locally finite inversive distance packings of the complex plane, but for the proof of conformality, all we need is the verification of the following specialized rigidity conjecture. In the conjecture, H is the constant 6-degree triangulation of the plane. The edges of H may be put into three equivalence classes depending to which edge of a fixed face τ a given edge is "parallel". A circle packing of $\mathbb{C}$ is *locally finite* provided each point of the plane has a neighborhood that meets only finitely many circles of the packing.

Conjecture 1. Let α, β, and γ be the inversive distances between respective pairs of three equi-radii circles in the plane whose centers are the vertices of a nondegenerate triangle. Let Θ be an inversive distance edge function for H that assigns the values α, β, and γ to the three respective edges of each face so that Θ is constant on each of the three equivalence classes. Then locally finite circle packings for (H, Θ) are unique up to Euclidean similarity, i.e., if $\mathcal{C}$ and $\mathcal{C}'$ are both locally finite circle packings for (H, Θ), then there is a similarity S such that $S(\mathcal{C}) = \{S(C) \colon C \in \mathcal{C}\} = \mathcal{C}'$.

If the conjecture is true, then any circle packing for (H, Θ) has $\mathbb{Z} \times \mathbb{Z}$-symmetry with fundamental domain the union of any two triangles formed

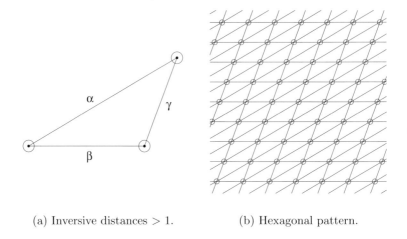

(a) Inversive distances > 1.   (b) Hexagonal pattern.

Fig. 7. Hexagonal rigidity.

by connecting neighboring circle centers and that meet along a common edge. This follows by placing three circles of equal radii in the plane such that the pairwise inversive distances are given by α, β, and γ. The plane then may be triangulated in the hexagonal pattern with isometric copies of the triangle obtained by connecting the centers of these three equi-radii circles. An example where $\alpha = 113.0346$, $\beta = 51.8889$, and $\gamma = 31$ appears in Fig. 7.

When each of α, β, and γ are no greater than unity, results of [6] verify the conjecture. The next theorem shows how normality of a quasi-conformal family with bounded dilatations and the conjecture are used to prove conformal convergence. We make the restriction in the theorem and its corollary that Φ never takes the value 0 so that the circle packings have no orthogonal neighboring circles.

Theorem 1. *If the sequence of discrete conformal mappings f_n exists, is quasi-conformal with bounded dilatations, and Conjecture 1 holds, then f_n converges uniformly on compact subsets of the interior of $|K|$ to a conformal mapping F of $|K|$ to a rectangle in the complex plane with $F(a) = 0$, $F(b) = 1$, and $F(c)$ and $F(d)$ in the upper half plane. Moreover, the maximum dilatation of f_n converges to unity uniformly on compact subsets of the complement of $\mathbf{V}$ in $|K|$.*

Proof. Suppose the hypotheses hold so that f_n is a sequence of quasi-conformal mappings of $|K|$ to the plane with a global bound κ on the quasi-conformal dilatations. Let μ be the conformal modulus of the conformal quadrilateral $|K|$ and let $F\colon |K| \to \mathcal{R}$ be the unique conformal mapping from $|K|$ to the rectangle in the plane with vertices 0, 1, and μi and $F(a) = 0$, $F(b) = 1$, $F(c) = 1+\mu i$, and $F(d) = \mu i$. As F is 1-quasi-conformal, each of the mappings $f_n \circ F^{-1}$ is κ-quasi-conformal. Theorem II 5.1 of [9] applies to show that the family of κ-quasi-conformal mappings $\mathcal{F} = \{f_n \circ F^{-1} : n = 1, 2, \ldots\}$ is normal in the interior of $\mathcal{R}$. Let w be any limit function of a sequence from $\mathcal{F}$, say $w = \lim f_{n(i)} \circ F^{-1}$ for a subsequence $f_{n(i)}$, where the limit is uniform on compact subsets of the interior of $\mathcal{R}$. By Theorem II 5.3 of [9], there are exactly three possibilities: the limit function w on the interior of $\mathcal{R}$ is a constant mapping, a mapping onto two distinct points, or a κ-quasi-conformal mapping. We show next that the first two possibilities do not occur.

Let $\mathcal{S}$ be the open infinite strip in the complex plane between the horizontal lines through $\pm\mu i$. Since the four corners and sides of $\mathcal{R}$ are mapped by $f_n \circ F^{-1}$ to the four corners and sides of the image rectangle $\mathcal{R}_n = f_n(|K|)$, the reflection principle for quasi-conformal mappings [9] may be iterated to produce a κ-quasi-conformal extension F_n of $f_n \circ F^{-1}$ to the domain $\mathcal{S}$, as well as a κ-quasi-conformal extension $\widetilde{w}$ of w. Since w is a limit function of a sequence from $\mathcal{F}$, $\widetilde{w}$ is a limit function of a sequence from $\widetilde{\mathcal{F}} = \{F_n : n = 1, 2, \ldots\}$. By Theorem II 5.3 of [9], there are exactly the same three possibilities for this function $\widetilde{w}$. Notice though that $\widetilde{w}$ is the identity on the set of integers, which are contained in the interior of $\mathcal{S}$, so the first two possibilities are ruled out. It follows that $\widetilde{w}$ is a κ-quasi-conformal mapping and, as w is the restriction of $\widetilde{w}$ to $\mathcal{R}$, so too is w.

The carrier $\mathcal{R}_n$ of $\mathcal{C}_n$ is a rectangle in the upper half plane with one side the unit interval. Theorem II 5.4 of [9] implies that the image of w is the kernel of the interiors of the rectangles $\mathcal{R}_{n(i)}$, and it is easy to see that such a kernel must be a rectangle with one side the unit interval. We show below that w is conformal, which immediately implies that this image rectangle $w(\mathcal{R})$ must be $\mathcal{R}$ itself and that w must be the identity mapping of $\mathcal{R}$ since it fixes the four corners. It follows that $f_{n(i)}$ converges uniformly on compact subsets of the interior of $|K|$ to F, the unique conformal mapping of $|K|$ to $\mathcal{R}$ with $F(a) = 0$ and $F(b) = 1$. As w is an arbitrary limit function of a sequence from $\mathcal{F}$, this argument shows that there is only one such limit function, namely the identity function on $\mathcal{R}$. As the collection $\mathcal{F}$ is a normal family of mappings, so that every infinite subset of $\mathcal{F}$ has a limit function, it follows that the sequence $f_n \circ F^{-1}$ itself converges uniformly on compact subsets of the interior of $\mathcal{R}$ to this identity function, or that the sequence f_n converges uniformly on compact subsets of $|K|$ to F. This completes the proof of convergence of the f_n to a conformal mapping of $|K|$ modulo the verification that w is in fact conformal. This will be accomplished next with the aid of Conjecture 1.

Let α, β, γ, H, and Θ be as in Conjecture 1 and for each n, let H_n be the subcomplex of H formed by n generations of the hexagonal grid about some fixed vertex v_0. Let $\sigma = \langle v_0, v_1, v_2 \rangle$ be a face of H containing v_0 and let Θ_n be the restriction of Θ to the edges of H_n. A proof using the rigidity of Conjecture 1 and the generalized ring lemma, similar to the proof of the hexagonal packing lemma of [10], shows that there is a sequence ε_n decreasing to zero such that, if $\mathcal{H}_n$ is any oriented circle packing for (H_n, Θ_n), and if τ_n is the triangle in $\mathbb{C}$ formed by connecting the centers of the circles in $\mathcal{H}_n$ corresponding to v_0, v_1, and v_2 and τ is the triangle formed by connecting the centers of the circles in the unique packing $\mathcal{H}$ corresponding to v_0, v_1, and v_2, then the vertex preserving affine map from τ_n to τ has dilatation at most $1 + \varepsilon_n$. This is a very strong statement concerning the shapes of the triangles τ_n as the constants ε_n do not depend on which packing for (H_n, Θ_n) is chosen, but merely on the fact that v_0 is "n-deep" within the complex H_n.

Let D be a compact subset contained in an open face σ of $|K|$. Let N be an arbitrary positive integer and choose n so large that each point z of D is centered in a simply connected neighborhood U_z formed by N generations of the hexagonal grid in $|K_n|$ that results from n hex-refinements of the face σ. The generalization of the hexagonal packing lemma of the previous paragraph guarantees that f_n has maximum dilatation at most $1 + \varepsilon_N$ on D, and since ε_N decreases to zero, the maximum dilatation converges to unity uniformly on D. This implies by Theorem II 5.3 of [9] that the dilatation of the limit mapping F at any point in the interior of a face of K is no more than $1 + \varepsilon_N$, for all N, and therefore F is conformal on the interiors of the faces of K. By removability of analytic arcs and isolated singularities, F is conformal on $|K|$. We emphasize here that this argument with the use of the generalized hexagonal packing lemma requires that our radius function R, from which the seed inversive distance function Φ_0 is calculated, be constant on the vertex set $\mathbf{V}$. One may run the algorithm with arbitrary variable radius function R, but the convergence generally will not be to a conformal mapping.

The last statement of the theorem requires a small modification to show uniform convergence when the compact set D hits edges, which we shall not present.   □

Corollary 1. *If all the edge lengths $|e|$ lie in the half-close interval $(\lambda, \sqrt{2}\lambda]$, for some positive constant λ and the Thurston conditions (T1) and (T2) hold, then the functions f_n exist and are quasi-conformal with bounded dilatations, and the sequence converges uniformly on compact subsets of the interior of $|K|$ to a conformal mapping F of $|K|$ to a rectangle in the complex plane with $F(a) = 0$, $F(b) = 1$, and $F(c)$ and $F(d)$ in the upper half plane. Moreover, the maximum dilatation of f_n converges to unity uniformly on compact subsets of the complement of $\mathbf{V}$ in $|K|$.*

Proof. If the initial radius function R_0 is chosen to have constant value $\lambda/\sqrt{2}$, then the initial inversive distance edge function Φ_0 takes values in the half-closed interval $(0, 1]$, since all the edge lengths $|e|$ lie in the interval $(\lambda, \sqrt{2}\lambda]$. Notice that the values of each Φ_n are the same as those for the initial inversive distance edge function Φ_0, so that the Φ_n values are in the unit interval. Thus the circle packings $\mathcal{C}_n$ are either tangency or nonorthogonal overlapping packings. The sequence f_n exists by existence-uniqueness results of [2] and [6] that cover the tangency and overlapping packing cases. Quasi-conformality of the f_n with bounded dilatations follows from the ring lemma of [10] for the tangency case and its generalization for the overlapping case. The verification of Conjecture 1 for the tangency case appears in [10] and for the overlapping case in [6]; see also [11]. Theorem 1 applies.   □

5 A Gallery of Quadrilaterals

Example 1. Our first examples are those of Figs. 1 and 2. In Fig. 1(a) all edges have unit length and the surface $|K|$ is an equilateral surface formed by gluing eight unit equilateral triangles along edges that meet at a common central vertex. By conformal symmetry at the central vertex, the central angles of all the triangles have measure $\pi/4$ in the conformal structure though they all have Euclidean measure $\pi/3$ in the piecewise flat structure. Notice that there are anti-conformal reflections across the diagonals from a to c through the center and from b to d through the center, as well as across the other two diagonals. Thus the dihedral group D_4, the symmetry group of the square, acts as a group of conformal symmetries of $|K|$. The only rectangles on which D_4 acts conformally are squares, so we know before running the inversive distance scheme that the conformal modulus of $|K|$ is 1 and $|K|$ is conformally equivalent to a square via a mapping taking the equilateral faces to congruent $(2, 4, 4)$ triangles formed by the diagonals and opposite edge bisectors of a square. `CirclePack` confirms this in Fig. 2(a). Since this is an equilateral surface, we used tangency packings with unit inversive distance function.

In Fig. 1(b) all edges have unit length except for the three boundary edges labeled by γ, each of which has edge length $\gamma = 2\sin\frac{\pi}{18} \approx 0.3473$. This makes the Euclidean angles opposite γ equal to $\pi/9$ so that the total Euclidean angle spanned opposite the three labeled sides is $\pi/3$, the same as the Euclidean angles of the equilateral triangles at that vertex. The total Euclidean angle sum around the central vertex is 2π, so the angles measured by the conformal structure at the central vertex agree with the Euclidean measures. In particular, a conformal mapping to a rectangle will map the faces so that the Euclidean angles at the central vertex are preserved. Again `CirclePack` confirms this in Fig. 2(b). This time the rectangle is not a square and the conformal modulus of the conformal quadrilateral $|K|$ is $\mu = 1.2031$. The fixed value we chose for the radius function R, from which the inversive distance edge function Φ is calculated by Equation 3, is $\gamma/2$. This makes the inversive distance values unity along the γ edges and 15.5817 otherwise.

In Fig. 1(c) all edges have unit length except for the four boundary edges labeled by δ, each of which has edge length $\delta = 2\sin\frac{\pi}{24} \approx 0.26105$. This makes the Euclidean angles opposite δ equal to $\pi/12$ so that the total Euclidean angle spanned opposite the four labeled sides is $\pi/3$. Thus the market share of these four angles totaled equals the market share of each of the other angles at the central vertex in the unit equilateral triangles. This means that

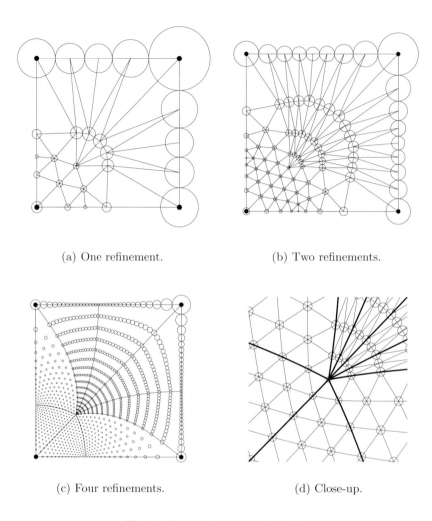

(a) One refinement.   (b) Two refinements.

(c) Four refinements.   (d) Close-up.

Fig. 8. Converging to conformality.

the conformal structure on $|K|$ measures the total angle spanned opposite the four labeled sides as $2\pi/5$ as well as the remaining four angles at the central vertex in the four equilateral triangles. In particular, the conformal structure measures each angle opposite δ as $\pi/10$ though the Euclidean measure is $\pi/12$. Also, $|K|$ has an anti-conformal reflection across the diagonal from a to c and, since squares are the only rectangles with a diagonal conformal symmetry, we know that the conformal modulus of $|K|$ is 1 and $|K|$ is conformally equivalent to a square. Again CirclePack confirms this in Fig. 2(c). The fixed value we chose for the radius function R is $\delta/2$, which makes the inversive distance values unity along the δ edges and 28.3477 otherwise. Fig. 8 shows the image rectangular packings at stages one, two, and four of the inversive distance iteration, with only the image of the original triangulation shown for the fourth stage packing, as well as a close-up of the central vertex from the stage four refinement.

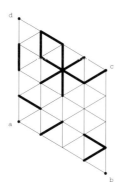

Fig. 9. Hexagonal grid: lengths of bold edges are 1.1; others are 1.4.

Example 2. The edge length assignments used in Fig. 9 for the complex K have $|e|$ equal to 1.1 for the bold edges and 1.4 otherwise. We approximate the conformal mapping of $|K|$ to a rectangle using three different choices for the initial radius function. The radius function $R(1)$ takes the constant value $1/\sqrt{2}$ where all neighboring circles overlap nontrivially. The second $R(2)$ takes the constant value $3/5$ where there is a mixture of overlapping and disjoint circles in the initial configuration. The third $R(3)$ takes the constant value $1/4$ where all circle pairs are disjoint. The inversive distance algorithm with any of the three seed radii should provide approximations that converge to the unique conformal mapping of $|K|$ to a rectangle of unit horizontal side length. Fig. 10 shows the fourth iterate of the inversive distance scheme applied with each of the three seed radii functions. The circle packings themselves with the images of the edges of the initial triangulation K darkened are shown, along with a close-up of a neighborhood of one of the vertices. This experimentation with `CirclePack` suggests that the convergence is independent of the initial constant radius value, as it should be. The ranges of the Φ values are 0.2100 to 0.9600 for $\Phi_{R(1)}$, 0.6806 to 1.7222 for $\Phi_{R(2)}$, and 8.6800 to 14.6800 for $\Phi_{R(3)}$.

Example 3. Corollary 1 confirms that the inversive distance scheme converges in case the piecewise flat metric is *equilateral*, i.e., when the edge length function $|-|$ takes the constant value 1. Then the metric surface $|K|$ is a union of equilateral triangles glued side-to-side and, when the radius function R takes the constant value $1/2$, the inversive distance function Φ_R takes the constant value 1. The rectangular packings are then tangency packings.

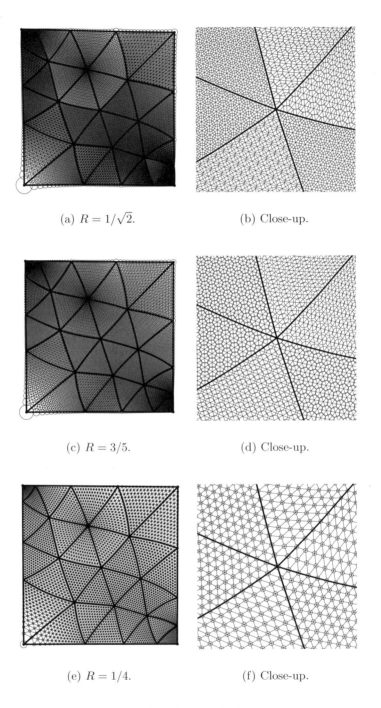

(a) $R = 1/\sqrt{2}$.   (b) Close-up.

(c) $R = 3/5$.   (d) Close-up.

(e) $R = 1/4$.   (f) Close-up.

Fig. 10. Hexagonal grid.

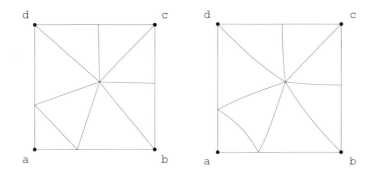

(a) A quadrilateral and its reflective triangulation.

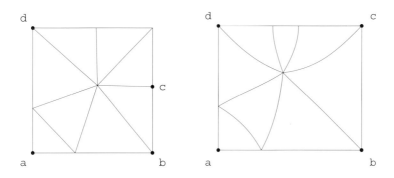

(b) A different corner point c.

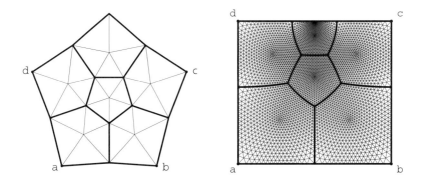

(c) A pentagonal packing and its reflective triangulation.

Fig. 11. Equilateral surfaces and their uniformizations.

Fig. 11 shows three examples of piecewise equilateral quadrilaterals and their uniformizations as rectangles. Each edge in the left-hand figures is given unit length, and four refinements are used to approximate the rectangular uniformizations on the right. An interesting feature of equilateral surfaces is that they have a *reflective* structure in which each face reflects across any interior edge to its companion face, see [4]. This reflection is an anti-conformal map and the whole surface is generated by fixing any one face and then reflecting across edges iteratively. This is obvious in the piecewise equilateral manifestation of the surface, and this translates into the following property of their rectangular conformal images. The image $\tau' = F(\tau)$ of any equilateral face τ of $|K|$ under the conformal mapping to a rectangle contains all the information about the rest of the map in the sense that the rest of the map and the image curvilinear triangulation of the rectangle can be recovered by anti-conformal reflections iterated starting with τ'. This suggests that, in principle, an arbitrary finite or even countably infinite amount of information can be represented in the shape of a single curvilinear triangle and then recovered by anti-conformal reflections. This is theoretically interesting and is illustrated in the next example.

Fig. 12(a) shows an eight-by-eight square with each subsquare divided into two triangles with either a right or left slash. The right slash encodes a zero and the left a one, and the rows encode the individual symbols of the expression 'vismath!'. The resulting triangulation is given an equilateral metric with all unit edge lengths and this surface is mapped conformally to a rectangle. The resulting reflective curvilinear triangulation is shown in Fig. 12(b) and the upper left-hand corner triangle is enlarged in Fig. 12(c). The whole triangulation in Fig. 12(b), and therefore the message 'vismath!', may be recovered from the lone triangle in Fig. 12(c) (or from any other triangle in the figure) by iterated anti-conformal reflection. Of course there is nothing to restrict our attention to finite triangulations. We might well triangulate the plane, prescribe that each face be a unit equilateral triangle, then conformally map the resulting piecewise equilateral surface to the plane $\mathbb{C}$ or to the unit disk. The image of any face then contains all the combinatorial information of the original triangulation. The interested reader might find the discussion of [3] enlightening.

Example 4. The left-hand graphic of Fig. 13(a) (Color Plate 1(a) on page 425) shows a three-dimensional rendering of the surface of a human cerebrum obtained from the Visible Man data from the National Library of Medicine. This example contains $52,360$ vertices and $103,845$ faces. Hurdal *et al* [8] flattened this mesh quasi-conformally using tangency packings where all inversive distances are set to unity. They then computed a textured bump map using a fake diffuse component for each circle using the surface normal in $\mathbb{R}^3$. The color for each circle was then scaled based on the diffuse value. In this way the fissures and sulci of the three-dimensional brain data can be represented in the flat mapping, see [8]. The results appear in the left-hand

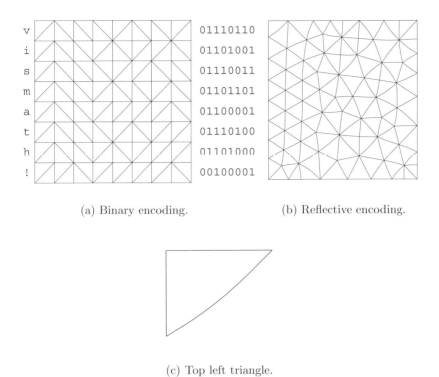

(a) Binary encoding.   (b) Reflective encoding.

(c) Top left triangle.

Fig. 12. Encoded 'vismath!'.

graphics of Figs. 13(b) and 13(c) (Color Plates 1(b) and 1(c) on page 425), where boundary data from the three-dimensional surface has been used to normalize the packing. One can see the dramatic effect bump map texturing has in these flattened images. In the right hand graphics of Fig. 13 (Color Plate 1 on page 425), we have isolated from this brain surface a quadrilateral region made up of 2943 faces with 1565 vertices, and mapped this subsurface conformally to a rectangle. We used the distances between neighboring vertices in the three-dimensional graphic of Fig. 13(a) (Color Plate 1(a) on page 425) to compute an edge-length function $|-|$ and then flattened using the inversive distance scheme. The first rectangular map, Fig. 13(b) (Color Plate 1(b) on page 425), is an inversive distance packing without the bump map texture, and the second, Fig. 13(c) (Color Plate 1(c) on page 425), is one with the bump map texture. This is a sample of ongoing work by a team of mathematicians and neuroscientists who are working to build a conformal flattening visualization tool for use in neuro-anatomical studies.

Another sample appears in Fig. 14 (Color Plate 2 on page 426) where we have conformally mapped two cerebellum images obtained from MRI scans to

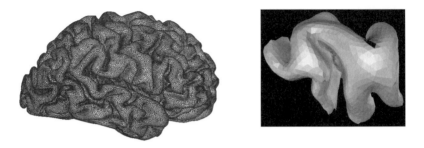

(a) Right hemisphere and subsurface.

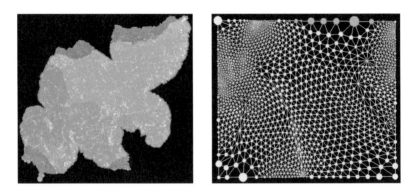

(b) Radii packing of hemisphere and inversive distance packing of a subsurface.

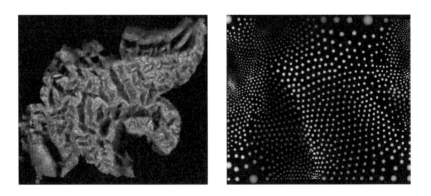

(c) Packing with bump map texture.

Fig. 13. Quasi-conformally mapping of the human brain to a planar domain.

a disk. The top two images show the cerebellum from two different subjects. The middle two images show a mapping to a disk. The bottom two images correspond to a close-up view of the disk mapping to highlight some of the detail in the central regions of the mappings. The color coding identifies regions of interest to neuro-anatomists with the orange regions indicating areas of PET activation when the subjects perform the same tasks.

6 Implementation: Practical Experimental, Computational, and Theoretical Issues

Implementation of the inversive distance scheme for approximating conformal mappings requires the development of a computational engine that computes oriented circle packings for given inversive distance data (K, Φ). Ken Stephenson has built such an engine in his program `CirclePack`. Its packing algorithm for tangency packings uses a refinement of Thurston's original idea in [12] as well as modern numerical schemes for fast approximation of transcendental functions. The reader may consult [5] for the latest detailed account of optimal packing algorithms. The packing algorithm generalizes to cover arbitrary inversive distance packings, though now there is no guarantee of convergence as there is in the tangency case. In fact, as we know of examples of inversive distance data (K, Φ) that have no circle packing realization, any such seed data for `CirclePack` would fail to converge. The packing algorithm is based on monotonicity results for the change in angle sums about vertices as the radius of a single circle is changed while preserving inversive distances. Again the interested reader is directed to [5] for details.

The reader might ask how practical it is to get really close approximations to the conformal mapping of $|K|$ to a rectangle since, obviously, the number of vertices grows exponentially as hex-refinement is iterated. The good news is that the experimental evidence suggests very fast convergence of the inversive distance scheme. Indeed, in all examples we have yet encountered, the difference between the fourth and fifth iteration is so small as to be unnoticeable. This points to a theoretical issue whose resolution would be very valuable for validating this experimental observation, namely, that of deriving analytic estimates on the quasi-conformal dilatations of the approximating mappings f_n. Sharp enough estimates might explain the observed fast convergence and give an alternate, constructive proof of the convergence of the scheme. For convergence to the correct conformal mapping, the scheme requires a constant radius function for the seed, but of course an arbitrary nonconstant radius function that satisfies Inequality 2 provides inversive distance data for which a packing might exist. Good analytic bounds on dilatations might provide a method for choosing a variable radius function whose packing closely approximates $|K|$ without iteration.

It would be unwise to pretend that there are no practical implementation problems with the computational engine that computes a circle packing for

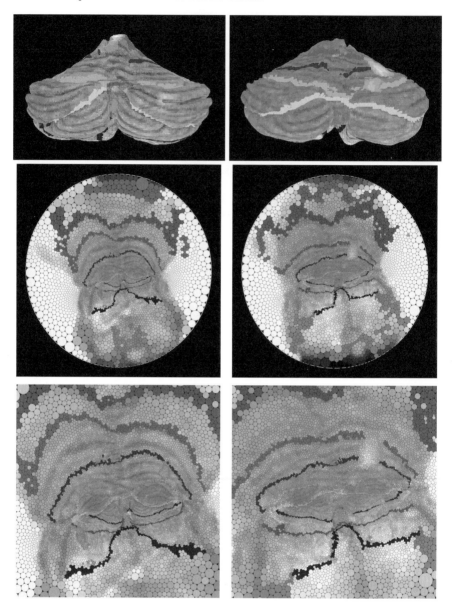

Fig. 14. Mapping two different cerebellum of the human brain.

given inversive distance data and with the resulting display. A particularly acute problem arises when there is a wide range of inversive distance values and the size of the complex K is large with, say, greater than 10^6 vertices. The resulting circle radii in the packing then have widely disparate values, which can lead to long computation times and numerical instabilities due to the high degree of numerical precision that is required. These large complexes, including those generated from a large number of hex refinements, can be difficult to visualize on a small computer screen. An advantage of conformal mapping for visualization of these large data sets comes into play here. Rather than map to a rectangle, we may map conformally to a disk, and then use Möbius transformations to bring various parts of the complex into focus near the disk center. This technique has proved extremely useful in developing a neuro-imaging tool.

Some improvements in the speed of the algorithm have been described by Collins and Stephenson in [5]. However there is room for considerable improvement for large data sets and for other investigations for optimizing the code. The algorithm for finding the circle packing for given inversive distance data is an iterative procedure, beginning with a specified vertex. The quasi-conformal results do not depend on the vertex chosen; however, it might be that nominating an alternate vertex would result in improved algorithm speed or faster convergence. Other computational experimental simulations may reveal additional insights into algorithm improvements. Since the algorithm is an iterative procedure, it seems to lend itself well to parallelization, which is an area of current research by Stephenson.

The computational issues described here become problems only for large complexes or data sets. For complexes such as the ones presented in this paper the algorithm is stable, robust and fast.

References

1. Beardon, A.F., (1983): The Geometry of Discrete Groups. Springer-Verlag, Berlin
2. Beardon, A.F., Stephenson, K. (1990): The uniformization theorem for circle packings. Indiana U. Math. J., **39**, 1383–1425
3. Bowers, P.L., Stephenson, K. (1997): A "regular" pentagonal tiling of the plane. Conformal Geom. Dynam., **1**, 58–86
4. Bowers, P.L., Stephenson, K. (to appear): Uniformizing dessins and Belyĭ maps via circle packings. Memoirs AMS
5. Collins, C., Stephenson, K. (to appear): A circle packing algorithm. Computational Geometry: Theory and Application
6. He, Z-X., (1999): Rigidity of infinite disk patterns. Ann. of Math., **149**, 1–33
7. Hurdal, M.K., Bowers, P.L., Stephenson, K., Sumners, D.L., Rehm, K., Schaper, K., Rottenberg, D.A. (1999): Quasi-conformally flat mapping the human cerebellum. In: Taylor, C., Colchester, A. (eds) Medical Image Computing and Computer-Assisted Intervention–MICCAI'99. Springer Berlin **1679**, 279–286

8. Hurdal, M.K., Kurtz, K.W., Banks, D.C. (2001): Case study: interacting with cortical flat maps of the human brain. In: Proceedings Visualization 2001, IEEE, Piscataway, NJ, 469-472, 591
9. Lehto, O., Virtanen, K.I., (1970): Quasiconformal Mappings in the Plane, 2nd ed. Springer-Verlag Berlin
10. Rodin, B., Sullivan, D. (1987): The convergence of circle packings to the Riemann mapping. J. Diff. Geo., **26**, 349–360
11. Schramm, O. (1991): Rigidity of infinite (circle) packings. J. AMS, **4**, 127–149
12. Thurston, W. (notes): The geometry and topology of 3-manifolds. Available from Princeton U. Math. Dept.

Discrete Differential-Geometry Operators for Triangulated 2-Manifolds

Mark Meyer[1], Mathieu Desbrun[1,2], Peter Schröder[1], and Alan H. Barr[1]

[1] Caltech, 1200 East California Boulevard, Pasadena, CA 91125, USA.
{*mmeyer,ps*} *@cs.caltech.edu, barr@gg.caltech.edu*
[2] USC, 3737 Watt Way, PHE 434, Los Angeles, CA 90089, USA.
desbrun@usc.edu

Summary. This paper proposes a unified and consistent set of flexible tools to approximate important geometric attributes, including normal vectors and curvatures on arbitrary triangle meshes. We present a consistent derivation of these first and second order differential properties using *averaging Voronoi cells* and the mixed Finite-Element/Finite-Volume method, and compare them to existing formulations. Building upon previous work in discrete geometry, these operators are closely related to the continuous case, guaranteeing an appropriate extension from the continuous to the discrete setting: they respect most intrinsic properties of the continuous differential operators. We show that these estimates are optimal in accuracy under mild smoothness conditions, and demonstrate their numerical quality. We also present applications of these operators, such as mesh smoothing, enhancement, and quality checking, and show results of denoising in higher dimensions, such as for tensor images.

1 Introduction

Despite extensive use of triangle meshes in Computer Graphics, there is no consensus on the most appropriate way to estimate simple geometric attributes such as normal vectors and curvatures on discrete surfaces. Many surface-oriented applications require an approximation of the first and second order properties with as much accuracy as possible. This could be done by polynomial reconstruction and analytical evaluation, but this often introduces overshooting or unexpected surface behavior between sample points. The triangle mesh is therefore often the only "reliable" approximation of the continuous surface at hand. Unfortunately, since meshes are piecewise linear surfaces, the notion of continuous normal vectors or curvatures is non trivial.

It is fundamental to guarantee accuracy in the treatment of discrete surfaces in many applications. For example, robust curvature estimates are important in the context of mesh simplification to guarantee optimal triangulations [13]. Even if the quadric error defined in [9] measures the Gaussian curvature on an infinitely subdivided mesh, the approximation becomes rapidly unreliable for sparse sampling. In surface modeling, a number of other techniques are designed to create very smooth surfaces from coarse meshes, and use discrete curvature approximations to measure the quality of the current approximation (for example, see [19]). Accurate curvature normals are also essential to the problem of surface denoising [5, 11] where good estimates of mean curvatures and normals are the key to undistorted smoothing. More generally, discrete operators satisfying appropriate discrete versions of continuous properties would guarantee reliable numerical behavior for many applications using meshes.

1.1 Previous Work

Several expressions for different surface properties have been designed. For instance, we often see the normal vector at a vertex defined as a (sometimes weighted) average of the normals of the adjacent faces of a mesh. Thürmer and Wüthrich [27] use the incident angle of each face at a vertex as the weights, since they claim the normal vector should only be defined very locally, independent of the shape or length of the adjacent faces. However, this normal remains consistent only if the faces are subdivided linearly, introducing vertices which are not on a smooth surface. Max [16] derived weights by assuming that the surface locally approximates a sphere. These weights are therefore exact if the object is a (even irregular) tessellation of a sphere. However, it is unclear how this approximation behaves on more complex meshes, since no error bounds are defined. Additionally, many meshes have local sampling adapted to local flatness, contradicting the main property of this approach. Even for a property as fundamental as the surface normal, we can see that several (often contradictory) formulæ exist.

Taubin proposed the most complete derivation of surface properties, leading to a discrete approximation of the curvature tensors for polyhedral surfaces [25]. Similarly, Hamann [12] proposed a simple way of determining the principle curvatures and their associated directions by a least-squares paraboloid fitting of the adjacent vertices, though the difficult task of selecting an appropriate tangent plane was left to the user. Our paper is closely related to these works since we also derive all first and second order properties for triangulated surfaces. However, many of the previous approaches do not preserve important differential geometry properties (invariants) on $\mathcal{C}^0$ surfaces such as polyhedral meshes.

In order to preserve fundamental invariants, we have followed a path initiated by Federer, Fu, Polthier, and Morvan to name a few [8, 22, 24, 18, 26]. This series of work proposed simple expressions for the total curvatures, as well as the Dirichlet energy for triangle meshes, and derived discrete methods to compute minimal surfaces or geodesics. We refer the reader to the overview compiled by Morvan [18]. Note also the tight connection with the "Mimetic Discretizations" used in computational physics by Shashkov, Hyman, and Steinberg [14, 15]. Although it shares a lot of similarities with all these approaches, our work offers a different, unified derivation that ensures accuracy

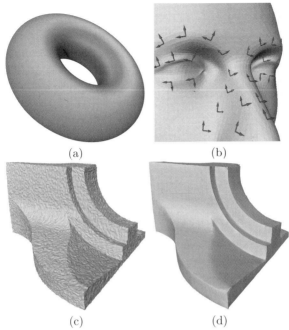

Fig. 1. *Some applications of our discrete operators: (a) mean curvature plot for a discrete surface, (b) principal curvature directions on a triangle mesh, (c-d) automatic feature-preserving denoising of a noisy mesh using anisotropic smoothing.*
and tight error bounds, leading to simple formulæ that are straightforward to implement.

Contributions

In this paper we define and derive the first and second order differential attributes (normal vector $\mathbf{n}$, mean curvature κ_H, Gaussian curvature κ_G, principal curvatures κ_1 and κ_2, and principal directions $\mathbf{e}_1$ and $\mathbf{e}_2$) for piecewise linear surfaces such as arbitrary triangle meshes. We present a unified framework for deriving such quantities resulting in a set of operators that is consistent, accurate, robust (in both regular and irregular sampling) and simple to compute.

The remainder of this paper is organized as follows. Details of why a *local spatial average* of these attributes over the immediate 1-ring neighborhood is a good choice to extend the continuous definition to the discrete setting is first given in Section 2. We then present a formal derivation of these quantities for triangle meshes using the mixed Finite-Element/Finite-Volume paradigm in Sections 3, 4 and 5. The relevance of our approach is demonstrated by showing the optimality of our operators under mild smoothness conditions. We demonstrate the accuracy and the use of these operators in different applications, including the smoothing and enhancement of meshes in Section 6. In Section 7, we generalize some of these operators to any 2-manifold or 3-manifold in an arbitrary dimension embedding space, offering tools for smoothing vector fields and volume data. Conclusions and perspectives are given in Section 8.

2 Defining Discrete Operators

In this section, we describe a general approach to define a number of useful differential quantities associated with a surface represented by a discrete triangle mesh. We begin with a review of several important quantities from differential geometry. This is followed by a technique for extending these quantities to the discrete domain using spatial averaging. Concluding this section is a general framework, used in the remaining sections, for deriving first and second order operators at the vertices of a mesh.

2.1 Notions from Differential Geometry

Let S be a surface (2-manifold) embedded in $\mathbb{R}^3$, described by an arbitrary parameterization of 2 variables. For each point on the surface S, we can locally approximate the surface by its tangent plane, orthogonal to the *normal vector* $\mathbf{n}$. Local bending of the surface is measured by *curvatures*. For every unit direction $\mathbf{e}_\theta$ in the tangent plane, the normal curvature $\kappa^N(\theta)$ is defined as the curvature of the curve that belongs to both the surface itself and the plane containing both $\mathbf{n}$ and $\mathbf{e}_\theta$. The two *principal curvatures* κ_1 and κ_2 of the surface S, with their associated orthogonal directions $\mathbf{e}_1$ and $\mathbf{e}_2$ are the extremum values of all the normal curvatures (see Figure 2(a)). The *mean curvature* κ_H is defined as the average of the normal curvatures:

$$\kappa_H = \frac{1}{2\pi} \int_0^{2\pi} \kappa^N(\theta) d\theta. \tag{1}$$

Expressing the normal curvature in terms of the principal curvatures, $\kappa^N(\theta) = \kappa_1 cos^2(\theta) + \kappa_2 sin^2(\theta)$, leads to the well-known definition: $\kappa_H = (\kappa_1 + \kappa_2)/2$. The *Gaussian curvature* κ_G is defined as the product of the two principle curvatures:

$$\kappa_G = \kappa_1 \kappa_2. \tag{2}$$

These latter two curvatures represent important local properties of a surface. Lagrange noticed that $\kappa_H = 0$ is the Euler-Lagrange equation for surface area minimization. This gave rise to a considerable body of literature on minimal surfaces and provides a direct relation between surface area minimization and mean curvature flow:

$$2\kappa_H \, \mathbf{n} = \lim_{diam(\mathcal{A}) \to 0} \frac{\nabla \mathcal{A}}{\mathcal{A}}$$

where $\mathcal{A}$ is a infinitesimal area around a point P on the surface, $diam(\mathcal{A})$ its diameter, and ∇ is the gradient with respect to the (x, y, z) coordinates of P. We will make extensive use of the mean curvature normal $\kappa_H \mathbf{n}$. Therefore, we will denote by $\mathbf{K}$ the operator that maps a point P on the surface to the vector $\mathbf{K}(P) = 2\kappa_H(P) \mathbf{n}(P)$. $\mathbf{K}$ is also known as the Laplace-Beltrami operator for the surface S. Note that in the remainder of this paper we will make no distinction between an operator and the value of this operator at a point as it will be clear from context. Gaussian curvature can also be expressed as a limit:

$$\kappa_G = \lim_{diam(\mathcal{A}) \to 0} \frac{\mathcal{A}^{\mathcal{G}}}{\mathcal{A}} \qquad (3)$$

where $\mathcal{A}^{\mathcal{G}}$ is the area of the image of the Gauss map (also called the spherical image) associated with the infinitesimal surface $\mathcal{A}$. The above definitions, as well as many more details, can be found in various sources on Differential Geometry [10, 4].

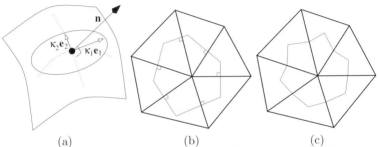

Fig. 2. *Local regions: (a) an infinitesimal neighborhood on a continuous surface patch; (b) a finite-volume region on a triangulated surface using Voronoi cells, or (c) Barycentric cells.*

2.2 Discrete Properties as Spatial Averages

Most of the smooth definitions described above need to be reformulated for $\mathcal{C}^0$ surfaces. We can consider a mesh as either the limit of a family of smooth surfaces, or as a linear (yet assumedly "good") approximation of an arbitrary surface. We define properties (geometric quantities) of the surface at each vertex as *spatial averages* around this vertex. If these averages are made *consistently*, and given a few assumptions such as non-degeneracy of the triangle mesh, a property at a given vertex will converge to the pointwise definition as the local sampling increases. Thus, by using these spatial averages, we extend the definition of curvature or normal vector from the continuous case to discrete meshes. Moreover, this definition is appropriate when, for example, geometric flows must be integrated over time on a mesh as a vertex will be updated according to the average behavior of the surface around it. Therefore, the piecewise linear result of the flow will be a correct approximation of the smoothed surface if the initial triangle mesh was a good approximation of the initial surface. Since we make no assumption on the smoothness of the surface, we will restrict the average to be within the immediately neighboring triangles, often referred as the 1-ring or star neighborhood. For example, we define the discrete Gaussian curvature, $\widehat{\kappa}_G$, at a vertex P as:

$$\widehat{\kappa}_G = \frac{1}{\mathcal{A}} \iint_{\mathcal{A}} \kappa_G \; dA$$

where $\mathcal{A}$ is a properly selected area around P. Note however that we will not distinguish between the (continuous) pointwise and the (discrete) spatially averaged notation, except when there may be ambiguity.

2.3 General Procedure Overview

The next three sections describe how we derive accurate numerical estimates of the first and second order operators at any vertex on an arbitrary mesh. We first restrict the averaging area to a family of special local surface patches denoted $\mathcal{A}_M$. These regions will be contained within the 1-ring neighborhood of each vertex, with piecewise linear boundaries crossing the mesh edges at their midpoints (Figures 2(b) and (c)). We show that this choice guarantees correspondences between the continuous and the discrete case. We then find the precise surface patch that optimizes the accuracy of our operators, completing the operator derivation. These steps will be explained in detail for the first operator, the mean curvature normal operator, $\mathbf{K}$, and a more direct derivation will be used for the Gaussian curvature operator κ_G, the two principal curvature operators κ_1 and κ_2, and the two principal direction operators $\mathbf{e}_1$ and $\mathbf{e}_2$. All these operators take a vertex $\mathbf{x}_i$ and its 1-ring neighborhood as input, and provide an estimate in the form of a simple formula that we will frame for clarity.

3 Discrete Mean Curvature Normal

We now provide a simple and accurate numerical approximation for both the normal vector, and the mean curvature for surface meshes in 3D.

3.1 Derivation of Local Integral using FE/FV

To derive a spatial average of geometric properties, we use a systematic approach which mixes finite elements and finite volumes. Since the triangle mesh is meant to visually represent the surface, we select a linear finite element on each triangle, that is, a linear interpolation between the three vertices corresponding to each triangle. Then, for each vertex, an associated surface patch (so-called finite volume in the Mechanics literature), over which the average will be computed, is chosen. Two main types of finite volumes are common in practice, see Figure 2(b-c). In each case, their piecewise linear boundaries connect the midpoints of the edges emanating from the center vertex and a point within each adjacent triangle. For the point inside each adjacent triangle, we can use either the barycenter or the circumcenter. The surface area formed from using the barycenters is denoted $\mathcal{A}_{\text{Barycenter}}$ while the surface area using the circumcenters is recognized as the local Voronoi cell and denoted $\mathcal{A}_{\text{Voronoi}}$. In the general case when this point could be anywhere, we will denote the surface area as $\mathcal{A}_{\text{M}}$.

We now wish to compute the integral of the mean curvature normal over the area $\mathcal{A}_{\text{M}}$. Since the mean curvature normal operator, also known as Laplace-Beltrami operator, is a generalization of the Laplacian from flat spaces to manifolds [4], we first compute the Laplacian of the surface with respect to the *conformal space* parameters u and v. As in [7] and [22], we use the current surface discretization as the conformal parameter space, that is, for each triangle of the mesh, the triangle itself defines the local surface metric. With such an induced metric, the Laplace-Beltrami operator simply turns into a Laplacian $\Delta_{u,v}\mathbf{x} = \mathbf{x}_{uu} + \mathbf{x}_{vv}$ [4]:

$$\iint_{\mathcal{A}_{\text{M}}} \mathbf{K}(\mathbf{x})\ dA = -\iint_{\mathcal{A}_{\text{M}}} \Delta_{u,v}\mathbf{x}\ du\ dv. \qquad (4)$$

Using Gauss's theorem as described in the Appendix, the integral of the Laplacian over a surface going through the midpoint of each 1-ring edge of a triangulated domain can be expressed as a function of the node values and the angles of the triangulation. The integral of the Laplace-Beltrami operator thus reduces to the following simple form:

$$\iint_{\mathcal{A}_{\text{M}}} \mathbf{K}(\mathbf{x})dA = \frac{1}{2}\sum_{j\in N_1(i)} (cot\ \alpha_{ij} + cot\ \beta_{ij})\ (\mathbf{x}_i - \mathbf{x}_j), \qquad (5)$$

where α_{ij} and β_{ij} are the two angles opposite to the edge in the two triangles sharing the edge $(\mathbf{x}_i, \mathbf{x}_j)$ as depicted in Figure 3(a), and $N_1(i)$ is the set of 1-ring neighbor vertices of vertex i.

Note that this equation was already obtained by minimizing the Dirichlet energy over a triangulation in [22]. Additionally, it is exactly the formula established in [5] for the gradient of surface area for the entire 1-ring neighborhood. This confirms, in the discrete setting, the area minimization nature of the mean curvature normal as derived by Lagrange. We can therefore express our previous result using the following general formula, valid for *any triangulation*:

$$\iint_{\mathcal{A}_{\text{M}}} \mathbf{K}(\mathbf{x})dA = \nabla\mathcal{A}_{\text{1-ring}}. \qquad (6)$$

where $\mathcal{A}_{\text{1-ring}}$ is the 1-ring area around a vertex P, and ∇ is the gradient with respect to the (x, y, z) coordinates of P.

Notice that the formula results in a zero value for any flat triangulation, regardless of the shape or size of the triangles of the locally-flat (zero curvature) mesh since the gradient of the area is zero for any locally flat region.

Although we have found an expression for the integral of the mean curvature normal independent of which of the two finite volume discretizations is used, one finite volume region must be chosen in order to provide an accurate estimate of the spatial average. We show in the next section that Voronoi cells provide provably tight error bounds under reasonable assumptions of smoothness.

3.2 Voronoi Regions for Tight Error Bounds

We now show that using Voronoi regions provides provably tight error bounds for the discrete operators by comparing the local spatial average of mean curvature with the actual pointwise value. Given a C^2 surface tiled by small patches $\mathcal{A}_i$ around n sample points $\mathbf{x}_i$, we can define the error E created by local averaging of the mean curvature normal compared to its pointwise value at $\mathbf{x}_i$ as:

$$
\begin{aligned}
E &= \sum_i \iint_{\mathcal{A}_i} \left\| \mathbf{K}(\mathbf{x}) - \mathbf{K}(\mathbf{x}_i) \right\|^2 dA \\
&\leq \sum_i \iint_{\mathcal{A}_i} C_i^2 \|\mathbf{x} - \mathbf{x}_i\|^2 \, dA \\
&\leq C_{max}^2 \sum_i \iint_{\mathcal{A}_i} \|\mathbf{x} - \mathbf{x}_i\|^2 \, dA,
\end{aligned}
$$

where C_i is the Lipschitz constant of the Beltrami operator over the smooth surface patch $\mathcal{A}_i$, and C_{max} the maximum of the Lipschitz constants. The Voronoi region of each sample point *by definition* minimizes $\|\mathbf{x} - \mathbf{x}_i\|$ since they contain the closest points to each sample, thus minimizing the bound on the error E due to spatial averaging [3]. Furthermore, if we add an extra assumption on the sampling rate with respect to the curvature such that the Lipschitz constants from patch to patch vary slowly with a ratio ϵ, we can actually guarantee that the Voronoi cell borders are less than $O(\epsilon)$ away from the optimal borders. As this still holds in the limit for a triangle mesh, we use the vertices of the mesh as sample points, and pick the Voronoi cells of the vertices as associated finite-volume regions. This will guarantee optimized numerical estimates and, as we will see, determining these Voronoi cells requires few extra computations.

3.3 Voronoi Region Area

Given a non-obtuse triangle P, Q, R with circumcenter O, as depicted in Figure 3(b), we must now compute the Voronoi region for P. Using the properties of perpendicular bisectors, we find : $a + b + c = \pi/2$, and therefore, $a =$

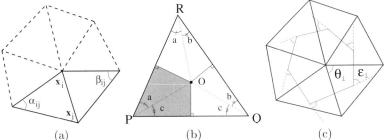

Fig. 3. *(a) 1-ring neighbors and angles opposite to an edge; (b) Voronoi region on a non-obtuse triangle; (c) External angles of a Voronoi region.*

$\pi/2 - \angle Q$ and $c = \pi/2 - \angle R$. The Voronoi area for point P lies within this triangle if the triangle is non-obtuse, and is thus: $\frac{1}{8}(|PR|^2 cot\angle Q + |PQ|^2 cot\angle R)$. Summing these areas for the whole 1-ring neighborhood, we can write the non-obtuse Voronoi area for a vertex $\mathbf{x}_i$ as a function of the neighbors $\mathbf{x}_j$:

$$\mathcal{A}_{\text{Voronoi}} = \frac{1}{8} \sum_{j \in N_1(i)} (cot\ \alpha_{ij} + cot\ \beta_{ij}) \|\mathbf{x}_i - \mathbf{x}_j\|^2. \tag{7}$$

Since the cotangent terms were already computed for Eq. (5), the Voronoi area can be computed very efficiently. However, if there is an obtuse triangle among the 1-ring neighbors or among the triangles edge-adjacent to the 1-ring triangles, the Voronoi region either extends beyond the 1-ring, or is truncated compared to our area computation. In either case our derived formula no longer stands.

3.4 Extension to Arbitrary Meshes

The previous expression for the Voronoi finite-volume area does not hold in the presence of obtuse angles. However, the integral of the Laplace-Beltrami operator given in equation (6) holds even for obtuse 1-ring neighborhoods - the only assumption used is that the finite-volume region goes through the midpoint of the edges. It is thus *still valid even in obtuse triangulations.* Therefore, we could simply divide the integral evaluation by the barycenter finite-volume area in lieu of the Voronoi area for vertices near obtuse angles to determine the spatial average value. We use a slightly more subtle area, to guarantee a perfect tiling of our surface, and therefore, optimized accuracy as each point on the surface is counted once and only once. We define a new surface area for each vertex $\mathbf{x}$, denoted $\mathcal{A}_{\text{Mixed}}$: for each non-obtuse triangle, we use the circumcenter point, and for each obtuse triangle, we use the midpoint of the edge opposite to the obtuse angle. Algorithmically, this area around a point $\mathbf{x}$ can be easily computed as detailed in Figure 4. Note that the derivation for the integral of the mean curvature normal is still valid for this mixed area since the boundaries of the area remain inside the 1-ring neighborhood and go through the midpoint of each edge. Moreover, these mixed areas tile the surface without overlapping. This new cell definition is equivalent to a local adjustment of the diagonal mass matrix in a finite element framework in order to ensure a correct evaluation. The error bounds are not as tight when local angles are more than $\pi/2$, and therefore, numerical experiments are expected to be worse in areas with obtuse triangles.

$\mathcal{A}_{\mathrm{Mixed}} = 0$
For each triangle T from the 1-ring neighborhood of $\mathbf{x}$
   If T is non-obtuse,   `// Voronoi safe`
      `// Add Voronoi formula (see Section 3.3)`
      $\mathcal{A}_{\mathrm{Mixed}}+ =$ Voronoi region of $\mathbf{x}$ in T
   Else   `// Voronoi inappropriate`
      `// Add either area(T)/4 or area(T)/2`
      If the angle of T at $\mathbf{x}$ is obtuse
         $\mathcal{A}_{\mathrm{Mixed}}+ =$ area$(T)/2$
      Else
         $\mathcal{A}_{\mathrm{Mixed}}+ =$ area$(T)/4$

Fig. 4. *Pseudo-code for region $\mathcal{A}_{Mixed}$ on an arbitrary mesh*

3.5 Discrete Mean Curvature Normal Operator

Now that the mixed area is defined, we can express the mean curvature normal operator $\mathbf{K}$ defined in Section 2.1 using the following expression:

$$\mathbf{K}(\mathbf{x}_i) = \frac{1}{2\mathcal{A}_{\mathrm{Mixed}}} \sum_{j \in N_1(i)} (cot\, \alpha_{ij} + cot\, \beta_{ij})\ (\mathbf{x}_i - \mathbf{x}_j) \tag{8}$$

From this expression, we can easily compute the mean curvature value κ_H by taking half of the magnitude of this last expression. As for the normal vector, we can just normalize the resulting vector $\mathbf{K}(\mathbf{x}_i)$. In the special (rare) case of zero mean curvature (flat plane or local saddle point), we simply average the 1-ring face normal vectors to evaluate $\mathbf{n}$ appropriately.

It is interesting to notice that using the barycentric area as an averaging region results in an operator very similar to the definition of the mean curvature normal by Desbrun *et al.* [5], since $\mathcal{A}_{\mathrm{Barycenter}}$ is a third of the whole 1-ring area $\mathcal{A}_{\text{1-ring}}$ used in their derivation - however, our new derivation uses *non-overlapping* regions and is therefore more accurate. At this time, we are not aware of a proof of convergence for this operator. However, our tests have shown no divergence as we refine a mesh, as long as we do not degrade the mesh quality (the triangles must not degenerate). We will give more precise numerical results in Section 6.1 showing the improved quality of our new estimate.

4 Discrete Gaussian Curvature

In this section, the Gaussian curvature κ_G for bivariate (2D) meshes embedded in 3D is studied. We will demonstrate that a derivation similar to the above is easily obtained.

4.1 Expression of the Local Integral of κ_G

Similar to what was done for the mean curvature normal operator, we first need to find an exact value of the integral of the Gaussian curvature κ_G over a finite-volume region on a piecewise linear surface. From Eq. (3), we could compute the integral over an area A_M as the associated spherical image area (also called the image of the Gauss map). Instead, we use the *Gauss-Bonnet theorem* [4, 10, 1] which proposes a very simple equality, valid over any surface patch. Applied to our local finite-volume regions, the Gauss-Bonnet theorem simply states:

$$\iint_{A_M} \kappa_G \, dA = 2\pi - \sum_j \epsilon_j$$

where the ϵ_j are the external angles of the boundary, as indicated in Figure 3(c). Note that this simplified form results from the fact that the integral of geodesic curvature on the piece-wise linear boundaries is zero. If we apply this expression to a Voronoi region, the external angles are zero across each edge (since the boundary stays perpendicular to the edge), and the external angle at a circumcenter is simply equal to θ_j, the angle of the triangle at the vertex P. Therefore, the integral of the Gaussian curvature (also called total curvature) for non-obtuse triangulations is: $2\pi - \sum_j \theta_j$. This result is still valid for the mixed region and is proven using a similar geometric argument. This result was already proven by Polthier and Schmies [24], who considered the area of the image of the Gauss map for a vertex on a polyhedral surface. Therefore, analogous to Eq. (6), we can now write for the 1-ring neighborhood of a vertex $\mathbf{x}_i$:

$$\iint_{A_M} \kappa_G dA = 2\pi - \sum_{j=1}^{\#f} \theta_j$$

where θ_j is the angle of the j-th face at the vertex $\mathbf{x}_i$, and $\#f$ denotes the number of faces around this vertex. Note again that this formula holds for any surface patch A_M within the 1-ring neighborhood whose boundary crosses the edges at their midpoint.

4.2 Discrete Gaussian Curvature Operator

To estimate the local spatial average of the Gaussian curvature, we use the same arguments as in 3.2 to claim that the Voronoi cell of each vertex is an

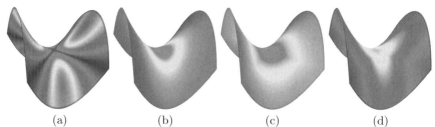

(a)   (b)   (c)   (d)

Fig. 5. *Curvature plots of a triangulated saddle using pseudo-colors: (a) Mean, (b) Gaussian, (c) Minimum, (d) Maximum.*

appropriate local region to use for good error bounds. In practice, we use the mixed area $\mathcal{A}_{\text{Mixed}}$ to account for obtuse triangulations. Since the mixed area cells tile the whole surface without any overlap, we will satisfy the (continuous) Gauss-Bonnet theorem: the integral of the discrete Gaussian curvature over an entire sphere for example will be equal to 4π *whatever the discretization used* since the sphere is a closed object of genus zero. This result ensures a robust numerical behavior of our discrete operator. Our *Gaussian curvature discrete operator* can thus be expressed as:

$$\kappa_G(\mathbf{x}_i) = (2\pi - \sum_{j=1}^{\#f} \theta_j)/\mathcal{A}_{\text{Mixed}} \tag{9}$$

Notice that this operator will return zero for any flat surface, as well as any roof-shaped 1-ring neighborhood, guaranteeing a satisfactory behavior for trivial cases. Note also that *convergence conditions* (using fatness or straightness) exist for this operator [8, 26], proving that if the triangle mesh does not degenerate, the approximation quality gets better as the mesh is refined. We postpone numerical tests until Section 6.1.

5 Discrete Principal Curvatures

We now wish to robustly determine the two principal curvatures, along with their associated directions. Since the previous derivations give estimates of both Gaussian and mean curvature, the only additional information that must be sought are the principal directions since the principal curvatures are, as we are about to see, easy to determine.

5.1 Principal Curvatures

We have seen in Section 2.1 that the mean and Gaussian curvatures are easy to express in terms of the two principal curvatures κ_1 and κ_2. Therefore, since both κ_H and κ_G have been derived for triangulated surfaces, we can define the *discrete principal curvatures* as:

$$\kappa_1(\mathbf{x}_i) = \kappa_H(\mathbf{x}_i) + \sqrt{\Delta(\mathbf{x}_i)} \qquad (10)$$

$$\kappa_2(\mathbf{x}_i) = \kappa_H(\mathbf{x}_i) - \sqrt{\Delta(\mathbf{x}_i)} \qquad (11)$$

with: $\Delta(\mathbf{x}_i) = \kappa_H^2(\mathbf{x}_i) - \kappa_G(\mathbf{x}_i)$ and $\kappa_H(\mathbf{x}_i) = \frac{1}{2}\|\mathbf{K}(\mathbf{x}_i)\|$.

Unlike the continuous case where Δ is always positive, we must make sure that κ_H^2 is always larger than κ_G to avoid any numerical problems, and threshold Δ to zero if it is not the case (an extremely rare occurrence).

5.2 Mean Curvature as a Quadrature

In order to determine the principal axes at a vertex, we will first show that the mean curvature from our previous expression can be interpreted as a quadrature of normal curvature samples:

$$\kappa_H(\mathbf{x}_i) = \frac{1}{2}\left(2\kappa_H(\mathbf{x}_i)\mathbf{n}\right) \cdot \mathbf{n} = \frac{1}{2}\mathbf{K}(\mathbf{x}_i) \cdot \mathbf{n}$$

$$= \frac{1}{4\mathcal{A}_{\text{Mixed}}} \sum_{j \in N_1(i)} (cot\ \alpha_{ij} + cot\ \beta_{ij})\ (\mathbf{x}_i - \mathbf{x}_j) \cdot \mathbf{n}$$

$$= \frac{1}{4\mathcal{A}_{\text{Mixed}}} \sum_{j \in N_1(i)} (cot\ \alpha_{ij} + cot\ \beta_{ij}) \frac{\|\mathbf{x}_i - \mathbf{x}_j\|^2}{\|\mathbf{x}_i - \mathbf{x}_j\|^2}(\mathbf{x}_i - \mathbf{x}_j) \cdot \mathbf{n}$$

$$= \frac{1}{\mathcal{A}_{\text{Mixed}}} \sum_{j \in N_1(i)} \left[\frac{1}{8}(cot\ \alpha_{ij} + cot\ \beta_{ij})\ \|\mathbf{x}_i - \mathbf{x}_j\|^2\right] \kappa_{i,j}^N, \qquad (12)$$

where we define:

$$\kappa_{i,j}^N = 2\frac{(\mathbf{x}_i - \mathbf{x}_j) \cdot \mathbf{n}}{\|\mathbf{x}_i - \mathbf{x}_j\|^2}.$$

This $\kappa_{i,j}^N$ can be shown to be an estimate of the normal curvature in the direction of the edge $\mathbf{x}_i\mathbf{x}_j$. The radius R of the osculating circle going through the vertices $\mathbf{x}_i$ and $\mathbf{x}_j$ is easily found using the mean curvature normal estimate as illustrated in Figure 9(a). Since we must have a right angle at the neighbor vertex $\mathbf{x}_j$, we have $(\mathbf{x}_i - \mathbf{x}_j) \cdot (\mathbf{x}_i - \mathbf{x}_j - 2R\ \mathbf{n}) = 0$. This implies:

$$R = \|\mathbf{x}_i - \mathbf{x}_j\|^2/(2\ (\mathbf{x}_i - \mathbf{x}_j) \cdot \mathbf{n})$$

This proves that $\kappa_{i,j}^N$ is a normal curvature estimate in the direction of the edge $\mathbf{x}_i\mathbf{x}_j$ (as it is the inverse of the radius of the osculating circle). This expression was also used in the context of curvature approximation in [19] and [25].

Therefore, Eq. (12) can be interpreted as a quadrature of the integral from Eq. (1), with weights w_{ij}:

$$\kappa_H(\mathbf{x}_i) = \sum_{j \in N_1(i)} w_{ij} \; \kappa_{i,j}^N,$$

where the $w_{ij} = \frac{1}{\mathcal{A}_{\text{Mixed}}} \left[\frac{1}{8} (cot \; \alpha_j + cot \; \beta_j) \|\mathbf{x}_i - \mathbf{x}_j\|^2 \right]$ sum to one for each i on a non-obtuse triangulation.

5.3 Least-Square Fitting for Principal Directions

In order to find the two orthogonal principal curvature directions we can simply compute the eigenvectors of the curvature tensor. Since the mean curvature obtained from our derivation can be seen as a quadrature using each edge as a sample direction, we use these samples to find the best fitting ellipse, in order to fully determine the curvature tensor. In practice, we select the symmetric curvature tensor B as being defined by three unknowns a, b, c:

$$B = \begin{pmatrix} a & b \\ b & c \end{pmatrix}.$$

This tensor will provide the normal curvature in any direction in the tangent plane. Therefore, when we use the direction of the edges of the 1-ring neighborhood, we should find:

$$\mathbf{d}_{i,j}^T \; B \; \mathbf{d}_{i,j} \; = \; \kappa_{i,j}^N,$$

where $\mathbf{d}_{i,j}$ is the unit direction *in the tangent plane* of the edge $\mathbf{x}_i\mathbf{x}_j$. Since we know the normal vector $\mathbf{n}$ to the tangent plane, this direction is calculated using a simple projection onto the tangent plane:

$$\mathbf{d}_{i,j} = \frac{(\mathbf{x}_j - \mathbf{x}_i) - [(\mathbf{x}_j - \mathbf{x}_i) \cdot \mathbf{n}] \; \mathbf{n}}{\|(\mathbf{x}_j - \mathbf{x}_i) - [(\mathbf{x}_j - \mathbf{x}_i) \cdot \mathbf{n}] \; \mathbf{n}\|}$$

A conventional least-square approximation can be obtained by minimizing the error E:

$$E(a,b,c) = \sum_j w_j \left(\mathbf{d}_{i,j}^T \; B \; \mathbf{d}_{i,j} - \kappa_{i,j}^N \right)^2$$

Adding the two constraints $a + b = 2\kappa_H$ and $ac - b^2 = \kappa_G$, to ensure coherent results, turns the minimization problem into a third degree polynomial root-finding problem. Once the three coefficients of the matrix B are found, we find the two principal axes $\mathbf{e}_1$ and $\mathbf{e}_2$ as the two (orthogonal) eigenvectors of B. In practice, all our experiments have demonstrated that the non-linear constraint on the determinant is not necessary (reducing the problem to a linear system). An example of these principal directions is shown in Figure 1(b) (Color Plate 23(b) on page 439).

Although we could actually determine the principal curvatures (and thus the mean and gaussian curvatures) using an unconstrained least squares procedure, we use our operators to compute the curvatures and only use the least squares for the principal directions as the curvature values computed from the least squares are often *less* accurate in practice while the directions are fairly robust. A plausible interpretation for the bad numerical properties of a pure least squares approach is the hypothesis of elliptic curvature variation: although this is perfectly valid for smooth surfaces, this is somewhat arbitrary for coarse, triangulated surfaces. It seems therefore more natural to use our previous operators that rely on differential properties still valid on discrete meshes.

6 Results and Applications

With robust curvature estimates at our disposal, we demonstrate some useful applications such as quality checking for surface design and tools for smoothing and enhancement of meshes. We first demonstrate the numerical quality of our operators.

6.1 Numerical Quality of our Operators

We performed a number of tests to demonstrate the accuracy of our approach in practice. First, we compared our operators to the well-known second-order accurate Finite Difference operators on several discrete meshes approximating simple surfaces such as spheres, or hyperboloids, where the curvatures are known analytically. In order to do so, we used special surfaces defined as height fields over a flat, regular grid so that the FD operators can be computed and tested against our results. The table below lists some representative results:

%error	FD κ_H	[DMSB99] κ_H	our κ_H	FD κ_G	our κ_G
Sphere patch	0.20	0.17	0.16	0.4	1.2
Paraboloid	0.0055	0.0038	0.0038	0.01	0.02
Torus (irregular)	-	0.047	0.036	-	0.05

Table 1. *Comparison of our operators with Finite Differences. The error is measured in mean percent error compared to the exact, known curvature values. Dashes "-" indicate that the FD tests cannot be performed since the triangulation is irregular. The angles θ_j needed for the Gaussian curvature were computed using the C function* atan2, *instead of* acos *or* asin *since* acos *and* asin *would significantly deteriorate the precision of the results.*

Overall, the numerical quality of our operators is equivalent to FD operators for regular sampling. A major advantage of our new operators over FD operators is that these differential-geometry based operators can *still* be used on irregular sampling, with the same order of accuracy.

We also tested our operators against one of the most widely used curvature estimation techniques [25]. We tested several simple surfaces (spheres, parametric surfaces, etc.) to determine the effect of sampling on the operators. The surfaces were created with 258 points, quadrisected and reprojected to create surfaces of 1026, 4098 and 16386 points. In all cases, the average percent error of our operators did not exceed 0.07% for mean curvature and 1.3% for gaussian curvature. The previous method had average errors of up to 1.8% for mean curvature and exceeding 10% in some instances for gaussian curvature.

Finally, we tested the effects of irregularity on the operators. In irregular areas of the surfaces (such as the area joining two regions of different sampling rates), our operators performed with the same order of accuracy as in the fairly regular regions (less than 0.2% average error for mean curvature and below 1.8% average error for gaussian curvature in regions of mild irregularity). The accuracy of our operators decreases as the irregularity (angle and edge length dispersion) increases, but, in practice, the rate at which the error increases is low.

6.2 Geometric Quality of Meshes

Producing high quality meshes is not an easy task. Checking if a given mesh is appropriately smooth requires a long inspection with directional or point light sources to detect any visually unpleasant discontinuities on the surface. Curvature plots (see Figure 5 and Color Plate 24 on page 439), using false color to texture the mesh according to the different curvatures, can immediately show problems or potential problems since they will reveal the variation of curvatures in an obvious way. Figure 6 demonstrates that even if a surface (obtained by a subdivision scheme) looks very smooth, a look at the mean curvature map reveals flaws such as discontinuities in the variation of curvature across the surface. Conversely, curvature plots can reveal unsuspected details on existing scanned meshes, like the veins on the horse.

6.3 Denoising and Enhancement of a Mesh

If the quality of a mesh is not sufficient (due to noise resulting from inaccurate scans for instance), denoising and enhancements can be performed using our discrete operators.

Isotropic Shape Smoothing Just like Laplacian filtering in image processing, a mean curvature flow will disperse the noise of a smooth mesh appropriately by minimizing the surface area as reported in [5]. We implemented this implicit fairing technique using our new operators with success. However, since our mesh can represent a surface with sharp edges, we sometimes experienced a dilemma: how can one get rid of the noise by smoothing the surface, while preserving sharp edges to keep the underlying geometry intact? We would like to smooth a noisy cube, for example, without turning it into a cushion-like shape (Figure 7(a) and (b)). A possible solution is to manually spray-paint the desired value of smoothing on the vertices [5], making the preservation of sharp edges possible while suppressing noise. But it is

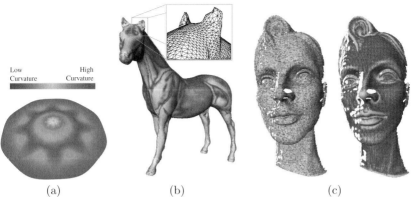

Fig. 6. *Mean curvature plots revealing surface details for: (a) a Loop surface from an 8-neighbor ring, (b) a horse mesh, (c) a noisy mesh obtained from a 3D scanner and the same mesh after smoothing. Our operator performs well on irregular sampling such as on the ear of the horse. Notice also how the operator correctly computes quickly varying curvatures on the noisy head while returning slowly varying curvatures on the smoothed version.*

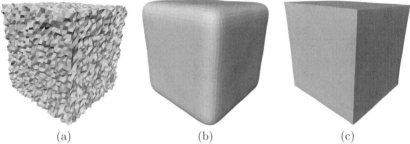

Fig. 7. *Cube: (a) Original, noisy mesh ($\pm 3\%$ uniform noise added along the normal direction). (b) Isotropic smoothing. (c) Anisotropic smoothing defined in Section 6.3.*

a rather time-consuming task for big meshes, and it will leave ragged edges on the vertices forced to a low smoothing amount.

Enhancement of Meshes We would like to automate the previous process, providing a way to smooth meshes while keeping clear features (like sharp edges) intact. This relates closely to the specific problem of image denoising, where clear features like object boundaries should be kept, while noisy, yet homogeneous regions should be smoothed. Different forms of anisotropic diffusion have shown good results for this problem in image processing [21], in flow visualization [23], and more recently on meshes too [2]. The underlying idea is to still diffuse the noise, but with an adaptive conductance over the image in order to preserve edges. We experimented with a simple technique to achieve similar results on meshes. Additionally, an enhancement procedure to help straighten edges has been designed.

An isotropic implicit curvature flow on regions uniformly noisy is desired, while special treatment must be applied for obvious edges and corners to prevent them from being smoothed away. In our previous work [6], we proposed a weighted mean curvature smoothing, where the weights are computed using the first fundamental form to preserve height field discontinuities. However, even if such an approach is appropriate for height fields, it does not capture enough information to perform adequately on a general mesh. The second fundamental form, i.e., local curvature, provides more information on the local variations of the surface, and therefore, will be more accurate for the weighting.

An Anisotropic Smoothing Technique Most of the meshes acquired from real object scans contain corners and ridges, which will be lost if isotropic denoising is used. Therefore, if these sharp edges are necessary features of a noisy mesh, the noise should be only directionally diffused in order to keep the characteristics intact. Presence of such features can be determined using the second-order properties of the surface. Indeed, in the case of an edge between two faces of a cube mesh, the minimum curvature is zero along the edge, while the maximum curvature is perpendicular to this edge. An immediate idea is to perform a weighted mean curvature flow that penalizes vertices that have a large ratio between their two principal curvatures. This way, clear features like sharp edges will remain present while noise, more symmetric by nature, will be greatly reduced.

We define the smoothing weight at a vertex $\mathbf{x}_i$ as being:

$$w_i = \begin{cases} 1 & \text{if } |\kappa_1| \leq T \text{ and } |\kappa_2| \leq T \\ 0 & \text{if } |\kappa_1| > T \text{ and } |\kappa_2| > T \text{ and } \kappa_1\kappa_2 > 0 \\ \kappa_1/\kappa_H & \text{if } |\kappa_1| = \min(|\kappa_1|, |\kappa_2|, |\kappa_H|) \\ \kappa_2/\kappa_H & \text{if } |\kappa_2| = \min(|\kappa_1|, |\kappa_2|, |\kappa_H|) \\ 1 & \text{if } |\kappa_H| = \min(|\kappa_1|, |\kappa_2|, |\kappa_H|) \end{cases}.$$

The parameter T is a user defined value determining edges. The general smoothing flow is then: $\partial \mathbf{x}_i/\partial t = -w_i \, \kappa_H(\mathbf{x}_i) \, \mathbf{n}(\mathbf{x}_i)$. As we can see, uniformly noisy regions (cases 1 and 5 in the weight definition given above) will be smoothed isotropically, while corners (case 2) will not move. For edges (cases 3 and 4), we smooth with a speed proportional to the minimum curvature, to be assured not to smooth ridges. The caveat is that this smoothing is no longer well-posed: we try to enhance edges, and this is by definition a very unstable process. Pre-mollification techniques have been reported successful in [23], and should be used in such a process. However, we have had good results by just thresholding the weights w_i to be no less than -0.1 to avoid strong inverse diffusion, and using implicit fairing to integrate the flow. As Figure 7 demonstrates, a noisy cube can be smoothed and enhanced into an almost perfect cube using our technique. For more complicated objects (see Figure 1(c-d) and Color Plate 23(c-d) on page 439), a pass of curve smoothing (also using implicit curvature flow) has been added to help straighten the edges.

7 Discrete Operators in nD

Up to this point, we defined and used our geometric operators for bivariate (2D) surfaces embedded in 3D. However, since they were derived in a dimensionless manner, our operators generalize easily to any embedding space dimensionality. The extension of our geometric operators to higher dimensional embedding spaces allows us to use the same smoothing technology used on meshes for vector fields or tensor images. To demonstrate the practical accuracy of our operator, we performed different smoothings on higher dimensional spaces. For instance, Figure 8 demonstrates how our operators can smooth a vector field, with or without preservation of features. As shown, anisotropic smoothing can indeed preserve significant discontinuities such as the boundary between the straight flow and the vortex, just as we preserved edges during mesh smoothing in 3D.

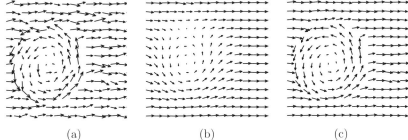

(a)   (b)   (c)

Fig. 8. *Vector field denoising: (a) Original, noisy vector field; (b) Smoothed using Beltrami flow; (c) Smoothed using anisotropic weighted flow to automatically preserve the vortex region.*

We can also extend the previous mean curvature normal operator, valid on triangulated surfaces, to tetrahedralized volumes which are 3-parameter volumes in an embedding space of arbitrary dimension (see [17] for a detailed derivation). This can be used, for example, on any MRI volume data (intensity, vector field or even tensor fields). For these 3-manifolds, we can compute the gradient of the 1-ring volume this time to extend the Beltrami operator. Once again, the cotangent formula turns out to be still valid, but this time for the dihedral angles of the tetrahedrons. This Beltrami operator can still be used to denoise volume data as it minimizes volume just as we denoised meshes through a surface area minimization.

8 Conclusion

A complete set of accurate differential operators for any triangulated surface has been presented. We consistently derived estimates for normal vectors and mean curvatures (Eq. (8)), Gaussian curvatures (Eq. (9)), principal curvatures (Eq. (10) and (11)), and principal directions (Section 5.3), and numerically showed their quality. Extended versions of our operator for surfaces and volumes in higher dimension embedding spaces have also been provided. Our operators perform as well as established methods such as Finite Differences in the regular setting and degrade gracefully as irregularity is increased. Moreover, we described how to use these simple, local operators to denoise arbitrary meshes or vector fields, including preservation and/or enhancement of features. These methods form a family of robust tools to help with processing noisy data, or simply to build a scale space out of a dataset to offer an adaptive description of the data. With little user interaction to select (and direct) the appropriate tools, noisy scanned meshes can be turned into high-quality meshes, vector fields can be smoothed to later segment the general flow, or MRI multi-valued images can be denoised. However, smoothing techniques do not deal well with large amounts of noise. Multiplicative noise, for example, can create large dents in a dataset, that only statistical techniques using local averages of n neighbors can try to deal with [20], often without guarantee of success. Yet we believe that, as in image processing, our framework can give rise to other anisotropic diffusion equations particularly designed for specific noise models.

We have confidence in the adequacy and efficiency of our simple discrete operators in other surface-based applications. The mean curvature normal operator for instance can be easily applied to function values on the surface, and it will define a Laplacian operator for the "natural" metric of the mesh. We are currently exploring other applications of these operators such as reparameterization, remeshing, geometry based subdivision schemes, and mesh simplification along the lines of [13].

Future work will try to answer some of the open questions. For instance, we are trying to determine what would be the minimum sampling rate of a continuous surface to guarantee that our discrete estimates are accurate within a given ϵ - laying the foundations for an irregular sampling theory. More generally, we would like to extend well known digital signal processing tools and theorems to digital geometry.

Acknowledgements This work was supported in part by the STC for Computer Graphics and Scientific Visualization (ASC-89-20219), IMSC - an NSF Engineering Research Center (EEC-9529152), an NSF CAREER award (CCR-0133983), NSF (DMS-9874082, ACI-9721349, DMS-9872890, and ACI-9982273), the DOE (W-7405-ENG-48/B341492), Intel, Alias—Wavefront, Pixar, Microsoft, and the Packard Foundation.

References

1. A. D. Aleksandrov and V. A. Zalgaller. *Intrinsic Geometry of Surfaces*. AMS, Rhode Island, USA, 1967.
2. U. Clarenz, U. Diewald, and M. Rumpf. Anisotropic Geometric Diffusion in Surface Processing. In *IEEE Visualization*, pages 397–405, 2000.
3. Qiang Du, Vance Faber, and Max Gunzburger. Centroidal Voronoi Tesselations: Applications and Algorithms. *SIAM Review*, 41(4):637–676, 1999.
4. Ulrich Dierkes, Stefan Hildebrandt, Albrecht Küster, and Ortwin Wohlrab. *Minimal Surfaces (I)*. Springer-Verlag, 1992.
5. Mathieu Desbrun, Mark Meyer, Peter Schröder, and Alan H. Barr. Implicit Fairing of Irregular Meshes using Diffusion and Curvature Flow. In *SIGGRAPH 99 Conference Proceedings*, pages 317–324, 1999.
6. Mathieu Desbrun, Mark Meyer, Peter Schröder, and Alan H. Barr. Anisotropic Feature-Preserving Denoising of Height Fields and Images. In *Graphics Interface'2000 Conference Proceedings*, pages 145–152, 2000.
7. G. Dziuk. An Algorithm for Evolutionary Surfaces . *Numer. Math.*, 58, 1991.
8. J. Fu. Convergence of Curvatures in Secant Approximations. *Journal of Differential Geometry*, 37:177–190, 1993.
9. Michael Garland and Paul S. Heckbert. Surface Simplification Using Quadric Error Metrics. In *SIGGRAPH 97 Conference Proceedings*, pages 209–216, August 1997.
10. Alfred Gray. *Modern Differential Geometry of Curves and Surfaces with Mathematica*. CRC Press, 1998.
11. Igor Guskov, Wim Sweldens, and Peter Schröder. Multiresolution Signal Processing for Meshes. In *SIGGRAPH 99 Conference Proceedings*, pages 325–334, 1999.
12. Bernd Hamann. Curvature Approximation for Triangulated Surfaces. In G. Farin *et al.*, editor, *Geometric Modelling*, pages 139–153. Springer Verlag, 1993.
13. Paul S. Heckbert and Michael Garland. Optimal Triangulation and Quadric-Based Surface Simplification. *Journal of Computational Geometry: Theory and Applications*, November 1999.
14. J. M. Hyman and M. Shashkov. Natural Discretizations for the Divergence, Gradient and Curl on Logically Rectangular Grids. *Applied Numerical Mathematics*, 25:413–442, 1997.
15. J. M. Hyman, M. Shashkov, and S. Steinberg. The numerical solution of diffusion problems in strongly heterogenous non-isotropic materials. *Journal of Computational Physics*, 132:130–148, 1997.
16. Nelson Max. Weights for Computing Vertex Normals from Facet Normals. *Journal of Graphics Tools*, 4(2):1–6, 1999.
17. Mark Meyer. *Differential Operators for Computer Graphics*. Ph.D. Thesis, Caltech, December 2002.
18. J.M. Morvan. On Generalized Curvatures. *Preprint*, 2001.
19. Henry P. Moreton and Carlo H. Séquin. Functional Minimization for Fair Surface Design. In *SIGGRAPH 92 Conference Proceedings*, pages 167–176, July 1992.
20. R. Malladi and J.A. Sethian. Image Processing: Flows under Min/Max Curvature and Mean Curvature. *Graphical Models and Image Processing*, 58(2):127–141, March 1996.

21. P. Perona and J. Malik. Scale-space and Edge Detection Using Anisotropic Diffusion. *IEEE Transactions on Pattern Analysis and Machine Intelligence*, 12(7):629–639, July 1990.
22. Ulrich Pinkall and Konrad Polthier. Computing Discrete Minimal Surfaces and Their Conjugates. *Experimental Mathematics*, 2(1):15–36, 1993.
23. T. Preußer and M. Rumpf. Anisotropic Nonlinear Diffusion in Flow Visualization. In *IEEE Visualization*, pages 323–332, 1999.
24. Konrad Polthier and Markus Schmies. Straightest Geodesics on Polyhedral Surfaces. In H.C. Hege and K. Polthier, editors, *Mathematical Visualization*. Springer Verlag, 1998.
25. Gabriel Taubin. Estimating the Tensor of Curvature of a Surface from a Polyhedral Approximation. In *Proc. 5th Intl. Conf. on Computer Vision (ICCV'95)*, pages 902–907, June 1995.
26. B. Thibert and J.M. Morvan. Approximations of A Smooth Surface with a Triangulated Mesh. *Preprint*, 2002.
27. Grit Thürmer and Charles Wüthrich. Computing Vertex Normals from Polygonal Facets. *Journal of Graphics Tools*, 3(1):43–46, 1998.

Appendix

Mean Curvature Normal on a Triangulated Domain

In this section, we derive the integral of the mean curvature normal over a triangulated domain. We begin by computing the integral of the Laplacian of the surface point $\mathbf{x}$ with respect to the conformal parameter space. Using Gauss's theorem, we can turn the integral of a Laplacian over a region into a line integral over the boundary of the region:

$$\iint_{\mathcal{A}_{\mathrm{M}}} \Delta_{u,v}\mathbf{x} \, du \, dv = \int_{\partial \mathcal{A}_{\mathrm{M}}} \nabla_{u,v}\mathbf{x} \cdot \mathbf{n}_{u,v} \, dl, \tag{13}$$

where the subscript u, v indicates that the operator or vector must be with respect to the parameter space.

Since we assumed our surface to be piecewise linear, its gradient $\nabla_{u,v}\mathbf{x}$ is constant over each triangle of the mesh. As a consequence, whatever the type of finite-volume discretization we use, the integral of the normal vector along the border $\partial \mathcal{A}_{\mathrm{M}}$ within a triangle will result in the same expression since the border of both regions passes through the edge midpoints as sketched in Figure 9(b). Inside a triangle $T = (\mathbf{x}_i, \mathbf{x}_j, \mathbf{x}_k)$, we can write:

$$\int_{\partial \mathcal{A}_{\mathrm{M}} \cap T} \nabla_{u,v}\mathbf{x} \cdot \mathbf{n}_{u,v} \, dl = \nabla_{u,v}\mathbf{x} \cdot [\mathbf{a} - \mathbf{b}]_{u,v}^{\perp} = \frac{1}{2}\nabla_{u,v}\mathbf{x} \cdot [\mathbf{x}_j - \mathbf{x}_k]_{u,v}^{\perp}$$

where $^{\perp}$ denotes a counterclockwise rotation of 90 degrees.

Since the function $\mathbf{x}$ is linear over any triangle T, using the linear basis functions B_l over the triangle, it follows:

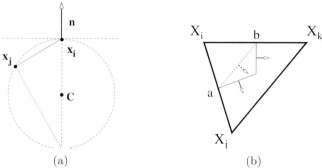

(a)   (b)

Fig. 9. *(a) Osculating circle for edge* $\mathbf{x}_i\mathbf{x}_j$*. (b) The integration of the surface gradient dotted with the normal of the region contour does not depend on the finite volume discretization used.*

$$\mathbf{x} = \mathbf{x}_i\ B_i(u,v) + \mathbf{x}_j\ B_j(u,v) + \mathbf{x}_k\ B_k(u,v)$$
$$\nabla_{u,v}\mathbf{x} = \mathbf{x}_i\ \nabla_{u,v}B_i(u,v) + \mathbf{x}_j\ \nabla_{u,v}B_j(u,v) + \mathbf{x}_k\ \nabla_{u,v}B_k(u,v)$$

Using the fact that the gradients of the 3 basis functions of any triangle T sum to zero and rearranging terms, the gradient of $\mathbf{x}$ over the triangle can be expressed as $\nabla_{u,v}\mathbf{x} = \frac{1}{2\mathcal{A}_T}[(\mathbf{x}_j-\mathbf{x}_i)\left([\mathbf{x}_i - \mathbf{x}_k]_{u,v}^{\perp}\right)^T+(\mathbf{x}_k-\mathbf{x}_i)\left([\mathbf{x}_j - \mathbf{x}_i]_{u,v}^{\perp}\right)^T]$, where ^{T} denotes the transpose. Note that $\nabla_{u,v}\mathbf{x}$ is an n x 2 matrix - n for the dimension of the embedding of $\mathbf{x}$ and 2 for the (u,v) space. The previous integral can then be rewritten as:

$$\int_{\partial\mathcal{A}\cap T} \nabla_{u,v}\mathbf{x}\cdot\mathbf{n}_{u,v}\ dl = \frac{1}{4\mathcal{A}_T}\left[([\mathbf{x}_i - \mathbf{x}_k]\cdot[\mathbf{x}_j - \mathbf{x}_k])_{u,v}\ (\mathbf{x}_j - \mathbf{x}_i)\right.$$
$$\left.+([\mathbf{x}_j - \mathbf{x}_i]\cdot[\mathbf{x}_j - \mathbf{x}_k])_{u,v}\ (\mathbf{x}_k - \mathbf{x}_i)\right].$$

Moreover, the area $\mathcal{A}_T$ is proportional to the sine of any angle of the triangle. Therefore, we can use the cotangent of the 2 angles opposite to $\mathbf{x}_i$ to simplify the parameter space coefficients and write:

$$\int_{\partial\mathcal{A}_M\cap T} \nabla_{u,v}\mathbf{x}\cdot\mathbf{n}_{u,v}\ dl = \frac{1}{2}[cot_{u,v}\ \angle(\mathbf{x}_k)(\mathbf{x}_j - \mathbf{x}_i) + cot_{u,v}\ \angle(\mathbf{x}_j)\ (\mathbf{x}_k - \mathbf{x}_i)].$$

Combining the previous equation with Eq. (4) and (13), using the current surface discretization as the conformal parameter space, and reorganizing terms by edge contribution, we obtain:

$$\iint_{\mathcal{A}_M} \mathbf{K}(\mathbf{x})dA = \frac{1}{2}\sum_{j\in N_1(i)} (cot\ \alpha_{ij} + cot\ \beta_{ij})\ (\mathbf{x}_i - \mathbf{x}_j)$$

where α_{ij} and β_{ij} are the two angles opposite to the edge in the two triangles sharing the edge $(\mathbf{x}_j, \mathbf{x}_i)$ as depicted in Figure 3(a).

Constructing Circle Patterns Using a New Functional

Boris A. Springborn

Institut für Mathematik, Sekr. 8-5, Technische Universität Berlin, Strasse des 17. Juni 136, 10623 Berlin, Germany. *springb@math.tu-berlin.de*

Summary. The problem of constructing circle patterns with specific properties is reduced to minimizing an explicitly given function of the (logarithmic) radii. Images of circle patterns produced in this way are shown. The implementation of the target function is discussed.

Keywords. circle packing, circle pattern

1 Introduction

Ever since Thurston, in the 1980's, called attention to the close relationship between packings and patterns of circles on the one hand and conformal maps on the other [12], the former have become a favorite object of mathematical research. It was shown that circle packings and circle patterns can be used to approximate conformal maps [9, 10]. This opens the possibility to develop numerical methods for conformal maps that are based on the construction of circle packings or patterns. Thurston's original algorithm to construct circle patterns was developed further by Stephenson and implemented in the program `circlepack` [4]. There are efforts to apply the approximation of conformal maps via circle patterns in medical imaging [5].

In the paper with Bobenko [2], we presented two new functionals for the construction of circle patterns. One belongs to Euclidean, the other to hyperbolic geometry. Building on methods that were developed by Colin de Verdière [3], we used the variational approach to prove existence and uniqueness results for circle patterns. We could also clarify the relationship between our functionals and the functionals found by Colin de Verdière [3], Brägger [1], and Rivin [8], and derive a functional which seems to be related to Leibon's [6]. For the practical construction of circle patterns, our functionals have obvious advantages over the others: They are given in closed form, whereas for Colin de Verdière's functionals only the derivatives are given in closed form. In principle, knowledge of the gradient is enough to minimize a function, but more efficient algorithms can be used if the function itself can be evaluated. Surprisingly, the evaluation of our functional is computationally not very expensive, as is shown in section 6. Moreover, the optimization of our functionals is not subject to any constraints. This is also true for Colin de Verdière's functionals, but the others require optimization under linear equality and inequality constraints.

Whereas the paper [2] is concerned with existence and uniqueness questions, the purpose of this paper is to give a brief but self-contained exposition of the Euclidean functional with a focus on the practical aspects. Examples of circle patterns that were constructed using the functional are presented. The last section describes how the functional can be evaluated numerically. For the sake of brevity we treat only the simplest case, circle patterns with no curvature which correspond to a cellular decomposition of the disc; and all combinatorial-topological subtleties are glossed over.

2 Circle Patterns

Figure 1 shows a cellular decomposition of a topological disc and a two pat-

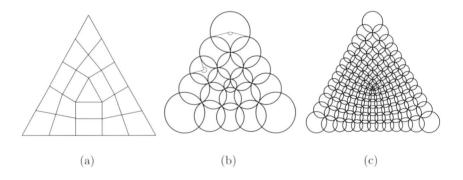

(a)	(b)	(c)

Fig. 1. A cellular decomposition of the disc and two circle patterns.

terns of circles. The cellular decomposition 1(a) corresponds to the pattern 1(b). The faces (2-cells) of the decomposition correspond to circles of the pattern and the vertices (0-cells) correspond to points where circles intersect. The pattern of figure 1(c) corresponds to an analogous cellular decomposition with more faces. The circle patterns are determined up to similarity by the following data:

- *The cellular decomposition,* or more precisely, its combinatorial type.
- *Intersection angles:* a function on the edges (1-cells) of the cellular decomposition prescribing the intersection angles of the circles corresponding to adjacent faces. In this example, all intersection angles are $\pi/2$.
- *Boundary conditions:* a function on the boundary faces of the cellular decomposition prescribing, for each circle, the total angle subtended by the

neighboring circles. In figure 1(b), this angle is indicated for two boundary circles. In this example, it is $5\pi/6$ for the corner circles and $3\pi/2$ for the other boundary circles. The boundary conditions were chosen such that the outer intersection points lie on the sides of an equilateral triangle.

The name of Neumann was put to this type of boundary conditions by Colin de Verdière. Prescribing instead the radii of the circles on the boundary constitutes the corresponding Dirichlet problem. It may also be treated with the functional below.

We are going to show how to construct a circle pattern from the above data using a variational approach. If, in addition to the cellular surface and the intersection angles, also the radii of the circles were known, then the circle pattern could be reconstructed in the following way: Start with an arbitrary face of the cellular decomposition and place the corresponding circle somewhere in the plane. Then add cyclically and one by one the circles corresponding to the neighboring faces. Continue in the obvious fashion until all the circles are placed. If the radii are correct, the circles will fit together and the whole pattern will be reconstructed. The problem is therefore reduced to finding the correct radii. It turns out to be more convenient to use the logarithmic radii as variables. We will derive a convex function of these logarithmic radii that achieves a minimum if the radii are correct.

3 Preliminaries

Consider a triangle with sides 1 and e^x and an angle θ between them, as shown in figure 2(a). The angle between the side of unit length and the third side is $\varphi = f_\theta(x)$, where

$$f_\theta(x) - \frac{1}{2i} \log \frac{1 - e^{x-i\theta}}{1 - e^{x+i\theta}}.$$

We always consider the principal branch of the logarithm function which takes values with imaginary part between $-\pi$ and π. If $\theta = \pi/2$, then $f_\theta(x) = \arctan e^x$. By rescaling the triangle, one sees that the third angle is $\varphi' = f_\theta(-x)$. Since the angles sum to π, it follows that

$$f_\theta(-x) = \theta^* - f_\theta(x), \tag{1}$$

where $\theta^* = \pi - \theta$. The derivative of f_θ is

$$f'_\theta(x) = \frac{\sin\theta}{2(\cosh x - \cos\theta)}. \tag{2}$$

It turns out that the integral is

$$\int_{-\infty}^{x} f_\theta(\xi)\, d\xi = \operatorname{Im} \operatorname{Li}_2(e^{x+i\theta}), \tag{3}$$

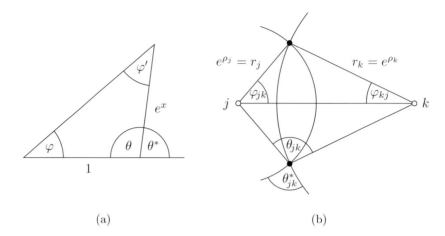

Fig. 2. Notation.

where Im denotes the imaginary part and Li_2 is the dilogarithm function [13]. Inside and on the unit circle, it is defined by the power series

$$\mathrm{Li}_2(z) = \sum_{k=1}^{\infty} \frac{z^k}{k^2}. \tag{4}$$

One deduces the integral representation

$$\mathrm{Li}_2(z) = -\int_0^z \frac{\log(1-\zeta)}{\zeta}\, d\zeta, \tag{5}$$

which reveals that the power series may be analytically continued to the complex plane with a cut from 1 to ∞ along the real axis. This is the principal branch of the dilogarithm, with which we will be concerned exclusively.

4 The Functional

Suppose that the intersection angles of a circle pattern are prescribed by a function θ (or equivalently $\theta^* = \pi - \theta$) on the edges of a cellular decomposition. (See figure 2(b).) The intersection angles may lie in the range $0 < \theta < \pi$. Suppose the boundary conditions are given by a positive function Φ on the boundary faces. It is convenient to extend Φ to a function on *all* the faces by assigning the value 2π to interior faces. Then the logarithmic radii $\rho = \log r$ have to satisfy the following equations. There is one equation for each face. The equation for a face j is

$$\sum_k 2f_{\theta_{jk}}(\rho_k - \rho_j) = \Phi_j, \tag{6}$$

where the sum is taken over all faces k meeting j in an edge, and θ_{jk} denotes the value of θ on the edge between the faces j and k.

Now we define the functional $S(\rho)$. For a function ρ on the faces, let

$$S(\rho) = \sum_{j \circ \!\!\bullet\!\! \circ k} \left(\operatorname{Im} \operatorname{Li}_2(e^{\rho_k - \rho_j + i\theta_{jk}}) + \operatorname{Im} \operatorname{Li}_2(e^{\rho_j - \rho_k + i\theta_{jk}}) - \theta_{jk}^*(\rho_j + \rho_k) \right)$$

$$+ \sum_{\circ j} \Phi_j \rho_j, \tag{7}$$

where the first sum is taken over all interior edges and j and k are the faces on either side. The second sum is taken over all faces j.

A function ρ on the faces satisfies equations (6) if and only if it is a critical point of S. Indeed, from equations and (1) and (3), one finds that

$$\frac{\partial S}{\partial \rho_j} = -\sum_k 2 f_{\theta_{jk}}(\rho_k - \rho_j) + \Phi_j, \tag{8}$$

where the sum is taken over all faces k meeting face j in an edge. Hence $\partial S / \partial \rho_j = 0$ if and only if equation (6) is satisfied.

Now, consider how the functional S behaves if all radii are scaled by a common factor. Thus, let ρ be a function on the faces, $h \in \mathbb{R}$, and $\tilde{\rho}_j = \rho_j + h$. Then

$$S(\tilde{\rho}) = S(\rho) + h \left(\sum_{\circ j} \Phi_j - \sum_{j \circ \!\!\bullet\!\! \circ k} 2\theta_{jk}^* \right),$$

where the first sum is taken over all faces j. The second sum is taken over all interior edges, and j and k are the faces on either side. Clearly, the functional S cannot have a critical point unless the term in brackets vanishes. Hence, the functional S is scale invariant if and only if

$$\sum_{\circ j} \Phi_j = \sum_{j \circ \!\!\bullet\!\! \circ k} 2\theta_{jk}^*, \tag{9}$$

and this is a necessary condition for the existence of a critical point. If equation (9) is not satisfied, then, under the gradient flow, all radii $r = e^\rho$ will either tend to zero or grow without bound. If, on the other hand, equation (9) is satisfied, so that S is scale invariant, then one may restrict the search for critical points to the set $\{\rho \mid \sum \rho_j = 0\}$.

Using equation (2) one sees that the second derivative of S is the quadratic form

$$S'' = \sum_{j \circ \!\!\bullet\!\! \circ k} \frac{\sin \theta_{jk}}{\cosh(\rho_k - \rho_j) - \cos \theta_{jk}} (d\rho_k - d\rho_j)^2,$$

where the sum is taken over all interior edges. Hence $S'' > 0$ unless $d\rho_j = d\rho_k$ for all pairs j, k of faces.

The following theorem summarizes the above results.

Theorem 2 ([2]). *A function ρ on the faces gives the logarithmic radii of a circle pattern with given data if and only if it is a critical point of the functional $S(\rho)$.*

The functional $S(\rho)$ is strictly convex on the set $\{\rho \mid \sum \rho_j = 0\}$. Critical points are hence minima. If a circle pattern exists for given data, it is unique up to similarity.

The question whether a circle pattern *exists* for given cellular decomposition, intersection angles, and boundary conditions is more delicate. One has to show that $S \to \infty$, if $\|\rho\| \to \infty$ while $\sum \rho_j = 0$. For a treatment of this problem, the reader is referred to [2].

For practical purposes, it may be sufficient to check the existence of a specific circle pattern experimentally. Indeed, if a circle pattern does *not* exist for given data, then some of the ρ will tend to $\pm\infty$ under the gradient flow. Conversely, if a numerical minimizer succeeds in approximating a minimum of $S(\rho)$, then this is a strong indication that the corresponding circle pattern really exists; especially if the radii do not differ greatly in magnitude and if they do not change much if the accuracy of the calculation is increased.

The functional $S(\rho)$ is so constructed that each partial derivative $dS/d\rho_j$ is the difference between the actual value of the total angle subtended by the neighboring circles and its nominal value of Φ_j, which is 2π for interior circles; see equation (8). Hence, if a numerical minimizer finds a solution ρ at which the gradient of $S(\rho)$ is almost zero, then the circles will fit together almost exactly when the pattern is reconstructed.

Currently, the method of choice to practically construct circle patterns seems to be Thurston's algorithm as developed further and implemented by Stephenson [12, 4]. Basically, it consists in adjusting, iteratively, the radius of each circle so that the neighboring circles (whose radii are fixed for the moment) fit around. This is equivalent to minimizing, iteratively, the functional $S(\rho)$ in each coordinate direction.

5 Examples

The circle patterns shown in figures 1(b) and 1(c) were created using the *Mathematica* software. It provides an implementation of the polylogarithms, but the evaluation is slow. Thus an `InterpolatingFunction` approximating Clausen's integral was computed once and for all from equation (10), and then formula (11) was used to compute the dilogarithm. The boundary conditions were chosen such that the outer intersection points lie on the sides of an equilateral triangle.

The circle patterns shown in figures 3, 4, and 6 were created using the *Java Tools for Experimental Mathematics*, which are currently being developed at the Sonderforschungsbereich 288 [11].

Figure 3 shows three circle patterns with square grid combinatorics and

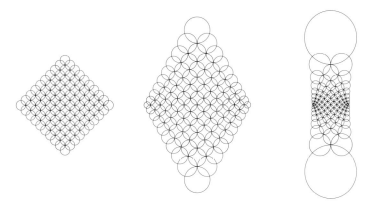

Fig. 3. Square grid combinatorics. $\theta = \pi/2$.

intersection angles $\pi/2$. The boundary conditions are given by $\Phi = 3\pi/2$ for all boundary circles except the ones in the corners, for which $\Phi = \pi$ for the pattern on the left, $\Phi = 5\pi/6$ and $7\pi/6$ for the pattern in the middle, and $\Phi = \pi/2$ and $3\pi/2$ for the pattern on the right.

Figure 4 shows circle patterns with hexagonal combinatorics. All inter-

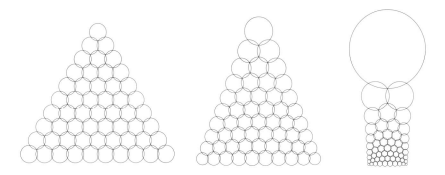

Fig. 4. Hexagonal combinatorics. $\theta = 2\pi/3$.

section angles are $\theta = 2\pi/3$. The boundary conditions were chosen such that the outer intersection points lie on the sides of a triangle. The triangle for the pattern on the right is degenerate. It has two right angles and one angle is zero.

We have been considering circle patterns in which the vertices of the cellular decomposition correspond to intersection points of circles. Thus, it is required that the intersection angles θ sum to 2π around each vertex. Often, a different type of pattern is considered, in which all vertices have three edges but the angles do not necessarily sum to 2π. At least the case where the sum is greater than 2π can be reduced to the type considered in this paper by adding circles as shown in figure 5(a). The angles for the new edges (see figure

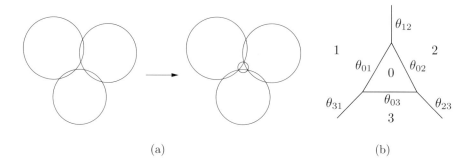

(a)   (b)

Fig. 5. Adding circles if $\sum \theta_{jk} > 2\pi$ around a vertex.

5(b)) are determined by the equations

$$\theta_{12} + \theta_{20} + \theta_{01} = 2\pi$$
$$\theta_{23} + \theta_{30} + \theta_{02} = 2\pi$$
$$\theta_{31} + \theta_{10} + \theta_{03} = 2\pi.$$

Hence $\theta_{01} = \pi - \frac{1}{2}(\theta_{12} - \theta_{23} + \theta_{31})$, etc. (Note that even if the condition $\theta_{12} + \theta_{23} + \theta_{31} > 2\pi$ is not satisfied, one may get values for $\theta_{01}, \theta_{02}, \theta_{02}$ which lie in the correct range $0 < \theta < \pi$. In this case, however, the circle pattern does not exist and the functional does not have a minimum.)

The circle patterns of figure 6 were constructed in this way. They are similar to those of figure 4. But here the intersection angles are $\theta = 3\pi/4$.

6 Numerical Evaluation of the Functional

How useful is the functional $S(\rho)$ for the practical construction of circle patterns? This depends primarily on how easily it can be computed numerically. Formula (7) is not so useful for this purpose since it involves the complex dilogarithm. However, the imaginary part of the dilogarithm function can be expressed in terms of a 2π-periodic real function, Clausen's integral $\text{Cl}_2(x)$. It can be defined by the imaginary part of dilogarithm on the unit circle:

$$\text{Cl}_2(x) = \text{Im} \, \text{Li}_2(e^{ix}). \tag{10}$$

From equation (4), its Fourier expansion is seen to be

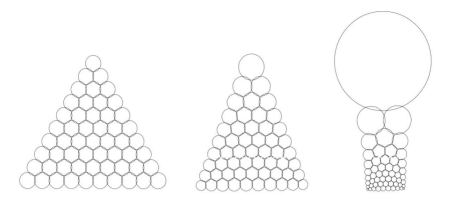

Fig. 6. Hexagonal combinatorics. $\theta = 3\pi/4$.

$$\mathrm{Cl}_2(x) = \sum_{k=1}^{\infty} \frac{\sin kx}{k^2}.$$

Substituting $\zeta = e^{i\xi}$ in equation (5), one obtains for $0 \leq x \leq 2\pi$ the equation

$$\mathrm{Cl}_2(x) = -\int_0^x \log\left(2\sin\frac{\xi}{2}\right) d\xi.$$

The imaginary part of the dilogarithm function is expressed in terms of Clausen's integral by the equation

$$\mathrm{Im\,Li}_2(e^{x+i\theta}) = \varphi x + \frac{1}{2}\mathrm{Cl}_2(2\varphi) - \frac{1}{2}\mathrm{Cl}_2(2\varphi + 2\theta) + \frac{1}{2}\mathrm{Cl}_2(2\theta), \qquad (11)$$

where $\varphi = f_\theta(x)$. If one wants to calculate the functional $S(x)$, the following formula is even more useful:

$$\mathrm{Im\,Li}_2(e^{x+i\theta}) + \mathrm{Im\,Li}_2(e^{-x+i\theta}) = px + \mathrm{Cl}_2(\theta^* + p) + \mathrm{Cl}_2(\theta^* - p) - \mathrm{Cl}_2(2\theta^*),$$

where $\theta^* = \pi - \theta$ and

$$p = 2\arctan\left(\tanh\left(\frac{x}{2}\right)\tan\left(\frac{\theta^*}{2}\right)\right). \qquad (12)$$

Clausen's integral $\mathrm{Cl}_2(x)$ was implemented in the following way using Chebyshev approximation. The function

$$h(x) = -\log\left(2\sin\frac{x}{2}\right) + \log x$$

is analytic at $x = 0$ and

$$\mathrm{Cl}_2(x) = \int_0^x h(\xi)\, d\xi - x(\log x - 1).$$

The function $h(x)$ was approximated in the interval from $-\pi$ to π by a Chebyshev series. This series was then integrated. The algorithms for fitting, integrating and evaluating Chebyshev series which were used are described in [7].

In principle, the evaluation of Clausen's integral is not more complicated or expensive than the evaluation of a trigonometric or exponential function. Note that, with p as in equation (12), $f_\theta(x) = \frac{1}{2}(\theta^* + p)$ and $f_\theta(-x) = \frac{1}{2}(\theta^* - p)$. Hence, when computing $S(\rho)$, the simultaneous computation of the gradient is (almost) free.

References

1. W. Brägger. Kreispackungen und Triangulierungen. *Enseign. Math.*, 38:201–217, 1992.
2. A. I. Bobenko and B. A. Springborn. Variational principles for circle patterns and Koebe's theorem. `arXiv:math.GT/0203250`, March 2002.
3. Y. Colin de Verdière. Un principe variationnel pour les empilements de cercles. *Invent. Math.*, 104:655–669, 1991.
4. Tomasz Dubejko and Kenneth Stephenson. Circle packing: experiments in discrete analytic function theory. *Experiment. Math.*, 4(4):307–348, 1995.
5. M. K. Hurdal, P. L. Bowers, K. Stephenson, et al. Quasi-conformally flat mapping the human cerebellum. In C. Taylor and A. Colchester, editors, *Medical Image Computing and Computer-Assisted Intervention—MICCAI '99*, volume 1679 of *Lecture Notes in Computer Science*, pages 279–286, Berlin, 1999. Springer.
6. G. Leibon. Characterizing the Delaunay decompositions of compact hyperbolic surfaces. *Geom. Topol.*, 6:361-391, 2002.
7. William H. Press, Saul A. Teukolsky, William T. Vetterling, and Brian P. Flannery. *Numerical recipes in C*. Cambridge University Press, Cambridge, second edition, 1992.
8. I. Rivin. Euclidean structures of simplicial surfaces and hyperbolic volume. *Ann. of Math.*, 139:553–580, 1994.
9. B. Rodin and D. Sullivan. The convergence of circle packings to the Riemann mapping. *J. Differential Geom.*, 26:349–360, 1987.
10. O. Schramm. Circle patterns with the combinatorics of the square grid. *Duke Math. J.*, 86(2):347–389, 1997.
11. Sfb 288. Java tools for experimental mathematics. `http://www.jtem.de`.
12. W. P. Thurston. The geometry and topology of three-manifolds. Electronic version 1.0 of 1997 currently provided by the MSRI under `http://www.msri.org/publications/books/gt3m/`.
13. L. Lewin. Polylogarithms and Associated Functions. North Holland, New York, 1981.

Constructing Hamiltonian Triangle Strips on Quadrilateral Meshes

Gabriel Taubin[1]

IBM T.J. Watson Research Center, P.O. Box 704, Yorktown Heights, NY 10598
taubin@us.ibm.com

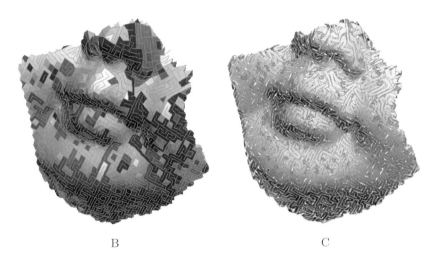

B   C

Fig. 1. Every connected manifold quadrilateral mesh without boundary can be represented as a single Hamiltonian generalized triangle strip cycle by splitting each face along one of its diagonals, and connecting the resulting triangles along the original mesh edges. (B) An arbitrary choice of face diagonals produces several cycles. (C) Cycles are then joined to form a single cycle by flipping diagonals.

Because of their numeric stability properties, quadrilateral meshes have become a popular representation for finite elements computations and computer animation. In this paper we address the problem of optimally representing quadrilateral meshes as generalized triangle strips (with one swap bit per triangle). This is important because 3D rendering hardware is optimized for rendering triangle meshes transmitted from the CPU to the GPU in the form of triangle strips. We describe simple linear time and space constructive algorithms, where each quadrilateral face is split along one of its two diagonals and the resulting triangles are linked along the original mesh edges. We show that with these algorithms every connected manifold quadrilateral mesh without boundary can be optimally represented as a single Hamiltonian generalized triangle strip cycle in multiple ways, and we discuss simple strategies to tailor the construction for transparent vertex caching.

1 Introduction

Because of their simplicity, triangle meshes (T-meshes) are one of the most widely used representations for 3D models in Computer Graphics, and the basic geometric primitive in the 3D rendering pipeline. Since in general a large number of triangles is required to faithfully describe the geometry of a complex surface, a bottleneck problem exists in the transmission of 3D models from the CPU to the GPU (graphics processing unit)[9], and also across other communication channels such as networks. This problem has motivated the search for more efficient encoding schemes for triangle meshes [22]. Instead of specifying each triangle by the coordinates of its three vertices, triangles can be sequentially organized forming *triangle strips* so that every pair of consecutive triangles share a *marching edge*. In this way, only the first triangle of a triangle strip requires the transmission of the coordinates of its three vertices, and each subsequent triangle is specified by only one vector of vertex coordinates.

Triangle strips (T-strips) are widely supported in graphics hardware. In a *sequential* T-strip of length n, composed of n triangles and specified by $n+2$ vertices $v_0, \ldots, v_{n+1}$, the corners of the i-th. triangle are the vertices v_i, v_{i+1}, and v_{i+2}. In a *generalized* T-strips one additional *marching bit* is transmitted for each triangle of the strip other than the last one (i.e., for each marching edge). This bit specifies whether the next triangle is attached to the left or the right edge of the last triangle opposite to the last marching edge. In the sequential triangle strips the marching bits alternate between left and right, and are implicitly specified. From now on, when we refer to a T-strip we will mean a generalized T-strip.

Quadrilateral meshes (Q-meshes) have become a popular representation in modelling and animation [28], in part because of Catmull-Clark Subdivision surfaces [5], and in finite element computations because of their superior numerical properties. In this paper we study the problem of optimally decomposing Q-meshes into T-strips by splitting each of the quadrilateral faces of the mesh along one of its two diagonals and connecting the resulting triangles. Since each T-strip of length n is specified by $n+2$ vertices, the absolute minimum representation cost of one vertex per triangle is achieved when the quadrilateral faces can be split so that all the triangles can be connected forming a single T-strip, with the last two vertices coinciding with the first two.

In this paper we show that this is always possible for connected manifold Q-meshes without boundary, and describe an efficient algorithm to do so. The resulting triangulation is called *Hamiltonian* because it corresponds to the existence of a Hamiltonian cycle in its dual graph.

Hardware support for T-strips requires a cache of size 2 in the graphics processing unit (GPU). Modern GPU's maintain a larger vertex FIFO cache to potentially achieve even higher performance. In the second part of the paper we discuss strategies to maximizing vertex locality in the construction of the Hamiltonian triangle strip to make better use of this cache.

1.1 Related Work

Although we are not aware of any previous solution to the problem that we address in this paper, a number of related problems have been studied. Whitney [27] proved that every planar triangulation without separating triangles has a Hamiltonian cycle. Tutte [25] extended this result to 4-connected planar graphs, and then other authors generalized the result in various ways [19, 23]. Arkin, et. al. [1] proved that every point set has a Hamiltonian triangulation, and showed that the problem of testing whether a triangulation is Hamiltonian or not is NP-complete. As a result, algorithms based on heuristics must be used to decompose triangle meshes into triangle strips. For example, Evans, et. al. [11] and Estkowski, et. al. [10] present algorithms for constructing triangle strips from triangulated models. Bose and Toussaint [4] studied the problem of constructing quadrangulations of point sets, and obtained an alternate method of computing Hamiltonian path triangulations. The problem of generating a quadrangulation of a set of points out of a triangulation has received considerable attention both in the mesh generation and the computational geometry literature. See for example, Ramaswami, et.al. [17] and related references. King et.al. proposes an algorithm to compress quadrilateral meshes [15] where the mesh faces are implicitly connected forming a tree. Velho, et.al. [26] present a refinement scheme that produces a hierarchy of triangle strip decompositions of a triangle mesh. Dafner, et. al. [8] constructs, in a way similar to our method, a Hamiltonian path in the regular grid defined by the pixels of an image for compression purposes.

2 Graphs and Meshes

In this section we introduce some definitions, notation, and facts to make the paper self-contained. We use this material in subsequent sections to formulate our main results more precisely. It can be skipped during a first reading. Just remember that in the rest of the paper, when we refer to a Q-mesh we always mean a connected manifold quadrilateral mesh without boundary (see definitions below).

2.1 Graphs

A *graph* $G = (V, E)$ is composed of a set V of *vertices* and a set E of *edges*. In addition, an *incidence* map $e \mapsto I(e) = \{v_1, v_2\}$ associates each edge with an *unordered pair* of vertices. The two vertices are the *ends* of the edge. Every edge *joins* or *connects* its ends and is *incident* to its ends. Two vertices

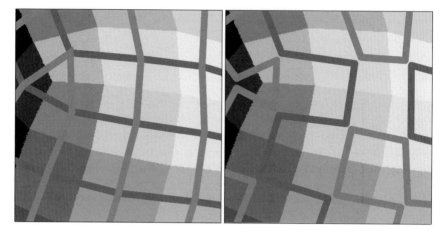

Fig. 2. Not every Eulerian circuit corresponds to a Hamiltonian triangulation. Edges that are opposite to each other on a face cannot be contiguous in the Eulerian circuit. We used two different colors to visualize the circuit intersections (or lack of), but both red and green dual edges belong to the same Eulerian circuit. The Eulerian circuit on the left does not correspond to any Hamiltonian triangulation.

connected by an edge are *adjacent* or *neighbors*. Two vertices not connected by an edge are *independent*. Two edges sharing exactly one end are *adjacent*. An edge is a *loop* if its two ends are the same. Two edges are *parallel* if they have the same ends. A graph is *simple* if it has no loops and no parallel edges. All the graphs in this paper will be simple. Since the incidence map is one-to-one, every edge of a simple graph is identified with the pair of its ends, and we write $e = \{v_1, v_2\}$.

A *walk* of length n in a graph G is an alternating sequence of vertices and edges $W = (v_1, e_1, \ldots, v_i, e_i, v_{i+1} \ldots, e_n, v_{n+1})$, possibly with repetitions, such that each edge connects the two vertices next to it in the sequence, i.e., $I(e_i) = \{v_i, v_{i+1}\}$. The first element of the sequence is the *beginning*, and the last one is the *end* of the walk. The beginning and end of the walk are the *ends* of the walk. A walk is *closed* if its two ends coincide, and *open* if not. A *trail* is a walk in which all the edges are different. A trail is *Eulerian* if it contains all the edges of the graph. A *path* is a walk with distinct vertices. In particular, every path is a trail. A *circuit* is a closed trail. A *cycle* is a circuit of length n with exactly n different vertices (a closed path). A *Hamiltonian* path (respectively cycle) contains all the vertices of the graph. A graph containing a Hamilton path or cycle is *Hamiltonian*. A graph is k-connected if between any pair of distinct vertices there are k edge-disjoint paths.

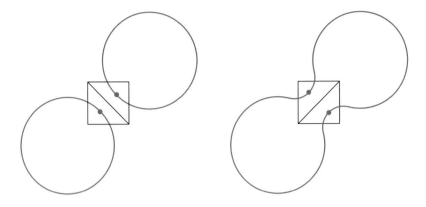

Fig. 3. If the two triangles generated by splitting one quadrilateral face belong to different cycles, flipping the diagonal joins the two cycles into a single one.

2.2 Meshes

A *polygon mesh* (P-mesh) is defined by the position of the vertices (geometry); by the association between each face and its sustaining vertices (connectivity); and optional colors, normals and texture coordinates (properties).

The connectivity of a P-mesh M is defined by the incidence relationships existing among its V vertices, E edges, and F faces. A *face* with n *corners* is a sequence of $n \geq 3$ different vertices. All the cyclical permutations of its corners are considered identical. Every face *joins* or *connects* its corners and is *incident* to its corners. An *edge* e is an un-ordered pair $e = \{v_1, v_2\}$ of different vertices that are consecutive in one or more faces of the mesh. The *graph* of a mesh is the graph defined by the V mesh vertices as graph vertices, and the E mesh edges as graph edges. The meshes considered in this paper have neither *isolated* vertices (not contained in any face) nor multiple connected faces (faces with holes).

We classify the vertices of a P-mesh as *boundary, regular,* or *singular*. A *boundary* mesh edge has exactly one, a *regular* mesh edge has exactly two, and a *singular* mesh edge has three or more incident faces. The *dual graph* of a P-mesh is the graph defined by the mesh faces as graph vertices, and the regular mesh edges as graph edges. The *edge star* of a vertex v is the set $E(v)$ of incident edges. The *vertex star* of a vertex v is the set $V(v)$ of adjacent vertices. The *face star* of a vertex v is the set of faces $F(v)$ incident to the vertex. A mesh vertex is a *boundary* vertex if its face star defines an open path in the dual graph. A mesh vertex is *regular* if its face star defines a cycle in the dual graph. A *singular* vertex is a vertex that is neither regular not boundary. In this paper a mesh with no singular edges is *manifold*. It is *manifold without boundary* if in addition all its edges are regular. Note that

```
BasicHamiltonianSplit (M = (V, E, F))
    # step 1
    split each face along one of its diagonals
    # step 2
    for each face f ∈ F
        if split triangles belong to different cycles
            flip diagonal of f
return
```

Fig. 4. Algorithm to split the faces of a Q-mesh along face diagonals so that the resulting T-mesh is Hamiltonian. In the first step, the face diagonals can be chosen in an arbitrary fashion.

this is not the usual definition of manifold, which are also required not to have singular vertices. Our manifolds are allowed to have *singular vertices*.

Two faces are *connected* if they are the ends of a path in the dual graph. This equivalence relation defines a partition of the set of faces into *connected components*. Note that as a consequence of our non-standard definition of manifold, faces belonging to different components may share singular vertices. A mesh with only one connected component is *connected*. An algorithm based on Tarjan's fast union-find data structure [20] can be used to generate the connected components. From now on, all the meshes are connected manifold without boundary.

3 Hamiltonian Paths

With respect to the construction of Hamiltonian paths, the situation for Q-meshes is quite different than for triangle meshes. Hakimi et.al. [13] proved that every 4-connected planar triangulation on V vertices contains not only one, but at least $V/\log_2(V)$ distinct Hamiltonian cycles. The dual graph of a Q-mesh connected manifold without boundary is 4-connected but not necessarily planar, but we will see that nevertheless we have considerable freedom in the construction of Hamiltonian triangulations.

The existence of a Hamiltonian cycle in the dual graph of the mesh would certainly solve the problem we are addressing in this paper. In this case each quadrilateral face can be split into two triangles consecutive in the traversal order defined by the Hamiltonian cycle. After the quadrilateral faces are split, the resulting triangles are linked through an alternating sequence of new diagonal edges and original mesh edges. The remaining original edges become the boundary edges of the resulting Hamiltonian T-strip cycle.

Unfortunately, there exist non-Hamiltonian Q-meshes. This is true even though determining the existence a Hamiltonian cycle in the dual graph of a Q-mesh is linear-time solvable [6]. In this case the alternative is to structure the faces of the Q-mesh as a tree with long runs and few branching nodes. In

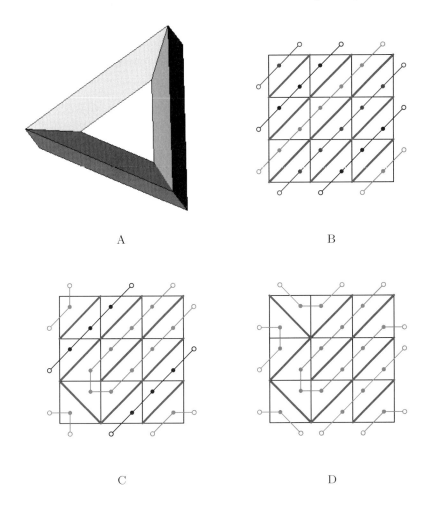

A

B

C

D

Fig. 5. The algorithm of figure 4 applied to a 3×3 torus (A). An arbitrary choice of diagonals produces a triangulation with three cycles (B). After one diagonal flip the number of cycles is two (C). After an additional flip we obtain a triangulation with a single Hamiltonian cycle (D). Remember that opposite boundary edges and vertices are identified. In particular, the four corners of the square correspond to the same mesh vertex.

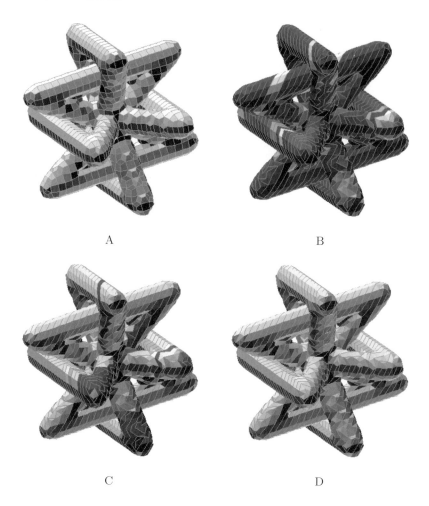

Fig. 6. The algorithm of figure 4 applied to a more complex mesh (A). A random choice of diagonals produces a triangulation with 73 cycles (B). After 30 diagonal flips the resulting triangulation still has 43 cycles (C). After 42 additional flips we obtain a triangulation with a single Hamiltonian cycle (D).

the algorithm introduced by King et.al. to compress the connectivity of Q-meshes [15], the quadrilateral faces are split along one diagonal while the dual graph is tree-traversed in depth-first order. In this algorithm the two triangles resulting from splitting a face are also consecutive in the tree traversal order. The tree can be cut into generalized triangle strips, but unfortunately, the number of branching nodes in the spanning tree that links the faces in the dual graph, where the cuts have to be made, could be arbitrarily large. This

```
PartitionEdges (M = (V, E, F), B)
    p_E = new Partition (E)
    for each face f = (v_0, v_1, v_2, v_3) ∈ F
        if b_f = 0
            p_E.join(e_01, e_12)
            p_E.join(e_23, e_30)
        else if b_f = 1
            p_E.join(e_12, e_23)
            p_E.join(e_30, e_01)
    return p_E


JoinCycles (M = (V, E, F), B, p_E, S)
    for each face f = (v_0, v_1, v_2, v_3) ∈ S
        if b_f = 0
            if p_E.find(e_01) ≠ p_E.find(e_23)
                b_f = 1
                p_E.join(e_01, e_23)
        else if b_f = 1
            if p_E.find(e_01) ≠ p_E.find(e_12)
                b_f = 0
                p_E.join(e_01, e_12)
    return


EfficientHamiltonianSplit (M = (V, E, F), B, S)
    # initialize edge partition
    p_E = PartitionEdges (M, B)
    # join cycles by flipping diagonals
    JoinCycles (M, B, p_E, S)
    return
```

Fig. 7. Efficient implementation of the algorithm illustrated in figure 4. B is a boolean array with one face bit b_f per face initialized by the calling function. The function `PartitionEdges` initializes the partition of the set of edges into the cycles defined by the array B. The function `JoinCycles` joins the multiple cycles into a single one by flipping some face bits corresponding to faces in the subset S of F. $S = F$ ensures success, but other choices will be discussed later. The mesh edge $e_{ij} = M.\text{getEdge}(v_i, v_j)$ connects vertices v_i and v_j.

problem is shared by a number of T-mesh connectivity encoding schemes [21, 24, 18]. As a result, we follow a different approach.

4 Basic Algorithm

Instead of a Hamiltonian cycle, our approach is based on the existence of an Eulerian circuit in the dual graph of the Q-mesh, which is guaranteed for every connected graph in which all the vertices have even order [12]. After the quadrilateral faces are split, the resulting triangles are linked through the original mesh edges. The new edges associated with the diagonal splits become the boundary edges of the resulting Hamiltonian T-strip cycle.

Note however, that not every Eulerian circuit on the dual graph of a Q-mesh defines a Hamiltonian triangulation, because edges that are opposite to each other on a face cannot be contiguous in the Eulerian circuit. Figure 2 (Color Plate 40 on page 447) illustrates this problem. We need a special algorithm to construct the Eulerian circuit taking into account the constraints imposed by the cyclical ordering of the edges around each face: if two mesh edges are not adjacent in the graph of the mesh, they cannot be contiguous in the Eulerian circuit.

The problem is to choose one out of the two diagonals of each quadrilateral face so that the result of splitting the faces along the chosen diagonals defines a single Hamiltonian T-strip cycle. If the diagonals are chosen at random, the graph defined by the triangles as graph vertices and the original mesh edges as graph edges is composed of a collection of disconnected cycles because each graph vertex is connected to exactly two other graph vertices. If the number of cycles in this graph is one, the problem is solved. If the number is larger than one, we only need to perform one additional step during which some diagonals are flipped. This step is based on the following observation illustrated in figure 3: if the two triangles generated by splitting one quadrilateral face belong to different cycles, splitting the same face along the other diagonal (flipping the diagonal) joins the two cycles into a single one. Since the number never increases, to construct a single cycle we just have to visit the faces of the mesh in an arbitrary order, and flip the diagonal of each face that joins different cycles. This simple algorithm is illustrated in figure 4. Figures 5 and 6 (Color Plate 41 on page 448) show examples of this algorithm applied to Q-meshes, with the state at an intermediate point where some diagonals have been flipped but not all. It is important to note that flipping a diagonal that joins two triangles that belong to the same cycle must be avoided, because doing so splits the cycle into two disconnected cycles.

5 Efficient Implementation

Figure 7 illustrates an efficient and more detailed implementation of the algorithm of figure 4. This implementation uses two data structures. The first one is a boolean array B with one bit b_f per face that indicates along which of the two diagonals the face must be split. If $b_f = 0$ the face $f = (v_0, v_1, v_2, v_3)$ is split along the (v_0, v_2) diagonal, and along the (v_1, v_3) diagonal if $b_f = 1$. This array is filled by the calling function and modified here so that splitting the faces along the corresponding diagonals generates a Hamiltonian triangulation. The second data structure is a set partition class based on Tarjan's fast Union-Find algorithm [20]. This set partition data structure efficiently implements the operations of membership (*find*) and union (*join*) of disjoint sets. It is used here to maintain a partition of the set of mesh edges into the cycles defined by the bits in the boolean array. It is neither necessary to maintain the cycles as linked lists, nor to explicitly split the faces into triangles. The Q-mesh is actually not modified by this algorithm, which returns just the boolean array. A subsequent traversal of the dual graph of the Q-mesh along the Eulerian circuit defined by the bits, can be used to create an explicit generalized T-strip representation. An additional hash table is used to efficiently implement the function $e_{ij} = M.\text{getEdge}(v_i, v_j)$ that locates an edge from its ends. But we consider this as part of the data structure used to represent the mesh connectivity.

Note that only the diagonals of faces belonging to the subset S of F, passed as an argument by the calling function as well, are considering for flipping. This subset must be chosen carefully to guarantee successful termination with a single cycle. A safe choice is $S = F$. Strategies to choose S containing a small number of faces will be discussed in the next section.

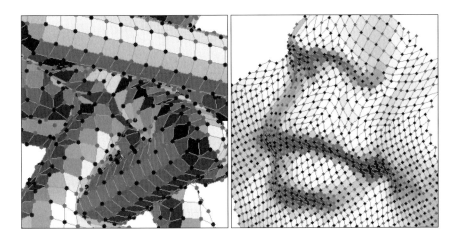

Fig. 8. Examples of 2-colorable Q-meshes.

6 Diagonal Graph Structure

The number of cycles produced by the algorithm described above, and as a consequence the number of diagonal flips necessary to link all the cycles into a single one, is in principle arbitrarily large. In this section we analyze the structure of the graph defined by the chosen diagonals in detail. This analysis will be used in the next section to develop strategies to tailor the construction of the array B and set S for applications to graphics cache optimization. In particular, we determine the minimum set of diagonals that, when considered for flipping during the second step of the algorithm, guarantee successful termination with a single cycle. We even show that in the most common cases there is actually no need to flip any diagonal.

The *diagonal graph* of a Q-mesh M is defined by the mesh vertices as graph vertices, and the diagonals of the quadrilateral faces as graph edges. If the Q-mesh has V vertices and F faces, then its diagonal graph has V vertices and $2F$ edges. The edges selected to split the faces by the algorithm described in the previous section define a maximal spanning subgraph D of the diagonal graph with V vertices and F edges.

6.1 Connected Components

We show here that D has at most two connected components. Later on we will see that we need different strategies to choose B and S for the two cases of one or two components. If we cut the Q-mesh through the chosen diagonals we obtain a single generalized triangle strip cycle. This T-mesh is topologically equivalent to a cylinder. Its boundary is a graph composed of two cycles. Let us call these cycles *left boundary* and *right boundary*. The Q-mesh vertices corresponding to left boundary T-mesh vertices and the diagonals that join pairs of these vertices form a subgraph of D that we denote D_L. This graph is clearly connected, because it can be obtained by clustering (identifying) vertices of a cycle, which is connected. A similar construction for the right boundary yields the connected subgraph D_R of D. Since D is the union of D_L and D_R, it follows that D has *at most* two connected components. But these two graphs may not be disjoint, in which case D has one connected component (is connected).

6.2 2-Coloring

A *2-coloring* of a graph is an assignment of one of two different colors (we will use red and black) to each vertex so that no edge has both ends of the same color. A graph is *2-colorable* if such assignment exists. A mesh is *2-colorable* if its graph is 2-colorable. Some examples of 2-colorable Q-meshes are shown in figure 8 (Color Plate 42 on page 448). Note that a mesh with a face with odd number of corners is not 2-colorable. In fact, a fundamental result in graph coloring is that a graph is 2-colorable if and only if it has no cycles of odd length [12].

We show now that D has two components if and only if the Q-mesh is 2-colorable. It is a lot easier to determine 2-colorability than to count connected components. For the necessity, since D has two connected components, paint red all the vertices of D_L and black all the vertices of D_R. In this case the mesh is 2-colorable because every mesh edge, being a marching edges of the T-strip, joins a vertex of D_L (red) and a vertex of D_R (black). For the sufficiency, let's assume that the Q-mesh is 2-colorable and D is connected.

Pick any edge of the mesh. Since D is connected, there is a path in D with the ends of the chosen edge as beginning and end. It follows that there is a path in the graph of the Q-mesh of even length with the same beginning and end (details left to the reader). If we add the original edge to this path we construct a cycle of odd length, which contradicts the 2-colorability.

```
Is2ColorableGraph (G = (V, E))
    T = SpanningTree (G)
    choose r ∈ V as the root of T
    # depth-first traversal
    for i ∈ V
        if depth_T(i) is even
            c_i = red
        else
            c_i = blue
    for e = (i, j) ∈ E
        if c_i = c_j
            return false
return true
```

Fig. 9. Algorithm to determine if a connected graph is 2-colorable or not.

Note that 2-colorability is independent of topological type. For example, verify that D has one connected component for the 3×3 torus of figure 5 and that the mesh is not 2-colorable (has a cycle of length 3). Do the same experiment with a 4×4 torus to verify that D has two connected components and that the mesh is 2-colorable.

It is important to point out that 2-colorable meshes are in widespread use in computer graphics, visualization, modelling and animation: it is not difficult to verify that Catmull-Clark meshes (even the non-manifold ones) and isosurfaces based on deformed cuberille Q-meshes (boundary mesh of set of voxels, regularized or not) are 2-colorable. It is also true that all planar Q-meshes (without handles) are 2-colorable (a proof is sketched in section 6.3). And a simple tree-traversal algorithm illustrated in figure 9, during which newly visited vertices are painted with alternating colors, can be used to determine if a connected graph is 2-colorable or not [12].

6.3 Spanning Forests

The *Euler characteristic* of a manifold mesh without border with V (non-singular) vertices, E edges, and F faces is the number $V - E + F$, which is a topological invariant. When a mesh is orientable, the Euler characteristic is also equal to $2 - 2G$, where G is the *genus*, or number of handles, of the mesh. A Q-mesh with V vertices, E edges, and F faces, has Euler characteristic

```
DiagonalSpanningForest (M = (V, E, F))
    B = new BooleanArray (F)
    S = new Set ()
    # build spanning forest in diagonal graph
    p_V = new Partition (V)
    for f = (v_0, v_1, v_2, v_3) ∈ F
        if p_V.find(v_0) = p_V.find(v_2)
            b_f = 0
            p_V.join(v_0, v_2)
        else if p_V.find(v_1) = p_V.find(v_3)
            b_f = 1
            p_V.join(v_1, v_3)
        else
            # collect cycle-producing faces in stack
            S.include (f)
            # assign temporary random value
    return (B, S)
```

Fig. 10. Algorithm to construct a spanning forest in the diagonal graph of a Q-mesh M with maximal number of edges, but such that at most one diagonal of each face is included in the forest. The faces without a diagonal in the forest are collected in a subset $S \subset F$ for subsequent processing. Inserting a diagonal of one of these faces in the forest creates a cycle.

$V - F$ because $E = 2F$ (each quadrilateral face is covered by four half-edges, and each edge is shared by two faces). A mesh of genus zero (with no handles) is *planar*.

Since D is composed of C (1 or 2) connected components, a spanning forest constructed in D with a maximal number of edges will contain C trees. Since the forest spans the diagonal graph, it contains $V - C$ edges, and so, the number of faces not split by a diagonal edge in the forest is $S = F - (V - C) = C + F - V$, which is typically a small number compared with the total number faces. In the orientable case this number is equal to $2G + (C - 2)$, i.e., $2G$ if the Q-mesh is 2-colorable, and $2G - 1$ if not. For example, for any planar Q-mesh ($G = 0$) the graph D is a forest composed of exactly two trees, and there is no need to flip diagonals. In particular, this proves that any planar Q-mesh is 2-colorable, and equivalently, any non-2-colorable Q-mesh is non-planar.

In the non-planar 2-colorable case we can avoid flipping diagonals as well. We first construct a spanning forest with a maximal number of edges in the diagonal graph, and collect the faces whose diagonals create cycles if inserted in the forest. Figure 10 illustrates an algorithm to do so, where the non-split faces are collected in the set S. Then we paint red all the vertices of one of the trees, and black all the vertices of the other tree. Finally, we construct D by inserting all the diagonals of the faces in S that join black vertices in the forest. At this point there is no need to run the algorithm of figure 7 because the set of diagonals already define a Hamiltonian cycle. The result of splitting the faces through this set of diagonals is a single cycle because cutting through the red tree produces a boundary cycle with as many vertices and edges as edges in the tree, and every triangle has either one or two vertices on this boundary. This cycle is one of the boundaries of the resulting T-strip. This is closely related to the Topological Surgery scheme for P-mesh connectivity compression [21], particularly as implemented in the MPEG-4 standard [16]. Figure 11 illustrates the relation between a tree of edges and the boundary cycle it produces for a general P-mesh.

In the non-2-colorable case D is connected and the spanning forest is composed of a single tree. Again, cutting the Q-mesh through the diagonals belonging to the spanning tree produces a new mesh with a single boundary cycle. This mesh is composed of the remaining quadrilateral faces collected in the set S, and a number of generalized T-strips, each starting and ending at an edge shared with a quadrilateral. There are $2S$ T-strips counting T-strips of length zero corresponding to edges shared by two quadrilateral faces. Each of these T-strips may start and end in different quadrilaterals, or in the same quadrilateral. In the later case, the two edges that connect the T-strip to the quadrilateral may be opposite or adjacent. If opposite, neither one of the two diagonals splits the mesh into disconnected parts. If adjacent, the diagonal that leaves the two edges on opposites sides is the only choice that keeps the mesh connected. The simplest strategy here is to revert to running the algorithm of figure 7 as a post-processing step. That is, we choose the diagonals of the faces collected in S at random, partition the edges of the mesh into connected components corresponding to the generated cycles, and sequentially, flip the diagonals that join different cycles. But here we only need to consider flipping the diagonals of the faces that belong to the set S.

7 Transparent Vertex Caching

Support for T-strips requires a cache of size 2 in the graphics processing unit (GPU). Since processing a cached vertex is much faster than an uncached one, more modern GPU's maintain a larger vertex FIFO cache. To make good use of this cache these *indexed T-strips* must be constructed maximizing vertex locality. In this section we discuss strategies to optimize the use of this cache in our construction.

The triangle ordering is called *rendering sequence* by Bogomjakov and Gotsman [3], who describe methods to construct *universal rendering sequences* for T-meshes that preserve locality at all scales. Hoppe [14] introduced the *transparent vertex caching* problem, and presented algorithms to optimize the decomposition of T-meshes into T-strips for a particular cache size. Several years earlier, Deering [9] presented a hardware-oriented geometry compression scheme based on an actively managed (non-FIFO) cache of

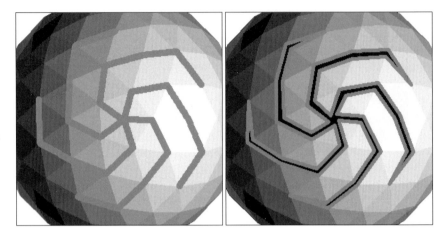

Fig. 11. Cutting through a tree, with green edges in the left picture, creates a boundary cycle, also with green boundary edges in the right picture. The black gap on the right picture is the back of the triangles on the other side of the sphere, which are not iluminated.

size 16. Chow [7] presented methods to decompose T-meshes into Deering's *generalized triangle meshes*, and so did Bar-Yehuda and Gotsman [2], who also showed that a cache of size $O(\sqrt{n})$ is necessary to minimize cache misses to zero.

A good rendering sequence minimizes the *average cache miss ratio* (acmr). This number, which measures the average number of cache misses per triangle, has a minimum value of about 0.5 (one cache miss per vertex) and a maximum of 3.0, which corresponds to the case when each T-strip is composed of a single triangle. In our case, since we can link all the triangles into a single T-strip, the actual maximum is slightly above 1.0.

7.1 Greedy Approach

As in the method proposed by Bogomjakov and Gotsman [3] for T-meshes, our scheme produces a rendering sequence independent of the cache size. Rather than an optimization-based procedure, we experimented with simple greedy schemes based on different face traversal algorithms and strategies to choose diagonals as the faces are visited. We implemented the traversal of faces in spiraling fashion, as used in most P-mesh connectivity encoding algorithms, and in random order, i.e., in the order the faces appear in the input file. When a face is visited during the traversal, one of its two diagonals is chosen. To maximize locality, we follow strategies that build spanning forests with lots of leaf nodes. This is illustrated in figure 12. A large number of leaf nodes implies a large number of branching nodes. The three most successful strategies were: 1) choosing the diagonal that locally maximizes the

Fig. 12. Maximizing vertex locality requires trees with large numbers of leafs and branching nodes.

vertex valence of the spanning forest constructed so far, 2) always choosing the diagonal across the traversal (in the spiraling traversal), and 3) randomly choosing one of the two diagonals as a function of a random face bit produced by a pseudo-random number generator. In the later case, each run produces a different result, but the results of different runs are very consistent. On the other hand, we have verified experimentally that a strategy that minimizes the number of leafs and branching nodes in the spanning forest produces much worse rendering sequence, often with $acmr \approx 1$.

7.2 Results

Figure 13 shows some of the models used in our experiments. The table in figure 14 shows the sizes and genus of these meshes, as well as the one shown in figure 1 (Color Plate 39 on page 447), and the $acmr$ we obtain for different cache sizes. The cube was obtained by recursive Catmull-Clark connectivity subdividision from a regular cube with six faces, and subsequent low-pass filtering. The bunny is not the original Stanford bunny, but a resampled version with quadrilateral faces and filled boundaries. The shape and spine meshes are isosurfaces. The skull and the head originally came from 3D scanned data and both had lots of boundaries. To remove the boundaries we generated the boundary surface of the solid resulting from extruding the mesh along the normal direction by a fixed amount such as average edge length. The resulting mesh is composed of two parallel copies of the original mesh connected along the corresponding boundaries by cycles of quadrilateral faces.

Roughly speaking, we obtain average $acmr$ of about 0.70 for cache size 16, 0.64 for cache size 32, and 0.60 for cache size 64, with very small deviation from these values, and very much independently of mesh size and regularity of the mesh. This performance values are comparable to those reported by Bogomjakov and Gotsman [3], whose algorithm for T-meshes is based on a complex combinatorial optimization procedure. Our results are also comparable with those produced by the greedy algorithm for T-meshes presented by Hoppe [14], which is tuned for a specific cache size.

It would be interesting to establish a theoretical *acmr* lower bound for Hamiltonian strips, and to see how much room is there for improvements using an additional combinatorial optimization step. Most probably, if such lower bound can be established, it will be larger than the 0.5 lower bound for arbitrary rendering sequences. However, given the good and consistent performance of the very simple strategies presented here, and the low computational cost, we do not see much practical need for further optimization.

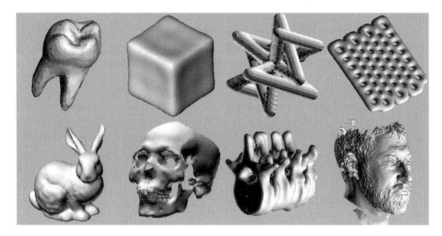

Fig. 13. Some Q-meshes used in our experiments.

8 Implementation and Complexity

We implemented all the algorithms described in previous sections in Java and integrated them into our interactive mesh processing tool. The different steps of the algorithm can be run independently of each other, or all together by pressing a single button. The user can interactively set all the parameters and options. A screen shot of this application is illustrated in figure 15. Because of this tight integration, and the additional operations performed for visualization purposes, it is difficult to measure running times with precision. Note that Hoppe [14] reports running times of up to 4 hours on meshes of about 100, 000 vertices for his optimization-based algorithm. Our greedy algorithms run at interactive rates for meshes of this size and larger. In fact, running times for our algorithms are comparable with rendering times. As

| mesh | | | | acmr | | |
| name | connectivity | | | cache size | | |
	V	F	G	16	32	64
toothQ	2,633	2,631	0	0.698	0.638	0.592
angelMouthQ	4,030	4,028	0	0.671	0.623	0.587
cube6146	6,146	6,144	0	0.685	0.619	0.583
shape51Q	6,912	6,938	14	0.714	0.641	0.602
g49plateQ	10,016	10,112	49	0.708	0.645	0.608
bunnyQ	20,758	20,756	0	0.673	0.614	0.579
skullQ	26,436	26,456	11	0.706	0.642	0.601
spine4Q	46,254	46,286	17	0.714	0.651	0.607
headscanQ	191,890	191,934	23	0.722	0.660	0.616
average				0.699	0.637	0.598

Fig. 14. Results corresponding to the meshes shown in figure 13.

mentioned before, our results are comparable with those produced by the Hoppe's greedy algorithm for T-meshes [14], and by the algorithm proposed by Bogomjakov and Gotsman [3], which is based on a complex combinatorial optimization procedure. But these last two algorithms solve a different problem. They both take an already triangulated mesh, not a quadrlateral mesh.

In terms of complexity, Tarjan's union-find algorithm [20] (used to maintain set partitions), and hash table access (used to represent mesh edges), correspond to steps with super-linear complexity. However, it is well known that the union-find algorithm requires linear storage and has linear complexity for practical purposes, and hash table access has expected linear complexity. The rest of the steps have linear space and time complexity.

9 Subdivision

Catmull-Clark Subdivision [5] is the method of choice to refine Q-meshes. Here each quadrilateral face is subdivided into four smaller quadrilateral faces. Suppose that we have chosen diagonals in the coarse mesh so that the resulting linked triangles form a Hamiltonian T-strip. The diagonal chosen to split each coarse face splits two of the corresponding fine faces as well. If the two parallel diagonals are chosen for the other two fine faces, when we link the resulting fine triangles trough the mesh edges we obtain two parallel cycles, and only one diagonal flip is sufficient to link them. For example, any of the fine faces not split by the coarse face diagonal is an acceptable choice here. Note that, although very simple, this procedure reduces the locality of the rendering sequence quite significantly. Consider two triangles that share a diagonal, and belong to the two original cycles. The distance along the joined Hamiltonian T-strip from one to the other may be up to one half the number of triangles in the strip. A better strategy is to first choose the diagonals

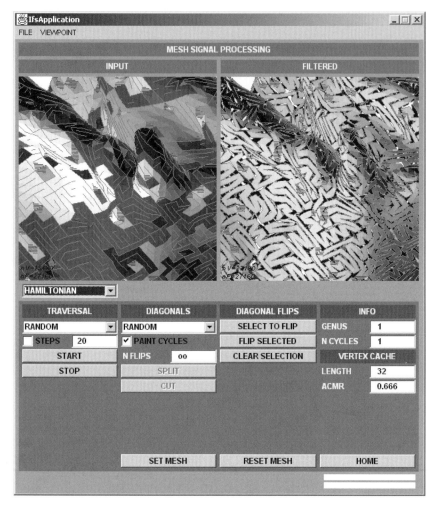

Fig. 15. Java implementation of algorithms as part of interactive mesh processing tool.

of the two fine faces not split by the original coarse diagonal orthogonal to the other diagonals, and then run the algorithm of figure 7 to join all these cycles into a single one. Figure 16 (Color Plate 43 on page 449) illustrates the concepts discussed in this section.

10 Borders

When applied to a manifold Q-meshes with border, the algorithm described above does not produce a Hamiltonian cycle. In fact, since all the original mesh edges become marching edges when the quadrilateral faces are split along the chosen diagonals, every boundary edge becomes the beginning or end of a T-strip. If the mesh has B boundary edges, the algorithm described in figures 4 and 7 will partition the mesh into $B/2$ T-strips, each one with one boundary edge as beginning, and another as end. One alternative here is to add boundary triangles to the mesh to *seal* the boundaries. We partition the boundary edges into pairs of adjacent edges. Three vertices are associated with each one of these pairs, the common vertex and the two other ends. One triangle with these vertices as corners is constructed and added to the mesh (note that we are ignoring the geometric problems that this procedure may create). The new edge that joins the two non-adjacent boundary vertices is regarded as a face diagonal. But this diagonal cannot be flipped. If all the boundary cycles of the mesh have even length, then the same algorithm will produce a single Hamiltonian T-strip. Note that all the 2-colorable Q-meshes fall into this category. A mesh with an even boundary cycle is not 2-colorable. In this case we can still connect pairs of adjacent boundary edges through an inserted triangle, and leave one open edge on each boundary of odd length.

11 Conclusions and Future Work

In this paper we presented very simple algorithms to represent any connected manifold quadrilateral mesh without boundary as a single Hamiltonian generalized triangle strip cycle in many different ways, by splitting each face along one of its two diagonals. We analyzed the structure of the graph of diagonals so produced, and developed practical strategies to choose a rendering sequence that makes good use of a graphics processing unit's FIFO vertex cache. There is potential for further optimization for cache utilization, and it will be interesting to investigate how much more can be gained by such a procedure. The algorithms presented in this paper do not extend in a straightforward manner to meshes with boundary and non-manifold meshes. It would be interesting to find ways to extend these ideas to this larger family of meshes. Converting T-meshes to Q-meshes is important for numerical simulations, but difficult to do. Very often the result of these procedures is a TQ-mesh, i.e., a mesh composed of a majority of quadrilateral faces, and a few triangular faces. We plan to extend our ideas to these meshes as well.

References

1. E.M. Arkin, M. Held, J.S.B. Mitchell, and S.S. Skiena. Hamiltonian triangulations for fast rendering. In J. Van Leeuwen, editor, *Algorithms-ESA '94*, volume 855 of *LNCS*, pages 36–47, Utrecht, NL, September 1994.
2. R. Bar-Yehuda and C. Gotsman. Time/space tradeoffs for polygon mesh rendering. *ACM Transactions on Graphics*, 15(2):141–152, April 1996.

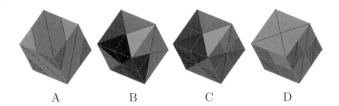

A   B   C   D

Fig. 16. Construction of Hamiltonian T-strip cycles on subdivision meshes. (A) Choosing all the diagonals of the coarse quadrilateral faces parallel to the diagonal chosen for the corresponding coarse face produces two parallel cycles with little vertex locality. (B) Flipping all the marching edges of these two cycles produces a large number of small cycles. (C) Some of these diagonals must be flipped back to link all these cycles into a single one. (D) The resulting Hamiltonian T-strip.

3. A. Bogomjakov and C. Gotsman. Universal rendering sequences for transparent vertex caching of progressive meshes. In *Proceedings, Graphics Interface, GI'2001*, pages 81–90, June 2001.

4. P. Bose and G.T. Toussaint. No quadrangulation is extremely odd. In *Proceedings, International Symposium on Algorithms and Computation*, Cairns, Australia, 1995.

5. E. Catmull and J. Clark. Recursively generated B-spline surfaces on arbitrary topological meshes. *Computer Aided Design*, 10:350–355, 1978.

6. N. Chiba and T. Nishizeki. The hamiltonian cycle problem is linear-time solvable for 4-connected planar graphs. *Journal of Algorithms*, 10(2):187–211, 1989.

7. M.M. Chow. Optimized geometry compression for real-time rendering. In *IEEE Visualization'97 Conference Proceedings*, pages 347–354, 1997.

8. R. Dafner, D. Cohen-Or, and Y. Matias Context-based Space Filling Curves. In *Eurographics 2000 Conference Proceedings*, 2000.

9. M. Deering. Geometric compression. In *Siggraph'95 Conference Proceedings*, pages 13–20, August 1995.

10. R. Estkowski, J. S. B. Mitchell, and X. Xiang. Optimal Decomposition of Polygonal Models into Triangle Strips. In *Proceedings, ACM Symposium on Computational Geometry*, 2002.

11. F. Evans, S. Skiena, and A. Varshney. Optimizing triangle strips for fast rendering. In *Proceedings, IEEE Visualization'96*, pages 319–326, 1996.

12. J. L. Gross and T. W. Tucker. *Topological Graph Theory*. Dover Publications, Inc., 2001.

13. S.L. Hakimi, E.F. Schmeichel, and C. Thomassen. On the number of hamiltonian cycles in a maximal planar graph. *Journal of Graph Theory*, pages 365–370, 1979.

14. H. Hoppe. Piecewise smooth subdivision surfaces with normal control. In *Siggraph'1999 Conference Proceedings*, pages 269–276, 1999.

15. A. King, D. Szymczak and J. Rossignac. Connectivity compression for irregular quadrilateral meshes. Technical Report GIT-GVU-99-36, Georgia Tech GVU, 1999.

16. ISO/IEC 14496-1 Information technology - Coding of audio-visual objects, Part 2: Visual / PDAM1 (MPEG-4 v.2), mar 1999.

17. S. Ramaswami, P. Ramos, and G. Toussaint. Converting triangulations to quadrangulations. In *Proceedings, Seventh Canadian Conference on Computational Geometry, CCCG'95*, 1995.
18. J. Rossignac. Edgebreaker: Connectivity compression for triangular meshes. *IEEE Transactions on Visualization and Computer Graphics*, 5(1):47–61, January-March 1999.
19. D.P. Sanders. On paths in planar graphs. *Journal of Graph Theory*, pages 341–345, 1997.
20. R.E. Tarjan. *Data Structures and Network Algorithms*. Number 44 in CBMS-NSF Regional Conference Series in Applied Mathematics. SIAM, 1983.
21. G. Taubin and J. Rossignac. Geometry Compression through Topological Surgery. *ACM Transactions on Graphics*, 17(2):84–115, April 1998.
22. G. Taubin and J. Rossignac. Course 38: 3d geometry compression. Siggraph'2000 Course Notes, July 2000.
23. R. Thomas and X. Yu. 4-connected projective-planar graphs are hamiltonian. *Journal of Combin. Theory Ser. B*, pages 114–132, 1994.
24. C. Touma and C. Gotsman. Triangle mesh compression. In *Graphics Interface Conference Proceedings*, Vancouver, June 1998.
25. W.T. Tutte. A theorem on planar graphs. *Trans. Amer. Math. Soc.*, pages 99–116, 1956.
26. L. Velho, L.H. de Figueiredo, and J. Gomes. Hierarchical generalized triangle strips. *The Visual Computer*, 15(1):21–35, 1999.
27. H. Whitney. A theorem on graphs. *Ann. Math.*, pages 378–390, 1931.
28. D. Zorin and P. Schröder. Course 23: Subdivision for modeling and animation. Siggraph'2000 Course Notes, July 2000.

Part II

Discrete Vector Fields and Topology

Visualizing Forman's Discrete Vector Field

Thomas Lewiner, Helio Lopes, and Geovan Tavares

Math&Media Laboratory, Department of Mathematics, Pontifical Catholic
University, Rio de Janeiro. Rua Marquês de São Vicente 225, Gávea, Rio de
Janeiro, Brazil, 22.453-900.
thomas.lewiner@polytechnique.org, {*lopes,tavares*}*@mat.puc-rio.br*

Summary. Morse theory has been considered to be a powerful tool in its applications to computational topology, computer graphics and geometric modeling.
Forman introduced a discrete version of it, which is purely combinatorial. This
opens Morse theory applications to a much larger scope.

The main objective of this work is to illustrate Forman's theory. We intend to
use some of Forman's concepts to visually analyze the topology of an object. We
present an algorithm to build a discrete gradient vector field on a cell complex as
defined in Forman's theory.

Keywords. Morse theory, Forman theory, vector field visualization

1 Introduction

Morse theory is a fundamental tool for investigating the topology of smooth
manifolds. Particularly for computer graphics, many applications have been
devised [10, 27, 16, 17]. Also in the new field of computational topology
[7, 29], Morse theory has been used to devise topology based algorithms and
data structures [8, 22]. The aim of this work is to visualize a similar tool for
discrete structures (see Fig. 1 and Color Plate 14 on page 434).

Fig. 1. The gradient vector field on a figure eight knot model [19].

Morse proved that the topology of a manifold is very closely related to the critical points of a real smooth map defined on it [25]. Morse theory is one of the most powerful tools to understand the topology of a manifold.

The recent results in Morse theory by Forman [12, 13] extended several aspects of this fundamental tool to cell complexes. This theory has already been used in a more theoretical context [4, 5]. The main goal of this work is to visually investigate topological aspects of a geometric or an abstract model, by using this theory.

The paper is organized as follows. In Section 2, we will briefly introduce the notion of cell complex, define discrete gradient vector field and its critical elements as defined in Forman's theory, and state a very nice result of Forman on homotopy. In Section 3 we will need some definitions of hypergraph theory, which are slightly different from the classical ones [3]. In Section 4, we will introduce our algorithm to build those gradient fields, trying to reach optimality. This algorithm is proven to give a minimal number of critical cells for the case of 2-manifolds [20]. Reaching the minimum in the general case is MAX SNP hard [11]. However, our algorithm shows to give a reasonable number of critical cells in quadratic time. We will illustrate some applications to visualization in the last section.

2 Basic Concepts

This section aims to give familiarity with Forman's theory. For a given cell complex, discrete Morse theory as introduced by Forman can be built on a class of discrete gradient vector field and its critical elements. We will define those notions in the following paragraphs.

Similarly to the classical Morse Theory, Forman proved that the topology of a cell complex is related to its critical elements in a very strong way. More precisely, a cell complex with a discrete gradient vector field V is homotopy equivalent to a complex composed of only the critical elements of V.

For a complete presentation of Forman's theory and its application, see [12, 13, 14, 15].

A *cell complex* is, roughly speaking, a generalization of the structures used to represent solid models: it is a consistent collection of cells (vertices, edges,...).

More formally, a *cell* $\alpha^{(p)}$ of dimension p is a set homeomorphic to the p-ball $\{x \in \mathbb{R}^p : \|x\| \leq 1\}$. When the dimension p of the cell is obvious, we will simply denote α instead of $\alpha^{(p)}$.

A *cell complex* K of dimension n is a collection of p-cells, $0 \leq p \leq n$, such that every intersection of two cells of K is also a cell of K. A complete introduction to cell complexes can be found in [24].

A p-cell $\alpha^{(p)}$ is a *face* of a q-cell $\beta^{(q)}$ ($p < q$) if $\alpha \in \beta$. We will use the notation $\alpha^{(p)} \prec \beta^{(q)}$, and say that α and β are *incident*.

In this paper, we will only consider finite cell complexes, i.e. complexes with a finite number of cells.

2.1 Forman's Discrete Gradient Vector Fields

Forman's theory relies on admissible functions on a cell complex, or equivalently their gradient vector field. We chose here to introduce the theory from the second point of view, although our construction of a vector field can be done in the same way to define discrete Morse function.

Definition 4 (Combinatorial vector field).
   A combinatorial vector field V defined on a cell complex K is a disjoint collection of pairs $\{\alpha^{(p)}, \beta^{(p+1)}\}$ of incident cells : $\alpha^{(p)} \prec \beta^{(p+1)}$.
   For such pairs, $V(\alpha) = \beta$ and $V(\beta) = 0$. If a cell σ does not belong to any pair, then $V(\sigma) = 0$.

   We will represent this paring with an arrow from $\alpha^{(p)}$ to $\beta^{(p+1)}$.
   A *non-trivial closed V-path* is an alternate sequence of r p- and $(p+1)$-cells $\alpha_0, \beta_0, \ldots, \alpha_r, \beta_r, \alpha_{r+1} = \alpha_0$ satisfying :

$$V(\alpha_i^{(p)}) = \beta_i^{(p+1)} \quad \text{and} \quad \beta_i^{(p+1)} \prec \alpha_{i+1}^{(p)} \neq \alpha_i^{(p)}.$$

Definition 5 (Discrete gradient vector field).
   A combinatorial vector field V will be called a discrete gradient vector field *if there is no non-trivial closed V-path.*

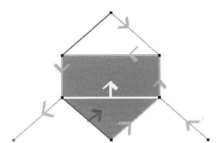

Fig. 2a. A cell complex with its discrete gradient vector field.

Fig. 2b. The Hasse diagram with the pairing.

In the example of Fig. 2a (Color Plate 15a on page 434), the discrete gradient vector field V is represented by arrows, from a cell of the complex to its image by V: from an edge to a face, and from a vertex to an edge.
   The corresponding Hasse diagram (Fig. 2b and Color Plate 15b on page 434) represents every cell by one node. The faces (2-cells) are aligned on top rank, the edges (1-cells) on the middle one and the vertices (0-cells) on the

bottom rank. A link between two nodes symbolizes that the corresponding cells are incident. We paired cells linked by a red line. Blue lines represent the incidences we selected for the spanning tree, as we will do in the algorithm (see Section 4).

The Hasse diagram is drawn as a directed graph with the AT&T software named `GraphViz - dot` [9].

Morse proved that the topology of a manifold is related to its critical elements. Forman gave an analogous result with the following definition for the critical cells.

Definition 6 (Critical Cells).
A cell α is critical *if it is not paired with any other cell, i.e.:*

$$V(\alpha) = 0 \qquad \text{and} \qquad \alpha \notin Im(V)$$

In the example of Fig. 2a (Color Plate 15a on page 434), critical cells are drawn in red: there is one critical vertex and one critical edge. In the Hasse diagram 2b (Color Plate 15b on page 434), red nodes represent those critical cells.

The number of critical cells is not a topological invariant, as it depends on the discrete gradient vector field defined. For example, an empty discrete vector field (i.e. no cells are paired) would have all its cells critical. Our algorithm is proven to give a minimal number of critical cells for the case of 2-manifolds [20]. Reaching the minimum in the general case is MAX SNP hard [11]. However, our algorithm gives a reasonable number of critical cells in quadratic time.

2.2 Homotopy Properties

Forman proved that a cell complex with a discrete gradient vector field V is *homotopy equivalent* to a complex built with exactly one cell for each critical element of V. In the example of Fig. 2a (Color Plate 15a on page 434), there is one critical vertex and one critical edge: the corresponding complex has the homotopy of a circle.

Homotopy equivalence means continuous deformation (see [1]), and Forman gave an explicit way of doing this deformation.

From homotopy theory, we know there is exactly one critical vertex (1-cell) per connected component of the complex. Starting from that vertex, we can follow the gradient to go from one cell to the incident ones, and from those to their paired cells, and so on as on figure... Critical cells are not paired, so this route stops at those, and forks at regular cells.

Forman proved that the inverse routes, without the critical parts, are deformation retracts, so do not alter the homotopy. The routes that end with a critical cell can also be retracted as above if we glue back the critical cell on the remaining cells. Thus, the critical cells represent the modification of the homotopy in this route.

Fig. 3. The inverse route of the gradient of last example

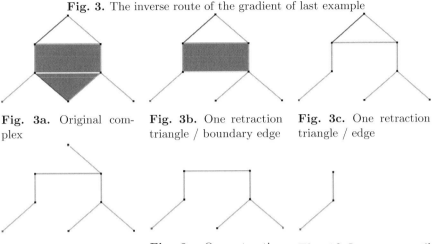

Fig. 3a. Original complex

Fig. 3b. One retraction triangle / boundary edge

Fig. 3c. One retraction triangle / edge

Fig. 3d. Passing below a critical cell

Fig. 3e. One retraction vertex / edge

Fig. 3f. Last steps until the critical vertex

In Fig. 3 we see this route at different steps. This corresponds to cutting a differentiable manifold at different heights, as in classical Morse theory [25], although Forman's theory is completely independent of the geometry of the complex.

3 Hypergraphs and Hypertrees

We need here to generalize the notion of graphs. For example, in the triangulation of a solid, an edge can be incident to more than one face. Thus the graph whose nodes are the triangles of the triangulation, and whose links joins triangles that share an edge would not be an ordinary graph: such links can have more than two end nodes.

We use here a slightly different structure of hypertrees than the classical ones. A complete introduction to hypergraphs can be found in [3] for more details.

Definition 7 (Hypergraph).
A hypergraph is a pair (N, L). N is the set of nodes. The elements of L are family of nodes, and are called hyperlinks.

We will classify hyperlinks into the *regular hyperlinks* (or shortly, *link*), which join two distinct nodes as in ordinary graphs, and the *non-regular hyperlinks*, which loop on one node or join three nodes or more.

We will give a hypergraph a simple orientation by distinguish one node in every non-regular hyperlink. We will call that node the *source node* of the hyperlink lk and write n_{lk}. The other nodes of lk will be called destination nodes of lk. The regular hyperlinks are not necessarily oriented.

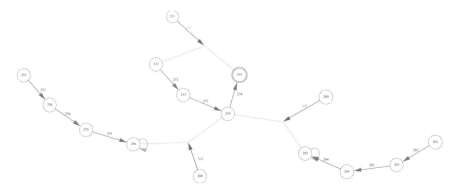

Fig. 4. A part of the dual hypertree resulting while processing a solid 2-sphere.

The graph of Fig. 4 (Color Plate 16 on page 434) represents a simply oriented hypergraph. Every regular link (in blue) has not yet a meaningful orientation (it represents the gradient vector field).

The non-regular hyperlinks are of two kinds: those incident to only one node (boundary links in orange), and those incident to more than two nodes (in green). In both cases, exactly one node is the origin of the non-regular hyperlink.

Definition 8 (Regular components).

The regular components *of a hypergraph* (N, L) *are the connected components of the ordinary graph* (N, R)*, where R is the set of regular hyperlinks.*

An hypercircuit in a simply oriented hypergraph is a sequence of hyperlinks $lk_1, lk_2, \ldots, lk_r$ where :

– lk_i and lk_{i+1} share a node, with the convention $lk_{r+1} = lk_1 : lk_i \cap lk_{i+1} \neq \emptyset$.
– if lk_{i+1} is a non-regular hyperlink, the shared node of lk_i and lk_{i+1} is the source node of $lk_{i+1} : lk_i \cap lk_{i+1} = n_{lk_i}$.

Definition 9 (Hypertree).

We will say that a simply oriented hypergraph (N,L) is a hypertree if the 3 conditions below are satisfied :

1. *Every regular component of* (N, L) *is an ordinary tree.*
2. *There is at most one source node in each regular component.*
3. (N, L) *has no hypercircuit.*

In Fig. 5 for example, we can see different regular component in blue. They are isolated or connected by a non-regular hyperlink (in green). Those hyperlinks in green form a kind of tree, respecting the above Def. 9.

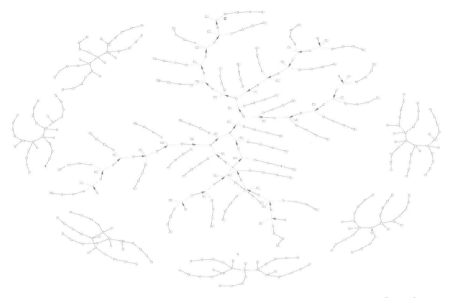

Fig. 5. The dual hypertree resulting while processing a model of $S^2 \times S^1$.

The regular component only have one source node. This source node is the one incident to a boundary link (orange loops) or to a non-regular hyperlink (on the dark green arrow side).

4 Algorithm

In this section we will introduce our algorithm to define a discrete gradient vector field for a given cell complex. This algorithm's validity and analysis will be published elsewhere.

The algorithm is optimal for surfaces [20], in the sense that it minimizes the number of critical cells. But the general case has been proven to be MAX SNP hard, i.e. any polynomial approximation can be arbitrarily far from the optimal. However, our algorithm shows to give a reasonable number of critical cells.

4.1 Outline

Let us consider a finite cell complex K of dimension n. The algorithm consists in the following steps :

1. In the first step, we select all n-cells, with some incident $(n-1)$-cells, as explained in Section 4.4.
   The algorithm optimality relies on this step, and its complexity is quadratic in the worst case. Elsewhere it has a linear complexity.

2. We then define the vector field for the selected cells as presented in Section 4.3.

   The cells of K not selected in the last step form again a complex K'. As every n-cell is selected during the first step, K' has dimension at most $n-1$.

3. So we repeat those steps until the unselected cells form a complex of dimension 1, i.e. a graph. At last, we build the vector field on that graph as explained in Section 4.2.

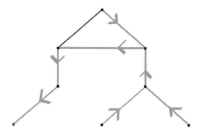

Fig. 6a. First step : selecting faces and edges in a spanning tree fashion.

Fig. 6b. Last step : processing the remaining vertex/edge graph.

Working again on the example of Fig. 2a (Color Plate 15a on page 434), we see in Fig. 4.1 the two steps of the algorithm. During the first step, the vector field is defined on a dual tree containing all faces (see 4.3 and 4.4). The unpaired vertices and edges form another cell complex, actually an ordinary graph. During the second and last step the vector field is defined on it.

4.2 Last Step: Construction on Graphs

An ordinary *graph* is a pair (N, L). N is a set which elements are called *nodes*. L is a family of pairs of nodes (i.e. duplicated edges are allowed), whose elements are called *links*. Such a graph is an ordinary *tree* if it is connected and contains no cycle.

We know from the topology of a graph that any graph is homotopy equivalent to a node with loops. From the vector field point of view, we have to pair every node except one, leaving the loops unpaired.

We build a spanning tree (with any of the classical methods) of the nodes of the graph. All the links which are not in the tree will remain unpaired, and thus be critical. We then choose a root node r for the tree, which will also be left unpaired. Then, every link $\{r, s\}$ incident to r will be paired with their other end node s. We repeat the process on all those nodes m, and so on until the leaves are reached. Finally, we have paired all the nodes and links of the spanning tree, except r.

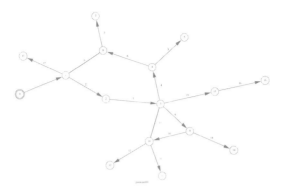

Fig. 7. The graph remaining after processing Hachimori's model of Poincare's homological sphere [18].

As the tree contains no cycle, the resulting vector field is admissible as a discrete gradient vector field. On Fig. 7, we can visualize the vector field on a graph resulting of the process of a model of Poincare's homological sphere [18]. There are two cycles in the graph, which remains unpaired and critical edges. The root node of the graph also remains unpaired, and is critical. Those graphs are drawn with the AT&T software named `GraphViz - neato` [26].

4.3 First Steps: Construction on Dual Hypertrees

As an extension of the case of a vertex spanning tree, we will define the vector field on dual hypertrees. A vertex spanning tree of a complex K has its nodes representing the vertices of K and its links representing some edges of K.

We will now consider a dual hypertree (N, L) extracted from K. Its nodes will represent the p-cells of K. A hyperlink representing a $(p-1)$-cell σ of K will join the nodes corresponding to the p-cells incident to σ. On the contrary of the previous case, the vector field will pair a node to a hyperlink. In the next section, we will present a procedure to choose the hyperlinks in such a way that (N, L) will be a hypertree.

We will process regular component per regular component. As (N, L) does not have any hypercircuit (Def. 9 condition 3), there is at least one component that has no destination node. If this component has a source node (and there is at most one by Def. 9 condition 2), we will denote it r. In the other case, r will denote an arbitrary node.

The component is an ordinary tree (Def. 9 condition 1). Thus we can pair each leaf node with its unique incident link. The unpaired elements of the component form again an ordinary tree, and we repeat the pairing until there is only the root node r left.

If the component had a source node, we pair that source node with its non-regular hyperlink. In the other case, we leave the root r unpaired and it will remain critical.

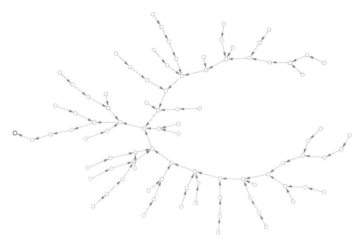

Fig. 8. A spanning tree tetrahedra/triangles processing Hachimori's model of Poincare's homological sphere [18].

For example on Fig. 8, the tree is processed from the leaves to the root, which is critical (in red).

As the hypertree does not contain hypercircuit, the resulting vector field will be admissible as a discrete gradient vector field.

The unpaired elements of (N, L) form again a dual hypertree, and we repeat the process above to pair all the hypertree except possibly the critical root nodes.

4.4 Selecting Cells of the Dual Hypertree

We use a greedy algorithm to select cells of the dual hypertrees. As in the construction of a spanning tree, we maintain an auxiliary structure that assigns to each cell its component number. This structure is similar to the union/find structure of [6].

The dual hypertree must contain all the cells of maximal dimension, say p, which will be represented by nodes. It must also contain some of the cells of dimension $(p - 1)$ represented by hyperlinks.

To reach optimality, we will try to select the maximum number of hyperlinks into the hypertree. For example, adding the hyperlink of the left side of Fig. 9 allows us to pair it with the node on the left. Thus, there will be less critical (unpaired) nodes.

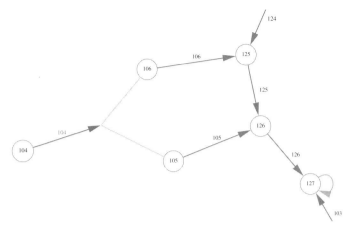

Fig. 9. Detail of a hyperlink insertion in the dual hypertree appearing with a solid torus model.

First, for each regular hyperlink, we test whether it loops inside a component or it joins two components. In the latter case, we select it for the hypertree and (lazily) update the auxiliary structure. At the end of this step, the connected components of the tree are the regular components of the final hypertree.

Then we process every *boundary cell*, i.e. a $(p - 1)$-cell that is incident to only one p-cell in the complex. We add to the hypertree at most one boundary cell per regular component (to respect Def. 9 condition 2). A regular component without boundary cell will be said *deficient* and in the other case *completed*.

Among the non-regular hyperlinks, we give priority to the boundary cell: it is the cheapest way to ensure the resulting hypergraph will not contain hypercircuit.

We finally process the non-regular hyperlinks. For each non-regular hyperlink lk left, we test if it is incident to a deficient component. We require the selected deficient components to have only one node of lk. We also check whether lk is incident to at least one completed component. In that case, we add lk to the hypertree and consider the deficient component as completed. As this has changed the configuration of deficient and connected component, we test again all non-regular hyperlinks.

As long as there are deficient components, we try until we cannot add non-regular hyperlinks. This is the bottleneck point where the algorithm is, in the worst case, quadratic in the number of those non-regular hyperlink.

5 Applications

5.1 Visualizing the Gradient Field of a Geometric Model

In differentiable Morse theory, the gradient vector field can be obtained with a Morse function by a derivative computation. For height functions, this leads to a very simple geometric interpretation of the vector field.

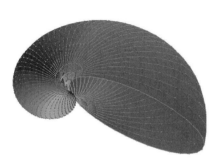

Fig. 10. A discrete gradient vector field on a shelf model.

Fig. 11. A discrete gradient vector field on a simple 2-sphere.

A similar result can be obtained by a purely combinatorial way, using Forman's theory, as in Fig. 10 (Color Plate 17 on page 435). The relation of classical Morse theory to geometry does not stand *as is* in the discrete theory. For example, the discrete gradient vector field can be disconnected from the geometry, even for simple models as for a sphere (see Fig. 11 and Color Plate 18 on page 435). This gives a real power of Forman's theory: all the above figures has been done without any geometrical test.

As mentioned in section 2.2, there is a natural way to go along the discrete gradient vector field: beginning with the critical vertex, following to the incident edge and their paired vertices, then continuing with a boundary edge to the faces and their incident edges.

The colors of the figure mark this route, following increasing the Hue component of the HSV decomposition: in green are the first visited faces, the route continue on with the blue and then purple faces, until the red ones.

Considering a classical Morse gradient field and following it in the same way, we would see cells of the same height drawn with the same color. The ordering of color (by the hue) gives the ordering of the heights. The color of the figures can be interpreted as a height function.

The discrete gradient vector field is also a powerful tool to understand the structure of a model. As above, edges drawn with the same color are at the same height. For example on figure 12 (model from the Math&Media Lab.,

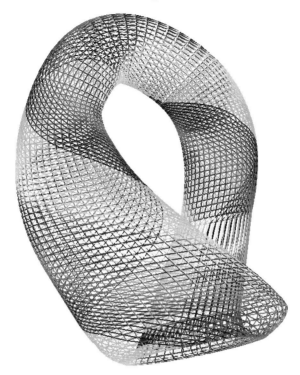

Fig. 12. The gradient vector field on edges/vertices of Klein bottle model.

created by Sinesio Pesco), we dawned the edges of a Klein bottle model to see the auto-intersection. Looking at the green edges for example, we see clearly a Möbius strip spiraling along the bottle. The discrete gradient vector field gives a more intuitive sense of the non-orientability of the Klein bottle.

5.2 Visualizing the Structure of an Abstract Complex

Many topological objects appear without a geometric model, or with a model in higher dimensions. Those structures are quite difficult to understand without visualization. Forman's theory points out a topologically consistent way of choosing significant cells. Those cells can be outlined in a Hasse diagram, as done in figures 13 and 14.

The Hasse diagram represents every cell by one node. Red nodes represent critical cells. Non-regular hyperlinks of the hypertrees are drawn in green. The cells of same dimension are displayed on the same row. The rows are ordered decreasingly on the dimension.

A link between two nodes symbolizes that the corresponding cells are incident one to the other. We linked by a red line paired cells. Blue lines represent the incidences we selected for the hypertrees and the final graph, as we did in the algorithm (see Section 4.4).

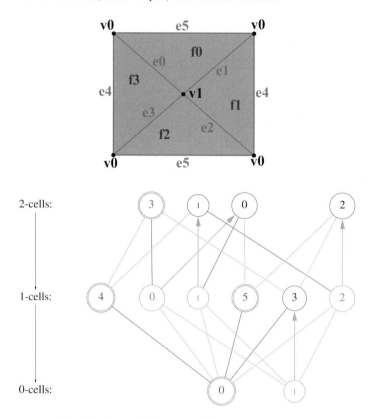

Fig. 13. A non-PL torus and its Hasse diagram.

Again, the Hasse diagram are drawn as a directed graph with the AT&T software named `GraphViz - dot` [9].

5.3 Topologically Controlling a Deformation

Morse theory studies the topology of an object by its critical points. Another way to analyze it is provided by the handlebody theory [25, 21]. This theory constructs an object by successively attaching handles to a disc. The addition of a critical point corresponds to a handle attachment. Forman provides a similar result as introduces in Section 2.2. This allows to describe a complex as glueing a few number of cells (the critical ones) *without any geometrical test*.

For example Fig. 15 (Color Plate 19 on page 435) represents a decomposition of a torus with 25600 cells in 4 critical cells: 1 face, 2 edges and 1 vertex. This result can be interpreted as follow. The first cell removed is the critical face, leading to a punctured torus (a red face in the meridian on Fig. 15a

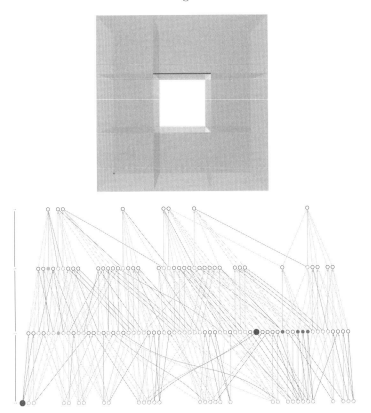

Fig. 14. A ring made of 8 cubes and its Hasse diagram.

and Color Plate 19 (a) on page 435). Then, the torus deformation along decreasing height (color) until it reaches one of the two critical edges, leading to Fig. 15b (Color Plate 19 (b) on page 435). Repeating the deformation until the second critical edge is reached, we get Fig. 15c (Color Plate 19 (c) on page 435), which is a disk. We continue retracting until having reduced all the faces (Fig. 15d and Color Plate 19 (d) on page 435), and continue reducing the remaining tree until reaching the critical vertex.

6 Future Works

We intended by this work to illustrate Forman's theory, and to use some of its concepts to visually analyze the topology of an object. We presented an explicit construction of a discrete gradient vector field. With this fundamental tool, we provided various ways of using it to visually extract topological information of a given combinatorial structure.

Fig. 15. A decomposition of a torus.

Fig. 15a. The gradient vector field.

Fig. 15b. First step: opening the torus along its meridian

Fig. 15c. Second step: opening the torus along its equator

Fig. 15d. Further steps continue on edges and vertices

An important application of this work to computer graphics would be in the field of geometric compression. The algorithm Grow&Fold of A. Szymczak and J. Rossignac [28] could be justified and enhanced by our algorithm to minimize the number of so-called "glue faces" in order to achieve a better encoding. This work has been done in an optimal way for the case of surfaces with handles in [23].

We plan to continue this work in three directions. Firstly, we intend to apply the auditory method designed by Axen and Edelsbrunner in [2], together with Forman's tools we provided. This would give a more sensitive way of studying higher-dimensional cell complexes. Secondly, we will try to develop graphical tools to capture as much as possible the topology 3-manifolds, where very hard mathematical problems remain unsolved. Finally, we look forward to produce a topologically consistent morphing based on mapping directly the discrete gradient field between two objects of the same homotopy type.

References

1. M.A. Armstrong (1979): Basic topology. McGraw-Hill, London.
2. U. Axen and H. Edelsbrunner (1998): Auditory Morse analysis of triangulated manifolds. In: Hege, H.C., Polthier, K. (ed) Mathematical Visualization, 223–236, Springer Berlin.
3. C. Berge (1970): Graphes et hypergraphes. Dunod, Paris.
4. M. Chari (2000): On discrete Morse functions and combinatorial decompositions. Discrete Math., **217**, 101–113.
5. M. Chari and M. Joswig (2001): Discrete Morse complexes. preprint.
6. C. Delfinado and H. Edelsbrunner (1993): An Incremental Algorithm for Betti Numbers of Simplicial Complexes. Proceedings of the 9th Annual Symposium on Computer Geometry, 232–239.
7. T.K. Dey, H. Edelsbrunner, and S. Guha (1999): Computational topology. In: Chazelle B., Goodman J.E., Pollack, R. (ed) Advances in Discrete and Computational Geometry, Contemporary mathematics **223**, American Mathematical Society, Providence.
8. T.K. Dey and S. Guha (2001): Algorithms for manifolds and simplicial complexes in euclidean 3-Space. preprint.
9. E. Koutsofios and S.C. North (1993): Drawing graphs with dot. Technical report, AT&T Bell Laboratories, Murray Hill, NJ. www.research.att.com/ñorth/graphviz/
10. H. Edelsbrunner, J. Harer, and A. Zomorodian (2001): Hierarchical Morse Complexes for Piecewise Linear 2-Manifolds. Proceedings of the 17th Annual Symposium on Computer Geometry, 70–79.
11. Ö. Eğecioğlu and T.F. Gonzalez (1996): A computationally intractable problem on simplicial complexes. Computational Geometry: Theory and Applications **6**, 85–98.
12. R. Forman (1995): A discrete Morse theory for cell complexes. In: Yau S.T. (ed), Geometry, Topology & Physics for Raoul Bott, International Press.
13. R. Forman (1998): Morse theory for cell complexes. Advances in Mathematics **134**, 90–145.
14. R. Forman (2001): Some applications of combinatorial differential topology. preprint.
15. R. Forman (2001): A user guide to discrete Morse theory. preprint.
16. J. Hart (1998): Morse theory for implicit surface modeling. In: Hege, H.C., Polthier, K. (ed) Mathematical Visualization 257–268, Springer Berlin.
17. J. Hart (1997): Morse theory for computer graphics. WSU Technical Report EECS-97-002. In: Hart, J., Ebert, D. (ed), New Frontiers in Modeling and Texturing.
18. M. Hachimori : Simplicial Complex Library. www.qci.jst.go.jp/~hachi/math/library/index_eng.html
19. R. Scharein : Knot-Plot. www.cs.ubc.ca/spider/scharein/
20. T. Lewiner, G. Tavares, and H. Lopes (2001): Optimal discrete Morse functions for 2-manifolds. To appear in Computational Geometry: Theory and Applications.
21. H. Lopes (1996): Algorithm to build and unbuild 2 and 3 dimensional manifolds. PhD. Thesis, PUC-Rio, Rio de Janeiro.

22. H. Lopes and G. Tavares (1997): Structure operators for modeling 3 dimensional manifolds. In: Hoffman, C., Bronsvort, W. (ed), ACM Siggraph Symposium on Solid Modeling and Applications, 10–18.

23. H. Lopes, J. Rossignac, A. Safanova, A. Szymczak, and G. Tavares (2002): Edgebreaker: a simple compression for surfaces with handles. To appear in 7th ACM Siggraph Symposium on Solid Modeling and Applications.

24. A. Lundell and S. Weingram (1969): The topology of CW complexes. Van Nostrand Reinhold, New York.

25. J. Milnor (1963): Morse theory. Princeton University Press, NJ.

26. S.C. North (1992): Neato User's Guide. Technical Report, AT&T Bell Laboratories, Murray Hill, NJ.
www.research.att.com/~north/graphviz/

27. Y. Shinagawa, T.L. Kunii, and Y.L. Kergosien (1991): Surface coding based on Morse theory. IEEE Computer Graphics and Applications **11**, 66–78.

28. A. Szymczak and J. Rossignac (2000): Grow & Fold: Compression of Tetrahedral Meshes. Computer–Aided Design **32**(8/9), 527–538.

29. G. Vegter (1997): Computational topology. In: Goodman, J.E., O'Rourke, J. (ed), Handlebook of Discrete Computational Geometry, 517–536, CRC Press.

Identifying Vector Field Singularities Using a Discrete Hodge Decomposition

Konrad Polthier and Eike Preuß

Technical University Berlin, Institute of Mathematics, MA 8-3, 10623 Berlin
polthier@math.tu-berlin.de, eike@sfb288.math.tu-berlin.de

Summary. We derive a Hodge decomposition of discrete vector fields on polyhedral surfaces, and apply it to the identification of vector field singularities. This novel approach allows us to easily detect and analyze singularities as critical points of corresponding potentials. Our method uses a global variational approach to independently compute two potentials whose gradient respectively co-gradient are rotation-free respectively divergence-free components of the vector field. The sinks and sources respectively vortices are then automatically identified as the critical points of the corresponding scalar-valued potentials. The global nature of the decomposition avoids the approximation problem of the Jacobian and higher order tensors used in local methods, while the two potentials plus a harmonic flow component are an exact decomposition of the vector field containing all information.

Keywords. vector field singularities, Hodge decomposition, discrete rotation and divergence potential function, discrete differential forms

1 Introduction and Related Work

Singularities of vector fields are among the most important features of flows. They determine the physical behavior of flows and allow one to characterize the flow topology [9][10]. The most prominent singularities are sinks, sources, and vortices. Higher order singularities often appear in magnetic fields. All these singularities must be detected and analyzed in order to understand the physical behavior of a flow or in order to use them as an ingredient for many topology-based algorithms [24][26]. Although feature analysis is an important area, only a few technical tools are available for the detection of singularities and their visualization.

Methods for direct vortex detection are often based on the assumption that there are regions with high amounts of rotation or divergence. See, for example, Banks and Singer [1] for an overview of possible quantities to investigate. The deficiencies of first-order approximations have been widely recognized, and, for example, higher-order methods try to overcome this problem [18]. The detection and visualization of higher-order singularities is an active research area where rather heavy mathematical methods have been employed [21].

The Jacobian $\nabla\xi$ of a differentiable vector field ξ in $\mathbb{R}^2$ or $\mathbb{R}^3$ can be decomposed into a stretching tensor S and a vorticity matrix Ω, the symmetric

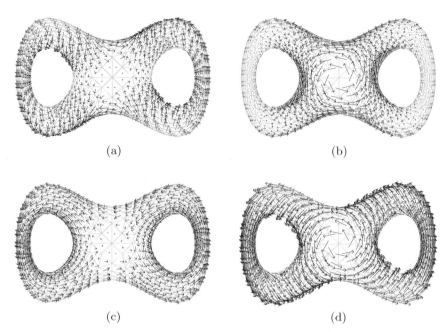

(a)   (b)

(c)   (d)

Fig. 1. Decomposition of a tangential vector field (d) on a pretzel in $\mathbb{R}^3$ in a rotation-free component (a) and a divergence-free component (b). The original vector field (d) was obtained by projection of a flow around the z-axis onto the tangential space of the curved, 2-dimensional surface. The harmonic component (c) belongs to an incompressible, rotation-free flow around the handles of the pretzel. See also Sect. 7.

and anti-symmetric parts of $\nabla \xi$. The eigenvalues of the diagonal matrix S correspond to the compressibility of the flow, and the off-diagonal entries of Ω are the components of the rotation vector. This matrix decomposition has classically also been used for discrete vector fields where the Jacobian is approximated by discrete difference techniques. The quality of this approach depends on the quality of the underlying grid and the accuracy of the vector field. For practical problems of vortex identification we refer to the case study of Kenwright and Haimes [12], and the eigenvector method in Sujudi and Haimes [22].

Another class of methods follows a geometric approach where geometric properties of streamlines and pathlines are investigated and put in relation to properties of the flow [19][20]. Tittgemeyer et al. [23] use a contraction mapping to detect singularities of displacement fields in magnetic resonance imaging. This helps in the understanding of pathological processes in a brain. Their method is applicable to any higher order singularities but fails to detect some critical points like centers of rotation or balanced saddle points.

Our approach uses a discrete version of the Hodge-Helmholtz decomposition of vector fields on curved surfaces M_h. We choose a global variational approach to compute the decomposition of a discrete vector field ξ which seems to be a novel approach to the detection and analysis of singularities of discrete vector fields. We compute two potentials u and w which determine the rotation-free and the divergence-free components of the flow. The remaining harmonic component v comprises the incompressible and irrotational component of the flow such that we have the exact decomposition (see Theorem 7)

$$\xi - \nabla u + \delta(w\omega) + v.$$

The potentials are obtained by a global variational approach where certain energy functionals are minimized in the set of scalar-valued functions on the surface M_h. The detection and analysis of vector field singularities is then transferred to the much simpler study of the critical points of the scalar valued potentials. In contrast to local methods, our approach avoids the approximation of the Jacobian matrix or higher order tensors from local information.

Although the Helmholtz decomposition [11] of smooth fields into a curl-free and divergence-free part is well-known in fluid dynamics [5], we have not found any application to the study of singularities of discrete vector fields. Discrete differential forms were introduced in differential geometry by Whitney [25] who invented the so-called Whitney forms. Whitney forms were brought to a new life in the pioneering work on discrete Hodge decompositions in computational electromagnetism by Bossavit [3][2] who applied them to the solution of boundary value problems. For simplicial complexes, Eckmann [8] developed a combinatorial Hodge theory. Dodziuk [7] showed that if a simplicial complex K is a smooth triangulation of a compact oriented Riemann manifold X then the combinatorial Hodge theory is an approximation of the Hodge theory of forms on X by choosing a suitable inner product on K.

Our discretization method has connections with weak derivatives used in finite element theory where the formal application of partial integration is used to shift the differentiation operation to differentiable test functions. In fact, the integrands of our discrete differential operators div_h and rot_h can be obtained from $\nabla\xi$ by formal partial integration with test functions. In contrast, our focus here is to emphasize the geometric interpretation of the discrete differentials, and to relate them with the discrete Hodge operator which also played a role in the discrete minimal surface theory in Polthier [14].

In Sect. 7 we apply our method to several test cases with artificial and simulated flows which are accurately analyzed. The simulated flow in the Bay of Gdansk reproduces similar results of Post and Sadarjoen [19], who applied different geometric methods.

2 Setup

In the following let M_h be a simplicial surface immersed in $\mathbb{R}^n$ (possibly with self-intersections and/or boundary), that is a surface consisting of planar triangles where the topological neighbourhood of any vertex consists of a collection of triangles homeomorphic to a disk (see [14] for an exact definition of a simplicial surface). We need the following finite element spaces, see the books [6][4] for an introduction.

Definition 10. *On a simplicial surface M_h we define the function space S_h of* conforming finite elements*:*

$$S_h := \left\{ v : M_h \to \mathbb{R} \mid v \in C^0(M_h) \text{ and } v \text{ is linear on each triangle} \right\}$$

S_h is a finite dimensional space spanned by the Lagrange basis functions $\{\varphi_1, .., \varphi_n\}$ corresponding to the set of vertices $\{p_1, ..., p_n\}$ of M_h, that is for each vertex p_i we have a function

$$\begin{aligned} \varphi_i : M_h &\to \mathbb{R}, \; \varphi_i \in S_h \\ \varphi_i(p_j) &= \delta_{ij} \quad \forall i, j \in \{1, .., n\} \\ \varphi_i \text{ is } &\text{linear on each triangle.} \end{aligned} \qquad (1)$$

Then each function $u_h \in S_h$ has a unique representation

$$u_h(p) = \sum_{j=1}^{n} u_j \varphi_j(p) \quad \forall \, p \in M_h$$

where $u_j = u_h(p_j) \in \mathbb{R}$. The function u_h is uniquely determined by its nodal vector $(u_1, ..., u_n) \in \mathbb{R}^n$.

The space of non-conforming finite elements includes discontinuous functions such that their use is still sometimes titled as a *variational crime* in the finite element literature. Nevertheless, non-conforming functions naturally appear in the Hodge decomposition and in the theory of discrete minimal surfaces [14].

Definition 11. *For a simplicial surface M_h, we define the space of* non-conforming finite elements *by*

$$S_h^* := \left\{ v : M_h \to \mathbb{R} \; \middle| \; \begin{array}{l} v_{|T} \text{ is linear for each } T \in M_h, \text{ and} \\ v \text{ is continuous at all edge midpoints} \end{array} \right\}$$

The space S_h^* is no longer a finite dimensional subspace of $H^1(M_h)$ as in the case of conforming elements, but S_h^* is a superset of S_h. Let $\{m_i\}$ denote the set of edge midpoints of M_h, then for each edge midpoint m_i we have a basis function

$$\begin{aligned} \psi_i : M_h &\to \mathbb{R} \quad \psi_i \in S_h^* \\ \psi_i(m_j) &= \delta_{ij} \quad \forall \, i, j \in \{1, 2, ..\} \\ \psi_i \text{ is } &\text{linear on each triangle.} \end{aligned} \qquad (2)$$

The support of a function ψ_i consists of the (at most two) triangles adjacent to the edge e_i, and ψ_i is usually not continuous on M_h. Each function $v \in S_h^*$ has a representation

$$v_h(p) = \sum_{\text{edges } e_i} v_i \psi_i(p) \quad \forall\, p \in M_h$$

where $v_i = v_h(m_i)$ is the value of v_h at the edge midpoint m_i of e_i.

We use the following space of vector fields on M_h

$$\Lambda_h^1 := \{ v \mid v_{|T} \text{ is a constant, tangential vector on each triangle} \}$$

which will later be considered in the wider setup of discrete differential forms. As common practice in the finite element context, the subindex h distinguishes this set from smooth concepts.

On each oriented triangle, we define the operator J that rotates each vector by an angle $\frac{\pi}{2}$ on the oriented surface.

Note, in this paper we use a slightly more general definition of the spaces S_h respectively S_h^*, namely we include functions which are only defined at vertices respectively at edge midpoints. For example, the (total) Gauß curvature is defined solely at vertices. Here for a given vector field ξ we will have $\mathrm{div}_h \xi \in S_h$ (respectively $div_h^* \xi \in S_h^*$) to be defined solely at a vertex. The motivation of this generalization is two-fold: first, a simplified notation of many statements, and, second, the fact that for visualization purposes one often extends these point-based values over the surface. For example, barycentric interpolation allows to color the interior of triangles based on the discrete Gauss curvature at its vertices. Caution should be taken if integral entities are derived.

3 Discrete Rotation

The rotation of a differentiable vector field on a smooth surfaces is at each point p a vector normal to the surface whose length measures the angular momentum of the flow. On a planar Euclidean domain with local coordinates (x, y), the rotation of a differentiable vector field $v = (v_1, v_2)$ is given by $\mathrm{rot}\, v = (0, 0, \frac{\partial}{\partial x} v_2 - \frac{\partial}{\partial y} v_1)$. In the discrete version of this differential operators, the (total) discrete rotation, we neglect the vectorial aspect and consider the rotation as a scalar value given by the normal component.

In the following we use a simplicial domain M_h which contains its boundary. The boundary is assumed to be counter clockwise parametrized. If $p \in \partial M_h$ is a vertex on the boundary then $\mathrm{star}\, p$ consists of all triangles containing p. If $m \in \partial M_h$ is the midpoint of an edge c then $\partial \mathrm{star}\, m$ does contain the edge c as well.

Definition 12. *Let $v \in \Lambda_h^1$ be a piecewise constant vector field on a simplicial surface M_h. Then the (total) discrete rotation* $\operatorname{rot} v$ *is a vertex based function in S_h given by*

$$\operatorname{rot}_h v(p) := \frac{1}{2} \int_{\partial\operatorname{star} p} v = \frac{1}{2} \sum_{i=1}^{k} \langle v, c_i \rangle$$

where c_i are the edges of the oriented boundary of the star of $p \in M_h$.

Additionally, the discrete rotation $\operatorname{rot}^* v$ at the midpoint m of each edge c is an edge-midpoint based function in S_h^* given by

$$\operatorname{rot}_h^* v(m) := \int_{\partial\operatorname{star} c(m)} v$$

where $\partial\operatorname{star} m$ is the oriented boundary of the triangles adjacent to edge c.

If the rotation of a vector field is positive on each edge of the link of a vertex then the vector field rotates counter clockwise around this vertex. Note that rot_h^* vanishes along the boundary. One easily shows the following Lemma.

Lemma 1. *Let p be a vertex of a simplicial surface M_h with emanating edges $\{c_1, ..., c_k\}$ with edge midpoints m_i. Then*

$$2\operatorname{rot}_h v(p) = \sum_{i=1}^{k} \operatorname{rot}_h^* v(m_i) \ .$$

Note, that $\operatorname{rot}_h^* v = 0$ at all edge midpoints implies $\operatorname{rot}_h v = 0$ on all vertices. The converse is not true in general.

Rotation-free vector fields are characterized by the existence of a discrete potential.

Theorem 3. *Let M_h be a simply connected simplicial surface with a piecewise constant vector field v. Then $v = \nabla u$ is the gradient of a function $u \in S_h$ if and only if*

$$\operatorname{rot}_h^* v(m) = 0 \quad \forall \ edge \ midpoints \ m \ .$$

Similarly, $v = \nabla u^*$ is the gradient of a function $u^* \in S_h^*$ if and only if

$$\operatorname{rot}_h v(p) = 0 \quad \forall \ interior \ vertices \ p.$$

Further, for a vertex $q \in \partial M_h$ the value $\operatorname{rot}_h \nabla u^*(q)$ is the difference of $u^*(q)$ at the two adjacent boundary triangles.

Proof. 1.) " $\Rightarrow$ ": Assume the orientation of the common edge $c = T_1 \cap T_2$ of two triangles leads to a positive orientation of ∂T_1. Then we obtain from the definition of rot_h^*

$$\operatorname{rot}_h^* v(m) = - \langle v_{|T_1}, c \rangle + \langle v_{|T_2}, c \rangle \ .$$

Let T_1 be a triangle with vertices $\{p_1, p_2, p_3\}$ and edges $c_j = p_{j-1} - p_{j+1}$. Assume $v = \nabla u_h$ is the gradient of a piecewise linear function $u_h \in S_h$. Let $u_j = u_h(p_j)$ be the function values at the vertices of T_1 then

$$\langle \nabla u_h, c_j \rangle = u_{j-1} - u_{j+1} \ .$$

The sum of the two scalar products $\langle \nabla u_h, c \rangle$ at the common edge of two adjacent triangles cancels because of the continuity of u_h and the reversed orientation of c in the second triangle.

" $\Leftarrow$ ": We construct a vertex spanning tree of M_h and orient its edges towards the root of the tree. Since v is rotation-free the scalar product of v with each oriented edge c_j is unique, and we denote it with $v_j := \langle v, c_j \rangle$. Now we construct a function u_h by assigning $u_h(r) := 0$ at the root of the spanning tree, and integrate along the edges of the spanning tree such that

$$u_h(p_{j_2}) - u_h(p_{j_1}) = v_j$$

if $c_j = p_{j_2} - p_{j_1}$. This leads to a function $u_h \in S_h$. On each triangle T we have $\nabla u_{|T} = v_{|T}$ since by construction we have $\langle \nabla u_h, c_j \rangle = u_h(p_{j_2}) - u_h(p_{j_1}) = v_j$ on each edge c_j.

We now show that the function u_h is independent of the choice of the spanning tree. It is sufficient to show that the integration is path independent on each triangle (which is clear) and around the link of each vertex. Around a vertex p denote the vertices of its oriented link with $\{q_1, ..., q_s\}$. Since we have the edge differences $u(p) - u(q_j) = v_j$ it follows that

$$\sum_{j=1}^{s} u(q_j) - u(q_{j+1}) = \sum_{j=1}^{s} u(q_j) - u(p) + u(p) - u(q_{j+1})$$

$$= \sum_{j=1}^{s} -v_j + v_{j|1} = 0 \ .$$

The function u_h depends solely on v and the integration constant $u_h(r)$.

2.) The second assumption follows from a similar calculation which we only sketch here.

" $\Rightarrow$ ": If v is the gradient of a function $u^* \in S_h^*$ then the assumption follows since at all interior vertices p we have

$$\int_{\gamma_p} \nabla u^* = \sum_{j=1}^{n} \langle \nabla u^*, m_j - m_{j-1} \rangle$$

$$= \sum_{j=1}^{n} u^*(m_j) - u^*(m_{j-1}) = 0$$

where γ_p is a polygon connecting the midpoints m_j of all edges emanating from p.

" $\Leftarrow$ " By assumption, the path integral of v along any closed curve γ crossing edges at their midpoints vanishes. Since v integrates to a linear function on each triangle, we obtain a well-defined function $u_h^* \in S_h^*$ similar to the procedure in 1.) by

$$u_h^*(p) := \int_\gamma v$$

where γ is any path from a base point $r \in M_h$ to p which crosses edges at their midpoints.

3.) The statement on the height difference of u^* at a boundary vertex q follows directly from the evaluation of rot $_h \nabla u^*(q)$.

The above theorem does not hold for non-simply connected surfaces since integration along closed curves, which are not null-homotopic, may lead to periods. Also note, that from $S_h \subset S_h^*$ follows $0 = $ rot $_h \nabla u = $ rot $_h^* \nabla u$ for any $u \in S_h$.

4 Discrete Divergence

In the smooth case the divergence of a field is a real-valued function measuring at each point p on a surface the amount of flow generated in an infinitesimal region around p. On a planar Euclidean domain with local coordinates (x, y), the divergence of a differentiable vector field $v = (v_1, v_2)$ is given by div $v = v_{1|x} + v_{2|y}$. The discrete version of this differential operators, the (total) discrete divergence, is obtained by a similar physical reasoning, that means we define the discrete divergence as the amount of flow generated inside the star p of a vertex p which is the total amount flowing through the boundary of star p.

At a boundary vertex p the discrete divergence must take into account the flow through the two boundary edges as well as divergence generated at all other edges emanating from p since the divergence at these interior edges has only been considered by half at interior vertices. The following definition also fulfills the formal integration by parts relation (3):

Definition 13. *Let $v \in \Lambda_h^1$ be a piecewise constant vector field on a simplicial surface M_h. Then the (total) discrete divergence div $_h : \Lambda_h^1 \to S_h$ of v is a vertex-based function given by*

$$\text{div }_h v(p) = \frac{1}{2} \int_{\partial \text{star } p} \langle v, \nu \rangle \, ds$$

where ν is the exterior normal along the oriented boundary of the star of $p \in M_h$. If $p \in \partial M_h$ then star p consists of all triangles containing p.

Additionally, we define the divergence operator div $_h^* : \Lambda_h^1 \to S_h^*$ based at the midpoint m of an edge c

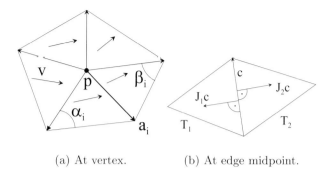

(a) At vertex.   (b) At edge midpoint.

Fig. 2. On the definition of discrete divergence.

$$\operatorname{div}_h^* v(m) = \int_{\partial \text{star } c} \langle v, \nu \rangle \, ds$$

where ∂star m is the oriented boundary of the triangles adjacent to edge c. If $m \in \partial M_h$ then ∂star m does not contain the edge c.

Note, the divergence div_h^* at an edge c common to two triangles T_1 and T_2 may equivalently be defined by $\operatorname{div}_h^* v(m) = \langle v, Jc_1 \rangle_{|T_1} + \langle v, Jc_2 \rangle_{|T_2}$ where the common edge has opposite orientation $c_1 = -c_2$ in each triangle. Let $\varphi_p \in S_h$ denote the Lagrange basis function associated to each vertex p of M_h. Then formally, the discrete divergence can obtained by applying Green's integration by parts

$$\operatorname{div}_h v(p) \; := \int_{\text{star } p} \text{"div } v\text{"} \cdot \varphi_p dx \tag{3}$$

$$= -\int_{\text{star } p} \langle v, \nabla \varphi_p \rangle \, dx + \int_{\partial \text{star } p} \langle v, \nu \rangle \, \varphi_p ds$$

although Green's formula does not hold in the discrete setting since v and φ_p are not differentiable on star p. On the right hand side the boundary integral vanishes since $\varphi_p = 0$ along ∂star p such that we obtain the same equation for $\operatorname{div}_h v(p)$ as in definition above.

The normalization of the divergence operator gives the following equality known from the smooth case:

$$\operatorname{div}_h \nabla u_h = \Delta_h u_h \tag{4}$$

using the discrete Laplace operator Δ_h in [13]. The same holds for the operators on S_h^*.

The discrete version of the Gauß integration theorem relates the divergence of a domain to the flow through its boundary.

Theorem 4. *Let M_h be a simplicial surface with boundary ∂M_h and piecewise constant vector field v. Then*

$$\sum_{p \in M_h} \operatorname{div}_h v(p) = \int_{\partial M_h} \langle v, \nu \rangle \tag{5}$$

where ν is the exterior normal along ∂M_h. Further, we have

$$\sum_{m \in M_h} \operatorname{div}_h^* v(m) = \int_{\partial M_h} \langle v, \nu \rangle \tag{6}$$

where m runs through the midpoints of all edges of M_h.

Proof. Sort and count edges, using that the integral along all edges of a single triangle vanishes.

For practical applications, we compute the explicit formula for the discrete divergence operator in terms of triangle quantities. Note the similarity with the formula of the discrete Laplace operator [13] where the influence of the domain metric solely appears in the cotangent factor.

Theorem 5. *Let $v \in \Lambda_h^1$ be a piecewise constant vector field on a simplicial surface M_h. Then the discrete divergence div_h of v is given at each vertex p by*

$$\operatorname{div}_h v(p) = -\frac{1}{2} \sum_{i=1}^{k} \langle v, Jc_i \rangle = \frac{1}{2} \sum_{i=1}^{k} (\cot \alpha_i + \cot \beta_i) \langle v, a_i \rangle \tag{7}$$

where J denotes the rotation of a vector by $\frac{\pi}{2}$ in each triangle, k the number of directed edges a_i emanating from p, and the edges c_i form the closed cycle of $\partial \mathrm{star}\, p$ in counter clockwise order. In the two triangles adjacent to an edge a_i we denote the vertex angles at the vertices opposite to a_i with α_i and β_i (see Fig. 2).

Proof. By definition Jc rotates an edge such that it points into the triangle, i.e. $Jc = -\nu|c|$ is in opposite direction of the outer normal of the triangle at c. Therefore, the representation of the discrete divergence operator follows from the representation of

$$Jc = \cot \alpha a + \cot \beta b$$

in each triangle with edges $c = a - b$, and sorting the terms around star p by edges.

Lemma 2. *The discrete rotation and divergence of a vector field $v \in \Lambda_h^1$ on a simplicial surface M_h relate by*

$$\operatorname{rot}_h Jv(p) = \operatorname{div}_h v(p),$$

respectively,

$$\operatorname{rot}_h^* Jv(m) = \operatorname{div}_h^* v(m)$$

where p is a vertex and m is the midpoint an edge of M_h.

Proof. The two relations of rot and div follow directly from definitions 13 and 12 of the differential operators.

The divergence at vertices and edges is related by the following lemma.

Lemma 3. *Let p be a vertex of simplicial surface M_h with emanating edges $\{c_1, ..., c_k\}$ with edge midpoints m_i. Then*

$$2\mathrm{div}\,_h v(p) = \sum_{i-1}^{k} \mathrm{div}\,_h^* v(m_i) \ .$$

Proof. On a single triangle we have $\mathrm{div}\,_h^* v(m_3) = -\mathrm{div}\,_h^* v(m_1) - \mathrm{div}\,_h^* v(m_2)$. Therefore, the right-hand side of the assumed equation is equal to

$$\int_{\partial \mathrm{star}\,p} \langle v, \nu \rangle$$

as assumed.

Divergence-free vector fields in $\mathbb{R}^2$ can be characterized by the existence of a discrete $2-$form ω with $\delta(\omega) = v$ (where δ is the co-differential operator, see also Def. 16) which is another justification of the discrete definition of div $_h$. Here we formulate the statement without the usage of differential forms which are introduced in the next section.

Theorem 6. *Let v be a piecewise constant vector field on a simply connected simplicial surface M_h. Then*

$$\mathrm{div}\,_h v(p) = 0 \quad \forall \ interior \ vertices \ p$$

if and only if there exists a function $u^* \in S_h^*$ with $v = J\nabla u^*$. Respectively,

$$\mathrm{div}\,_h^* v(m) = 0 \quad \forall \ edge \ midpoints \ m$$

if and only if there exists a function $u \in S_h$ with $v = J\nabla u$. In both cases, the function is unique up to an integration constant.

Proof. Using the relation between the discrete rotation and divergence of Lemma 2 the statement follows directly from the integrability conditions proven in Theorem 3.

5 Hodge Type Decomposition of Vector Fields

On each triangle we have a well-defined volume form ω from the induced metric of the triangle which can be expressed as $\omega = \nabla x \wedge \nabla y$ in local orthonormal coordinates (x, y) of the triangle, and for each vector v a one-form v which can expressed as $v_1 \nabla x + v_2 \nabla y$.

Definition 14. *The spaces of* discrete differential forms *on a simplicial sur-face M_h are defined piecewise per triangle T:*

$$\Lambda_h^0 := \big\{ u : M_h \to \mathbb{R} \mid u \text{ is continuous and } u_{|T} \text{ linear} \big\} \cong S_h$$

$$\Lambda_h^1 := \big\{ v \mid v_{|T} \text{ is a constant, tangential vector} \big\}$$

$$\Lambda_h^2 := \left\{ \begin{array}{c} w \mid \text{ on each simply connected region } D \\ w_{|D} = u\omega \text{ with a function } u \in \Lambda^0 \end{array} \right\} .$$

Additionally, we define the spaces $\Lambda^{0,*} \supset \Lambda^0$ and $\Lambda^{2,*} \supset \Lambda^2$ having functional representatives in S_h^*.

The space Λ_h^1 is the space of discrete vector fields on a polyhedral surface which are tangential and constant on each triangle. In the following we try to avoid too much formalism and, on simply connected domains, identify a $2-$form $w = u\omega$ with its function u without explicitly listing the volume form ω. Similarly, we identify vector fields with $1-$forms.

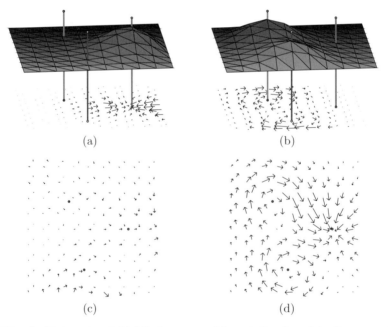

(a)   (b)

(c)   (d)

Fig. 3. Test vector field (d) decomposed in rotation-free (a), divergence-free (b) and harmonic component (c). The three vertical lines in (a) and (b) indicate the centers of the original potentials for comparision with the extrema of the calculated potential functions.

Definition 15. *On a simplicial surface M_h the Hodge operator $*$ is a map*

$$* : \Lambda_h^i \to \Lambda_h^{2-i}$$

such that locally

$$*u = u\omega \quad \forall u \in \Lambda_h^0 \text{ respectively } \Lambda_h^{0,*}$$
$$*v = Jv \quad \forall v \in \Lambda_h^1$$
$$*(u\omega) = u \quad \forall(u\omega) \in \Lambda_h^2 \text{ respectively } \Lambda_h^{2,*} .$$

The gradient operator ∇ used in S_h^* generalizes to two differential operators ∇ and δ on differential forms. We use rot_h^* respectively div_h in order to synchronize with the integrability condition of discrete vector fields in the following sequences.

Definition 16. *The* differential operator $\nabla : \Lambda_h^0 \to \Lambda_h^1 = \Lambda_h^{1,*} \to \Lambda_h^{2,*}$ *on a simplicial surface M_h is defined by*

$$\nabla u = \nabla u \quad \forall u \in \Lambda_h^0$$
$$\nabla v = \operatorname{rot}_h^* v \quad \forall v \in \Lambda_h^1$$
$$\nabla(u\omega) = 0 \quad \forall(u\omega) \in \Lambda_h^{2,*} .$$

The co-differential operator δ *is defined by* $\delta := *d* : \Lambda_h^{2,*} \to \Lambda_h^{1,*} = \Lambda_h^1 \to \Lambda_h^0$, *that is*

$$\delta u := 0 \quad \forall u \in \Lambda_h^0$$
$$\delta v := \operatorname{div}_h v \quad \forall v \in \Lambda_h^1$$
$$\delta(u\omega) := J\nabla u \quad \forall(u\omega) \in \Lambda_h^{2,*} .$$

Both operators are similarly defined on $\Lambda_h^{0,*}$ *respectively* Λ_h^2 *using* rot_h *respectively* div_h^*.

We remind that in the smooth situation for a vector field $v = (v_1, v_2)$, we have $\nabla v = (v_{2_{|x}} - v_{1_{|y}})\nabla x \nabla y$ and $\delta v = v_{1_{|x}} + v_{2_{|y}}$ on a planar Euclidean domain with coordinates (x, y).

Lemma 4. *Let $u \in \Lambda^0$ and $w \in \Lambda^2$ then*

$$\nabla^2 u(m) = 0 \text{ and } \delta^2 w(m) = 0 \text{ at each edge midpoint } m \in M_h .$$

Similarly, if $u \in \Lambda^{0,*}$ and $w \in \Lambda^{2,*}$ then

$$\nabla^2 u(p) = 0 \text{ and } \delta^2 w(p) = 0 \text{ at each interior vertex } p \in M_h,$$

Proof. Direct consequence of the corollaries of the previous section.

We now state a Hodge-type decomposition of $1-$forms (vector fields) on simplicial surfaces into a rotation-free, divergence-free, and a harmonic field.

Theorem 7. *Let M_h be a simplicial surface. Then any tangential vector field $\xi \in \Lambda^1(M_h)$ has a unique decomposition*

$$\xi = \nabla u + \delta(w\omega) + v \tag{8}$$

with $u \in \Lambda^0$, $w\omega \in \Lambda^2$ and harmonic component $v \in \Lambda^1$ with $\mathrm{div}_h v = \mathrm{rot}_h v = 0$, i.e. $\nabla v = 0$, $\delta v = 0$. Uniqueness of the decomposition follows from the normalization

$$\int_{M_h} u = 0, \ \int_{M_h} w\omega = 0 .$$

Since u and w are functions, ∇u is rotation-free and $\delta(w\omega)$ is divergence-free.

Proof. First, we derive the potential $u \in \Lambda^0$ of the rotation-free component of a given vector field ξ. We define the following quadratic functional F for functions in S_h

$$F(u) := \int_{M_h} (|\nabla u|^2 - 2\langle \nabla u, \xi \rangle) \tag{9}$$

which associates a real-valued energy to each function u_h. A quadratic functional has a unique minimizer which we denote with $u \in S_h$. As a minimizer, u is a critical point of the functional which fulfills at each vertex p the following minimality condition

$$0 \overset{!}{=} \frac{d}{du_p} F(u) = 2 \int_{\mathrm{star}\, p} \langle \nabla u - \xi, \nabla \varphi_p \rangle \tag{10}$$

where $\varphi_p \in S_h$ is the Lagrange basis function corresponding to vertex p. Formally, u solves the Poisson equation $\mathrm{div}_h \nabla u = \mathrm{div}_h \xi$ respectively $\Delta_h u = \delta\xi$.

To obtain the divergence-free component we define a similar functional

$$G(w) := \int_{M_h} (|\delta(w\omega)|^2 - 2\langle \delta(w\omega), \xi \rangle) \tag{11}$$

and compute the potential $w \in S_h$ as its unique minimizer which solves

$$0 \overset{!}{=} \frac{d}{dw_p} G(w) = 2 \int_{\mathrm{star}\, p} \langle \delta(w\omega) - \xi, J\nabla \varphi_p \rangle . \tag{12}$$

Formally, w solves $\mathrm{rot}_h \delta(w\omega) = \mathrm{rot}_h \xi$ resp. $\Delta_h w = \nabla \xi$.

The harmonic remainder is defined as $v := \xi - \nabla u - J\nabla w$. Using the above relations and the fact that for a 2-form $w\omega \in \Lambda_h^2$ we have $\mathrm{div}_h^* \delta(w\omega) = 0$ which implies $\mathrm{div}_h \delta(w\omega) = 0$, one easily verifies

$$\mathrm{div}_h v(p) = \mathrm{div}_h(\xi - \nabla u) - \mathrm{div}_h \delta(w\omega) = 0 .$$

And using the fact that for a function $u \in \Lambda_h^1$ we have $\mathrm{rot}_h^* \nabla u = 0$ which implies $\mathrm{rot}_h \nabla u = 0$, we obtain

$$\mathrm{rot}_h v(p) = \mathrm{rot}_h(\xi - \delta(w\omega)) - \mathrm{rot}_h \nabla u = 0 .$$

Theorem 7 was stated without proof as Theorem 2 in [17] where it was not made clear enough that the remaining component v is harmonic with respect to the *vertex based* operators div $_h$ and rot $_h$.

6 Decomposition Algorithm and Detecting Vector Field Singularities

For arbitrary tangential piecewise-constant vector fields on a simplicial (planar or curved) surface M_h. Let M_h again be a simplicial planar or curved surface and $\xi \in \Lambda_h^1$ a vector field on M_h. If M_h is a curved surface, ξ is given by a constant vector in each triangle that lies in the triangle's plane. We now apply the discrete Hodge-Helmholtz decomposition introduced in the previous sections to split the given vector field into a rotation-free, a divergence-free, and a remaining harmonic component. The first component is the gradient field of a function u, and the second component is the co-gradient field, i.e. the gradient rotated with J by 90 degree, of a second function w. The decomposition is obtained by directly computing the functions as minimizers of the energy functionals above.

The steps for a practical decomposition of a vector field are now straightforward. Assume, we want to compute the rotation-free component of a given vector field ξ on a simplicial surface M_h with boundary. We begin with an arbitrary initial function $u_0 \in S_h$. The functional F in (10) is a quadratic in ∇u, so the minimization problem has a unique solution for ∇u (independent of the initial function u_0), and two solutions u differ only in a constant vertical offset. Since the offset has no no effect on the critical points of u one can use any initial function u_0. In practice we usually start with zero values everywhere. Then we apply a standard conjugate gradient method to minimize the energy functional F by modifying the function values of u_0. As a result we obtain the a minimizer u of F. The same approach using the functional G is performed to compute the second potential w. Note that the minimization processes can be performed independently and simultaneously, e.g. on two different processors.

Furthermore, the minimization can be speed up rapidly by precomputing all terms which solely depend on the domain surface M_h, because only a few terms in the functionals and their gradients depend on the free variables. The evaluation of the gradient of the functional F in (10), for example, can be made rather efficient using the explicit representation of the discrete divergence operator given in (7):

$$\frac{d}{du_p} F(u) = 2 \int_{\text{star } p} \langle \nabla u - \xi, \nabla \varphi_p \rangle$$

$$= \sum_{i=1}^{k} (\cot \alpha_i + \cot \beta_i) \langle \nabla u - \xi, a_i \rangle \ .$$

Since the cotangent values belong to the triangulation M_h their non-linear computation can be done in a pre-processing step once before the conjugate gradient method starts. Precomputing the scalar products of the vector field ξ with the edges a_i is also possible. During runtime of the conjugate gradient method it computes repeatedly per vertex the k scalar products $\langle \nabla u, a_i \rangle$ and k scalar multiplications and one addition, where k is the number of triangles in the star of the vertex p.

The vector field components of ξ are easily derived by differentiating the potentials. For efficiency, one does not need to store the three vector field components explicitly since they are explicitly determined by the scalar-valued potentials u and w. Further, if one is interested only, say, in identifying the vortices of a vector field ξ, then it suffices to calculate w and to avoid the calculation of the full decomposition. Note that if one only stores the two potentials and the remaining harmonic component v of the vector field, then one is still able to fully reconstruct the original vector field ξ using the decomposition equation (8).

The following algorithms describes how to identify singularities of a given vector field ξ:

1. Calculate the potential u by minimizing the functional (9). The gradient of u is the rotation-free component of ξ.
2. Locate the local maxima and minima of the scalar valued function u over the two-dimensional surface, which are the centers of sinks or sources respectively. The maxima and minima can be automatically detected by searching for vertices p whose function value $u(p)$ is smaller or larger than the function values of all vertices on its link:

$$p \text{ is a sink} \Leftrightarrow u(p) < u(q) \ \forall \ q \in \partial \text{star} \, p$$
$$p \text{ is a source} \Leftrightarrow u(p) > u(q) \ \forall \ q \in \partial \text{star} \, p.$$

A similar algorithm determines first-order vortices of the vector field ξ:

1'. Calculate the potential w by minimizing the functional (11).
2'. Locate the local maxima and minima of the potential w on the surface M_h which are the centers of vortices. Vortex rotation direction is determined by the type of extremal point (maximum or minimum).

$$p \text{ is a vortex} \Leftrightarrow u(p) < \text{or} \ > u(q) \ \forall \ q \in \partial \text{star} \, p.$$

Local methods for vortex identification and feature analysis of discrete vector fields often try to approximate the Jacobian by discrete differences or by higher-order interpolation of the vector field. This approach often suffers from numerical or measured inaccuracies of the vector field which make it a delicate task to extract higher order data such as the Jacobian or even higher order differential tensors.

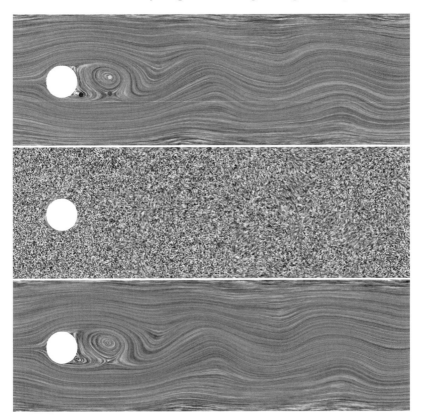

Fig. 4. Incompressible flow around a cylinder (bottom) with divergence-free (top) and harmonic component (middle). Two dots in the divergence-free part are the centers of rotation (white=clockwise, black=counter clockwise), and a third dot marks a saddle. All singularities were detected automatically by our method.

Our approach is global in the sense that u is rather independent of small local variations of the vector field which might have been introduced by numerical errors in the simulation, during the measurement, or by deficiency from a bad triangle mesh. This global approach is mainly due to the use of integrated values during the minimization of the functional.

7 Examples

The first example in Fig. 3 (Color Plate 27 on page 441) is an artificial vector field. The field is the sum of a gradient vector field and two co-gradient fields. Application of the Hodge decomposition leads to two potentials u and w with gradient ∇u and co-gradient $J\nabla w$ shown in the upper two images on the left and right. The location of the original, generating potentials are indicated by dots. The algorithm detects the singularities and clearly separates the source from the two vortices. The centers of the potentials may be varied at interactive speed since on a smaller grid the decomposition is done in real-time as shown at the web site [15].

In Fig. 4 the decomposition is applied to an incompressible flow around a cylinder from a CFD simulation. The rotation-free component of the incompressible flow vanishes as expected. The harmonic component shows the incompressible, non-rotational part of the flow.

The flow in the Bay of Gdansk in Fig. 5 (Color Plate 28 on page 442), a coastal region in Poland, is data from a simulation performed at WL | Delft Hydraulics using a curvilinear grid of $43 * 28 * 20$ nodes. The goal of the simulation was to investigate the flow patterns induced by wind and several inflows. In [19] Sadarjoen and Post derive geometric quantities from curvature properties of streamlines to find vortex cores and analyze their qualitative behaviour. We computed the potentials of the gradient and co-gradient components and easily recovered the vortices. We also detected some sinks and sources, that come from vertical flows in the originally three-dimensional vector field.

The harmonic component of a vector field corresponds to an incompressible, irrotational flow. On compact surfaces this harmonic component represents the non-integrable flows around the handles of the surface. The artificial vector field in Fig. 1 (Color Plate 26 on page 441) is obtained from the restriction of the tangent field of a rotation of 3−space onto the pretzel. Around each handle we see a well-distinguished harmonic flow. There are also two sinks and two sources at the upper side and the lower side.

8 Conclusions and Future Work

We present a method for feature detection that is mathematically well founded and that adopts the discrete nature of experimental data. It succeeds in the detection of first order singularities. The two potentials we compute seem to hold much more information than we discussed in this paper. Therefore it is interesting to extract, for example, magnitudes, axes and angular velocities of vortices and strengths of sources/sinks from them. Since our variational approach is based on integrated values it has a smoothing effect on the potentials u and w, which is a good property for the detection of higher order critical points of these potentials. Work in these directions has been done and will be presented in a future paper.

Acknowledgments

The numerical flows were contributed by different sources. The simulation data of the Bay of Gdansk with courtesy WL|Delft Hydraulics and the help of Frits Post and Ari Sadarjoen. The flow around a cylinder by Michael

Hinze at Technische Universität Berlin. The numerical computations and visualizations were performed in JavaView [16] and are available at `http://www.javaview.de`.

References

1. D. Banks and B. Singer. A predictor-corrector technique for visualizing unsteady flow. *IEEE Transactions on Visualization and Computer Graphics*, 1(2):151–163, June 1995.
2. A. Bossavit. Mixed finite elements and the complex of whitney forms. In J. Whiteman, editor, *The Mathematics of Finite Elements and Applications VI*, pages 137–144. Academic Press, 1988.
3. A. Bossavit. *Computational Electromagnetism*. Academic Press, Boston, 1998.
4. S. C. Brenner and L. R. Scott. *The Mathematical Theory of Finite Element Methods*. Springer Verlag, 1994.
5. A. J. Chorin and J. E. Marsden. *A Mathematical Introduction to Fluid Mechanics*. Springer Verlag, 1998.
6. P. Ciarlet. *The Finite Element Method for Elliptic Problems*. North-Holland, 1978.
7. J. Dodziuk. Finite-difference approach to the hodge theory of harmonic forms. *Amer. J. of Math.*, 98(1):79–104, 1976.
8. B. Eckmann. Harmonische Funktionen und Randwertaufgaben in einem Komplex. *Commentarii Math. Helvetici*, 17:240–245, 1944–45.
9. J. Helman and L. Hesselink. Representation and display of vector field topology in fluid flow data sets. *IEEE Computer*, 22(8):27–36, August 1989.
10. J. Helman and L. Hesselink. Visualizing vector field topology in fluid flows. *IEEE Computer Graphics & Applications*, 11(3):36–46, 1991.
11. H. Helmholtz. Über Integrale der Hydrodynamischen Gleichungen. *J. Reine Angew. Math.*, 55:25–55, 1858.
12. D. Kenwright and R. Haimes. Vortex identification - applications in aerodynamics: A case study. In R. Yagel and H. Hagen, editors, *Proc. Visualization '97*, pages 413 – 416, 1997.
13. U. Pinkall and K. Polthier. Computing discrete minimal surfaces and their conjugates. *Experim. Math.*, 2(1):15–36, 1993.
14. K. Polthier. Computational aspects and discrete minimal surfaces. In D. H. et al., editor, *Proceedings of the 2001 Clay Mathematics Institute Summer School on the Global Theory of Minimal Surfaces*. AMS, 2002.
15. K. Polthier, S. Khadem-Al-Charieh, E. Preuß, and U. Reitebuch. JavaView Homepage, 1998–2002. `http://www.javaview.de/`.
16. K. Polthier, S. Khadem-Al-Charieh, E. Preuß, and U. Reitebuch. Publication of interactive visualizations with JavaView. In J. Borwein, M. H. Morales, K. Polthier, and J. F. Rodrigues, editors, *Multimedia Tools for Communicating Mathematics*, pages 241–264. Springer Verlag, 2002. `http://www.javaview.de`.
17. K. Polthier and E. Preuß. Variational approach to vector field decomposition. In R. van Liere, F. Post, and et.al., editors, *Proc. of Eurographics Workshop on Scientific Visualization*. Springer Verlag, 2000.

18. M. Roth and R. Peikert. A higher-order method for finding vortex core lines. In D. Ebert, H. Hagen, and H. Rushmeier, editors, *Proc. Visualization '98*, pages 143 – 150. IEEE Computer Society Press, 1998.
19. I. A. Sadarjoen and F. H. Post. Geometric methods for vortex extraction. In E. Gröller, H. Löffelmann, and W. Ribarsky, editors, *Data Visualization '99*, pages 53 – 62. Springer Verlag, 1999.
20. I. A. Sadarjoen, F. H. Post, B. Ma, D. Banks, and H. Pagendarm. Selective visualization of vortices in hydrodynamic flows. In D. Ebert, H. Hagen, and H. Rushmeier, editors, *Proc. Visualization '98*, pages 419 – 423. IEEE Computer Society Press, 1998.
21. G. Scheuermann, H. Hagen, and H. Krüger. Clifford algebra in vector field visualization. In H.-C. Hege and K. Polthier, editors, *Mathematical Visualization*, pages 343–351. Springer-Verlag, Heidelberg, 1998.
22. D. Sujudi and R. Haimes. Identification of swirling flow in 3-d vector fields. Technical report, AIAA Paper 95-1715, 1995.
23. M. Tittgemeyer, G. Wollny, and F. Kruggel. Visualising deformation fields computed by non-linear image registration. *Computing and Visualization in Science*, 2002. to appear.
24. R. Westermann, C. Johnson, and T. Ertl. Topology preserving smoothing of vector fields. *Transactions on Visualization and Computer Graphics*, pages 222–229, 2001.
25. H. Whitney. *Geometric Integration Theory*. Princeton University Press, 1957.
26. T. Wischgoll and G. Scheuermann. Detection and visualization of closed streamlines in planar flows. *Transactions on Visualization and Computer Graphics*, 7(2):165–172, 2001.

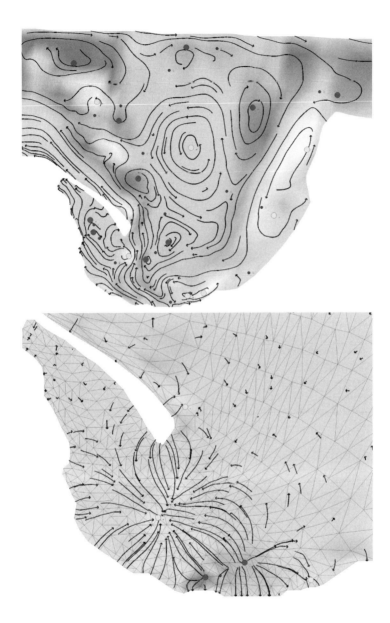

Fig. 5. Automatic identification of vector field singularities using a Hodge decomposition of a horizontal section of a flow in Bay of Gdansk. Rotation-free component (bottom) with sinks and sources, which come from vertical flows, and divergence-free component (top). The big dots indicate the location of sinks and sources (bottom) respectively vortices (top). The small dark dots mark saddle points (top and bottom). The bay is colorshaded by its discrete rotation (top) and divergence (bottom).

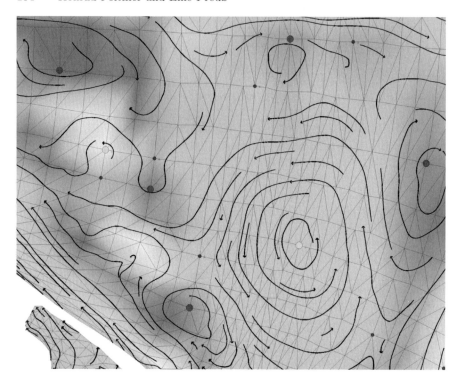

Fig. 6. Zoom into a region in Bay of Gdansk showing the divergence-free component of the Hodge decomposition with its integral curves. Big dots indicate locations of left and right rotating vortices, and small dark dots mark saddle points. The visualization of the vector field uses a vertex based, linear interpolated version of the element based vector field used for the decomposition, therefore sometimes there appear small discrepancies between visual centers of integral lines and indicated centers.

Searching for Knotted Spheres in 4-dimensional Space

Dennis Roseman

Department of Mathematics, University of Iowa, Iowa City, IA 52252, USA.
roseman@math.uiowa.edu

Summary. We search for knotted spheres in 4-dimensional space. This includes the generation of random examples, and experimental methods for the classification by topological type. Here we focus on a first step, to detect trivially knotted spheres.

Keywords. knotted spheres, knot theory

1 Background for Random Knots

The study of random knotted and linked circles in 3-dimensional space is of considerable interest in chemistry and biology. There are some interesting purely mathematical aspects. A good reference is the special issue of the Journal of Knot Theory and Its Applications (1994) Vol. 3, No. 3.

These mathematical problems have natural extensions to higher dimensions. Our goal is to investigate "random" surfaces in 4-dimensional space. Each surface is represented as a union of triangles and in our calculations. We take the number of triangles gives a measure of *computational complexity* of the surface. Some aspects of these problems are of interest in non-mathematical areas [26, 27, 28].

This paper discusses some first steps in a study of randomly generated spheres in R^4 with emphasis on computational aspects. Aspects of this program rely greatly on use of visualization, Section 6. To date most mathematical studies in this area have focused on asymptotic behavior—general statistical estimates of limits of behavior for large complexity. Our focus is on concretely constructing and examining random collections of objects each a union of small number of triangles. In the process, we will obtain a large collection of examples including all of the most basic knotted spheres with a small number of triangles.

Although the study of surfaces in 4-dimensional space is the natural analogue of knotted circles and links, there are some important differences, see Section 2.1. In particular it is a non-trivial task to generate random examples, Sections 3. Once we generate examples, we want to classify them by topological type. The first step of this process—detecting trivially knotted spheres—is discussed in Section 3.3. Other more advanced techniques are discussed in Section 5. Some motivation for this research is indicated in Section 7.

1.1 Basic Terminology

In this paper $\mathbb{R}^n$ denotes *n-dimensional space*; we will mostly consider $n \geq 4$. In $\mathbb{R}^4$ we use coordinates (x, y, z, w).

The *unit square in* $\mathbb{R}^3$ is defined: $Q^2 = [-1, 1] \times [-1, 1]$. Similarly the *unit cube in* $\mathbb{R}^3$ is $Q^3 = [-1, 1] \times [-1, 1] \times [-1, 1]$ and the *unit hypercube in* $\mathbb{R}^4$ is $Q^4 = [-1, 1] \times [-1, 1] \times [-1, 1] \times [-1, 1]$.

In a triangulation, the *order of a vertex* v is the number of edges which meet v. A *collection of line segments is self-avoiding* if any two distinct line segments, if they intersect, intersect in a common vertex. A *collection of triangles is self-avoiding* if any two distinct triangles, if they intersect, intersect either in a single common vertex or a single common edge.

We are interested in generating certain subsets of $\mathbb{R}^n$ which are unions of line segments and triangles. We refer to these as *geometric objects*. However we are interested in *topological* classification of these.

If $X, Y \subseteq \mathbb{R}^n$ we say *X and Y are topologically equivalent subsets* if there is a homeomorphism $h \colon \mathbb{R}^n \to \mathbb{R}^n$ such that $h(X) = Y$. For the subsets we will consider, this notion coincides with the more intuitive notion of isotopy. An *isotopy* of X and Y is a continuous one-parameter family of embeddings $h_t \colon X \to \mathbb{R}^n$ with h_0 the inclusion map and $h_1(X) = Y$. Roughly we can say that an isotopy is a flexible motion of X in $\mathbb{R}^n$ where, at time t, X is at position $h_t(X)$.

1.2 Review of Random Knotted Circles and Links in $\mathbb{R}^3$

Suppose we (somehow) define a notion of a random circle in $\mathbb{R}^3$ and use this to define the probability of a random knot. Here are some interesting naive questions. What is the probability that a randomly chosen circle is knotted [23]? More generally for each knot type k, what is the probability $p(k)$ that a randomly chosen circle has this knot type? What relations, if any are there between $p(k)$ and the known algebraic invariants (for example crossing number, etc). What relations, if any, are there between $p(k)$ and known geometric constructions. For example: If k is a connected sum of knots with types k_1 and k_2, is there a relation between $p(k)$, $p(k_1)$ and $p(k_2)$?

To study probabilities for knots the first step is to choose a method of representing a knot (which is an equivalence class of subsets) by a geometric object.

Most commonly one fixes an integer n and considers those knots represented as closed self-avoiding polygonal paths of n line segments. Sometimes one adds additional constraints. We might insist that all edges have the same length [18], or that the edge vectors have Gaussian distribution [9, 8], or that all edges are restricted to lie on the unit cubic lattice [22, 38].

Next one tries to answer the question: what knots can be constructed according the these constraints? After this one tries to find approximate values for the probability $p_n(k)$ that a random n-segment circular path has type k. One repeats for increasing values of n. The final challenge is to make sense of all the data this process produces.

A link is a collection of disjoint knotted circles. One could also ask the same questions about links as well and also explore higher dimensional analogues such as [37]. For brevity however we only consider the questions about knots in this paper.

2 Generating Random Collections of Objects in $\mathbb{R}^n$

Let X be either an N-gon or a triangulation of a sphere with N vertices and triangulation fixed for the discussion. In either case, denote the vertices by $V = \{v_1, \ldots, v_N\}$. We are interested in the collection of random images of piecewise-linear maps of X into $\mathbb{R}^n$.

Take an *ordered* collection of points $P = \{p_1, \ldots, p_n\}$ of $\mathbb{R}^n$, chosen "randomly" according to some criterion. The correspondence $v_i \leftrightarrow p_i$ determines a unique piecewise linear subset $\mathcal{I}_P \subseteq \mathbb{R}^n$. Namely, $\mathcal{I}_P$ contains the line segment between p_i and p_j if and only if there is an edge in X joining v_i and v_j. Similarly $\mathcal{I}_P$ contains the triangle with vertices p_i, p_j and p_k if and only if there is a triangle in X with vertices v_i, v_j and v_k.

For piecewise-linear circle in $\mathbb{R}^3$ it is not difficult to define a notion of a random and to generate examples of random N-gons that will represent circles [4, 8, 18, 9]. This is because such a random closed path will be self-avoiding with probability 1.

However this method does not easily generalize for generating random knotted spheres in $\mathbb{R}^4$. There are two major difficulties. First, a circle can be expressed in essentially one way as union of n line segments—as an n-gon. However for a given n, there are many combinatorially incompatible ways of expressing the sphere as the union of n triangles. The second problem is more vexing: there is a non-zero (often large) probability that the of triangles of $\mathcal{I}_P$ will have self-intersections.

2.1 Some Aspects of Dimension

To understand the some basic geometric issues we consider three simple examples which serve as local models of the general problems we will encounter.

1. In $\mathbb{R}^2$ consider the x-axis L_x and the the y-axis L_y. The intersection $L_x \cap L_y$ is the single point $\mathbf{0}$. Furthermore any line L in $\mathbb{R}^2$ close to L_y will also intersect L_x in a single point.
2. In $\mathbb{R}^3$ consider the x-axis L_x and the the y-axis L_y. The intersection $L_x \cap L_y$ is the single point $\mathbf{0}$. However, most lines L in $\mathbb{R}^3$ close to L_y will be disjoint from L_x.
3. In $\mathbb{R}^4$ consider the xy-coordinate plane L_{xy} and the zw-coordinate plane L_{zw}. Again $L_{xy} \cap L_{zw}$ is the single point $\mathbf{0}$. Furthermore any plane L in $\mathbb{R}^4$ close to L_{zw} will also intersect L_{xy} in a single point.

So it is clear that computationally 2-dimensional objects in $\mathbb{R}^4$ behave similar to 1-dimensional objects in $\mathbb{R}^2$. However, as subsets of R^n, for mathematical reasons, *topologically* 2-dimensional objects are in $\mathbb{R}^4$ are most similar to 1-dimensional objects in $\mathbb{R}^3$. There is piecewise linear knotting of circles only in R^3 and surfaces exhibit piecewise-linear knotting behavior only in R^4.

2.2 Knotted Surfaces

Classical knot theory is the study of ambient isotopy classes of compact 1-manifolds in $\mathbb{R}^3$. To eliminate wild knots, we assume these submanifolds are either smooth or equivalent to a closed polygonal path. If the sub-manifold is connected it is a *knotted circle*, otherwise it is a *link*.

A knotted circle k represents a *trivial knotting* if it is the boundary of an embedded piecewise linear disk $D \subseteq \mathbb{R}^3$. In this terminology, standard in knot theory, we say the knotted circle k is, in fact, *unknotted*. Standard results in geometric topology imply that k is unknotted if and only if it is topologically equivalent to a planar circle.

In one dimension higher, knot theory is the study of ambient isotopy classes of compact 2-manifolds in $\mathbb{R}^4$. To eliminate wild knots, we assume the submanifolds are given as a finite union of triangles [12].

A knotting of a surface Σ is a finite collection of (2-dimensional) flat polygons $\{\sigma_\alpha\}$ whose vertices are points of $\mathbb{R}^4$ whose union is homeomorphic to a surface. In this paper this surface will be a sphere. In any of these dimensions we will call the geometric submanifold a *knotting*, reserving the term *knot* for the topological equivalence class.

A knotted sphere Σ represents a *trivial knotting* if it is the boundary of an embedded piecewise linear ball $D \subseteq \mathbb{R}^4$. Standard results in geometric topology imply that Σ is unknotted if it is equivalent to a sphere which lies in a 3-dimensional hyperplane.

3 Methods of Generating Random Spheres in $\mathbb{R}^4$

3.1 Some Nice Triangulations

With so many possibilities for triangulation of a sphere, we need to narrow our considerations. For technical mathematical and computational reasons, it is best to have triangulations (more or less) uniform and to avoid triangulations that have vertices of high order or very low order. In addition very simple triangulations (such as the 4 faces of a tetrahedron) are avoided since they will not give interesting spheres in $\mathbb{R}^4$.

We are currently restricting our triangulations to be the icosahedron or its refinements. In Figure 1 we see the icosahedron on the left and the first refinement on the right. The refinement is obtained by dividing each of the equilateral triangles of the icosahedron into four equilateral triangles using midpoints of the edges.

The icosahedron has 20 triangles 12 vertices and 30 edges. The refined icosahedron has 80 triangles, 42 vertices and 120 edges. Some other nice triangulations for possible later study are the barycentric subdivision of the tetrahedron with 24 triangles, 14 vertices, and 36 edges and the refinement of this which has 96 triangles, 50 vertices and 144 edges.

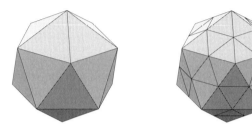

Fig. 1. The icosahedron and the refined icosahedron

By using stereographic projection from a vertex v_0 we may get a planar diagram to represent the 80 triangles of the refined icosahedron, Figure 2. The rays in this Figure which extend from the outermost pentagon are projections of those edges with vertex v_0

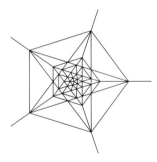

Fig. 2. Stereographic projection of triangulation of refined icosahedron.

3.2 Searching for Embedded Spheres

If we were only interested in embedded spheres we would produce random images, test for self-intersections and only consider examples with no self-intersections.

On the other hand we believe that generic images of spheres is a subject of interest especially in the case that these are images of an immersion (that is with no local knotting).

A smoothly generically immersed sphere in $\mathbb{R}^4$ which has n self-intersections will be called a *sphere with n kinks*. If n is not significant, we will call it a *kinky sphere*. The study of generic immersions of a disk in a 4-dimensional manifold is of great importance was initiated by Andrew Casson who called these "kinky disks". The study of these was fundamental to the proof of the 4-dimensional Poincaré conjecture [10].

The study of isotopy classes of kinky spheres is of some interest and we intend to pursue this in a later paper.

3.3 Knot Energy and Detecting Unknotted Spheres

Before we discuss details on generating examples of knotted spheres we should mention some techniques we use to analyze them. The obvious first step in this process is to identify those which are trivially knotted.

For this we have been using computations based on the concept of "knot energy" [2, 17], analogous to knot energy for circles [13, 19, 20, 11]. The idea is to model the subset as a self-repelling with fixed total area. We use the program Surface Evolver of Ken Brakke [3]. If a knotting of a surface is trivial one expects that under an energy minimizing flow, we would obtain an isotopy to a nearly round sphere that lies in a hyperplane.

Indeed we have had success with such flows as reported in [32, 36]. Those few examples had initial triangulations with thousands of triangles and it was known, for mathematical reasons that they were trivial knottings. In contrast we are now examining large groups of relatively simple examples of unknown knot type. This simplicity is not necessarily an advantage. Energy flow calculations in general behave much better with large data sets that approximate well a smooth submanifold.

One can use visualization tools to determine if an embedding which is near an energy (local) minimum is a trivial knotting—some projection into a hyperplane will be an embedding. We would like a quick computational test to see if a given knotting is "almost an embedding in some hyperplane". One such test which we have implemented with the control language of evolver is the following. Given a knotting with vertices at points $P = \{P_1, \ldots, P_n\}$ which is the union of oriented triangles $\{t_j\}$ we calculate the centroid c of P. Let ρ be the average distance from c to points of P. Let $\{T_j\}$ be the oriented tetrahedra T_j the cone from c to t_j and calculate an oriented normal n_j to T_j so that orientation of n_j followed by orientation for T_j gives the standard orientation for $\mathbb{R}^4$. Our hope is that the union of these tetrahedra $D = \cup_j T_j$ will be a disk with boundary σ. We check to see if these normals are pointing in the same direction, say $\mathbf{u}$. Let H be the hyperplane orthogonal to $\mathbf{u}$. The Surface Evolver keeps track of the area A_Σ of Σ and if we also find that $A_\Sigma = 4\pi\rho^2$ then projection of D to H will be an embedding. But even if the normals are more or less pointing in a consistent direction and $A_\Sigma \simeq 4\pi\rho^2$, we have a good indication that Σ might be close to passing such an unknotting test.

4 Random Icosahedra in $\mathbb{R}^4$

We generate our subsets, images of icosahedra (or refined icosahedra), with vertices $\{v_1, \ldots, v_N\}$ by taking N ordered (pseudo) random points $P = \{p_1, \ldots, p_N\}$ in the 4-cube Q^4, $N = 12$ or $N = 42$. Then we count the number of self-intersections of $\mathcal{I}_P$ denote this $n(\mathcal{I}_P)$. We would like to find examples where $n(\mathcal{I}_P) = 0$.

It takes very little time to generate thousands of examples of icosahedra in $\mathbb{R}^4$ with no self-intersections. Slightly more than 3% of randomly generated have no self-intersections. We do not know the maximum value of $n(\mathcal{I})$, but of the several hundreds of thousands of examples we have generated we have yet to find an example with $n(\mathcal{I}_P) > 40$.

Before starting out with our calculations, we attempted and failed to construct by hand an icosahedron in $\mathbb{R}^4$ which was not trivially knotted. Of the random icosahedra in $\mathbb{R}^4$ we have generated and successfully analyzed, all have been trivially knotted.

4.1 Refined Icosahedra

Since the refined icosahedron has 80 triangles, we expect randomly generated examples $\mathcal{I}_P$ higher values of $n(\mathcal{I}_P)$, and to have a lower success rate at finding examples with $n(\mathcal{I}_P) = 0$. We were unpleasantly surprised that we could find none at all. In fact, in a hundred thousand tries we did not find a single example with $n(\mathcal{I}_P) < 60$.

So our direct approach used for icosahedra seems impractical for refined icosahedron. One can try to be clever and build examples which avoid self-intersection. The problem is that results of such process do not seem to qualify as "random".

However, we have found a process we call *twiddling* which appears to have a great degree of randomness and at least is successful in generating a large number of examples of refined icosahedra with $n(\mathcal{I}_P) = 0$.

We begin with a speculation that almost every (perhaps every) collection of 42 random points of Q^4 can serve as vertices of some refined icosahedron with $n(\mathcal{I}_P) = 0$. If this is true then for some permutation P' of P we would have $n(\mathcal{I}_{P'}) = 0$

It is hopeless to systematically search all 42! possible permutations. There are 120 symmetries of an icosahedron, thus a total of $42!/120 \simeq 10^{49}$ distinct icosahedra corresponding to a single set of 42 points. This number is close to some estimates of the number of atoms in the earth.

Twiddling is based on the following idea. Suppose $\mathcal{I}_P$ is a refined icosahedron in $\mathbb{R}^4$ and that τ is a triangle of $\mathcal{I}_P$ with vertices p_i, p_j and p_k. There are five other icosahedra obtained by performing a permutation on P fixed except for these three vertices. Perhaps one of these permutations, call it P_1 will give fewer total intersections—that is $n(\mathcal{I}_{P_1}) < n(\mathcal{I}_P)$. If so, the plan is to turn our attention to $\mathcal{I}_{P_1}$ and search for a triangle of this which can similarly reduce our intersection number. The hope is that we can continue this process and end up with examples with no self-intersections.

Perhaps the easiest way to understand this process is to look at a process of obtaining a self-avoiding path in the plane from a randomly generated closed path. In Figure 3 in the upper left we see a closed path in the plane of 8 line segments and two self-intersection points. Think of this as the "original path". For each segment consider the path obtained by reversing the order of the endpoints of a single edge. This is a lower dimensional analogue of a twiddle. In Figure 3 the path second from the left, top row, shows the effect of reversing the longest horizontal edge. The third from the left on the top row shows the effect of reversing (only) the right-most vertical edge in the original. We continue in this manner showing the effects of reversing one edge of the original. This example also illustrates the fact that this process might not lead to a simplification. In this case all alterations result in an *increase* in the number of self-intersections.

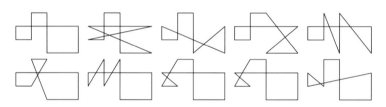

Fig. 3. An example of an analogue of twiddling for a path in the unit square.

Here is the process we are currently using. We call it *twiddling a refined icosahedron*. We begin with an ordered set P of 42 random points of Q^4. Call $\mathcal{I}_P$ the "initial data set".

1. randomize an ordering of the triangles of $\mathcal{I}_P$
2. in this order, try twiddling all 80 triangles. If for any triangle you find an improvement (lower total self-intersection) by twiddling, P' replace P with P' and continue with the remaining triangles
3. If we get do all 80 triangles and have been able to lower self-intersection number in this process, repeat steps 1 and 2 steps using this new $\mathcal{I}_P$; otherwise begin again with the initial data set and *new* randomization of orderings of triangles.

The interesting thing is that this process works reasonably well. We generate one initial data set and twiddle 100 times, then generate a new initial data set and twiddle this 100 times etc. This generates examples with zero self-intersections about 1.2% of the time. On a personal computer running all day (on non-optimized code), one can generate about one example per day. Considering that it currently takes much longer to analyze even one such example, this production rate is satisfactory.

It seems that there is sufficient randomness in the process to allow it to be at least considered a contender as a random process.

In Figure 4 we see projections of four refined icosahedra obtained in this way each having zero self-intersections. To get some idea of how random twiddling is, the four examples of Figure 4 are all obtained from the same initial data set. To date we have found many examples of data sets which when twiddled 100 times give more than one success—in none of these cases have we found any that are identical.

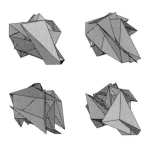

Fig. 4. Four spheres with no self-intersections. These are all obtained from the same data set by twiddling

4.2 Refined Icosahedra Can Knot in $\mathbb{R}^4$

Motivating our interest in refined icosahedra in $\mathbb{R}^4$ is the fact there are nontrivially knotted examples. We outline how to construct one.

One of the simplest knotted spheres is the spun trefoil [1]. Using the description in [35] one can take a knotted path in half space with M edges embed each in one of K pages in $\mathbb{R}^4$. If we are careful in our choice of coordinates for the knotted path we will be able to connect up the arcs in adjacent pages and not introduce self-intersections thus giving an piecewise linear embedding equivalent to the spun knot.

By choosing appropriate coordinates for the knot pictured in Figure 5 we have been able to get such an example using only three pages. In this triangulation of the sphere the endpoints of the arc give rise to vertices of order

3. Otherwise the rectangle joining corresponding line segments in adjacent pages is divided diagonally into two triangles, resulting in a triangulation shown in Figure 6a with $3 + 4 \times 6 + 3 = 30$ triangles.

We need one more step to complete our proof. The refined icosahedron in Figure 2 has more than enough triangles—but is this triangulation compatible with that shown in Figure 6a? Figure 6 shows that it is. On the right in this figure note the bold subgraph (this includes the non-visible vertex v_0 from which stereographic projection was taken). There is a piecewise linear map from the triangulation on the right to the one on the left that takes the bold graph to the bold graph. The shading in the figure helps to see this map—the shaded region on the right is sent to the shaded region on the left.

Fig. 5. A knotted arc of six segments used to produce a spun knot.

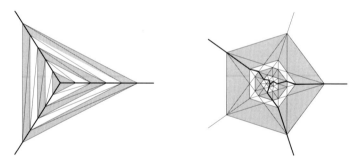

Fig. 6. Relating the two triangulations.

5 Sorting Knottings into Equivalence Classes

We have outlined above a program for generating a large number of knottings of a sphere into 4-space. The most challenging problem is how to sort these into equivalence classes. To show that two knottings are *not* equivalent we generally use algebraic techniques although in special situations, one can use topological methods such in Section 5.3. To show two embeddings *are* equivalent we must use topological methods.

Some of these topics in this section require some familiarity with Reidemeister moves and the fundamental group as applied to knot theory. Reference [7] is an excellent introduction; some more advanced standard references are [5, 29, 16].

5.1 Basic Topological Tools and Constructions

The two fundamental methods for the analysis of higher dimensional knots are slicing and projection. The *projection of a knotting* Σ [39, 30] is the subset $\Sigma_* = \pi(\Sigma)$ under the projection $\pi(x, y, z, w) = (x, y, z)$ together with information of the relative values of the w-coordinates of points $d \in \Sigma$ such that $D = \pi^{-1} \circ \pi(d) \cap \Sigma$ consists of more than one point. Generically D will consist of one, two or three points—in the case of two points, one point will be "under"; the other "over"; at triple points we can label these "under", "between", and "over". The set $D \subseteq S^2$ is called the *double point set of Σ with respect to the projection π*. For D we encode relative height relations at triple points and obtain what appears to be a diagram for a classic link projection. This is called a *projection link* [33].

For *slicing* we consider hyperplanes $H_{w_o} = \{(x, y, z, w) \in R^4 \colon w = w_0\}$ then analyze the knot by examination of the sets $\Sigma_w = \Sigma \cap H_w$ [12] which can be viewed as a sequence of classical knots and links connected by level sets of isolated non-degenerate critical points.

Suppose two knottings are isotopic then such an isotopy can be described as a finite sequence of elementary moves, see [33, 6]. In classical knot theory this are known as Reidemeister moves. Since there are no general algorithms for producing such realizing sequences, even in the case of classical knots, such a calculation must be at least in part interactive. Indeed in classical knot theory equivalence classes are often discovered by means such as physical manipulation of string on a table.

In [31] we described a topological drawing program TOPDRAWER to analyze and manipulate projections of classical knots. One of our current efforts is the generalization of this program to analyze and manipulate projections of knotted spheres. The code for TOPDRAWER is object oriented and much of the code can be reused in this project.

There are two main steps to this project. In the first we obtain data structure of projection link. Here we need only encode data of dimension at most 2—this will enable investigation of special moves as shown in [33, 15]. In the second step we extend this to include 3-dimensional information of the $R^3 - \Sigma$ —this will enable investigation of the generalized Reidemeister moves.

5.2 Basic Tools of Algebraic Topology

In classical knot theory one has many algebraic invariants which can distinguish distinct some knot types. Taken together they provide excellent practical tools for distinguishing most knottings. But as far as is currently known, they fall short of being able to distinguish *any* two given knottings. That is, there may exist knots which are not equivalent and yet have identical invariants of all known kinds.

For brevity, we focus discussion on one algebraic invariant, the knot group. The *knot group* of a knotting $K \subseteq R^n$ is defined to be the fundamental group $\pi_1(R^n - K)$. This invariant and its associated invariants have been used by most researchers of knotted surfaces such at Artin's first paper [1] and the work of Fox [12]. Hillman's book [21] summarizes many important topics.

In classical knot theory one can easily get a presentation of the knot group by using a planar projection of a piecewise linear circle together with the "under–over" information at the crossing points of the projection. In a similar way one can derive a presentation of the knot group of a surface using the projection $\Sigma_* \subseteq R^3$ [39, 30]. More simply, one can obtain a presentation from the data of a projection link.

A basic theorem of classical knot theory is that a piecewise linear circle k is unknotted if and only if $\pi_1(R^n - K)$ is isomorphic to the integers Z. It is not difficult to show that the knot group of the unknotted sphere is isomorphic to Z. However it is not known if a piecewise linear knotting with knot group Z must be a trivial knotting! (It is known that such a sphere must bound a *topological* sub-ball of R^4 but not known if it is a trivial knotting in our sense of bounding a *piecewise linear* sub-ball.) Thus the only methods available to prove a given knotting represents a trivial knotting are geometric.

5.3 Local Knotting

One problem we must address is that of local knotting. This has no analogue in classical knot theory.

Let Σ be a piecewise linear knotting of a sphere in R^4 and let v be a vertex of order n. Let S be a small round 3-sphere with center v and which contains only one vertex of Σ, namely v. Let $k_v = S \cap \Sigma$. Then k_v is the union of n arcs of great circles. We call such a union a *piecewise geodesic path*. Clearly k_v is a circle topologically. We say that Σ is *locally knotted at* v if k_v is a non-trivial knotted circle in S; otherwise we say Σ is *locally unknotted at* v

Smooth surfaces in R^4 are locally unknotted. In fact, a piecewise linear surface is topologically equivalent to a smooth surface if and only if it is locally unknotted.

For an example of local knotting take a non-trivial piecewise linear knotting $k \subseteq R^3 \subseteq R^4$ let $p = (0,0,0,1)$ and $q = (0,0,0,-1)$. Let C_+ be the union of all line segments in R^4 from p to a point of k; let C_- be the union of all line segments in R^4 from q to a point of k. Let $\Sigma_k = C_+ \cup C_-$. This will be a piecewise linear sphere called the *suspension of* k, see [1]. It is locally knotted at p and q.

An unknotted sphere is locally unknotted at every vertex. In particular, if a sphere is locally knotted at some vertex then it is not a trivial knotting. We will sketch a proof that local knotting cannot occur for icosahedra, but can occur for refined icosahedra. We outline a method of calculation that will determine if a refined icosahedron in R^4 is locally unknotted.

Suppose k is a polygonal circle in R^3 with N edges. Suppose e and e' are consecutive edges of k. These determine a triangle called a *peripheral triangle* T. Any of the $N - 4$ edges of k which do not intersect either e or e' will be called a *non-adjacent edge with respect to* T. We say that a *peripheral triangle is free* if it does not intersect any non-adjacent edge.

Using results and ideas in the proofs of [24, 25] one can show that any knotting with 5 edges must be a trivial knot. Also, any 6 edge knot k with no free peripheral triangle is equivalent to a trefoil knot; otherwise k is a trivial knotting.

These results can be reformulated in terms of piecewise geodesic curves and geodesic triangles in a sphere. Thus, if Σ is a knotted surface in R^4 with every vertex of order 5 or less, then Σ is locally unknotted. In particular an image of an octahedron is locally unknotted. Also, if Σ is a knotted surface in R^4 with every vertex of order 6 or less, then if Σ is locally knotted, the local knot must be a trefoil. By adopting the peripheral triangle test one can determine which of these cases hold.

The process of refinement will always give a triangulation with 12 vertices of order 5, the rest of order 6.

Locally knotted refined icosahedra exist. One can take a suspension of a six edge trefoil—this will have 8 vertices, and 12 faces. As we have done with other examples one can find a piecewise linear embedding of this triangulation into the refined icosahedron triangulation showing the realizability of this example.

By looking at random embeddings of a cone with hexagonal base, computer simulations indicate that approximately 0.03% of such embeddings are locally knotted at the cone point. Since there are 30 vertices of order 6 in a refined icosahedron, can roughly estimate that a random image would have at least one point where it is locally knotted approximately 8% of the time. If we restrict our attention to 0-kink images we would expect the percent of locally knotted examples to be less. We examined about 100 of the examples obtained from twiddling and found *no* examples which were locally unknotted. Clearly we need many more examples to get an idea of the percent of locally knotted examples, but this data seems to indicate that the twiddling process, in attempting to reduce self-intersections by local changes has produced, as a by-product, a filter of local knottedness.

6 The Role of Visualization

Visualization plays an essential role in assisting our analysis. The tools we use are not new but the visualization problems we encounter are new. Most visualization efforts of knotted surfaces in the past have involved inspection of examples about which some mathematical properties were known and whose coordinates were calculated by the user. In our current research we are trying to understand a completely unknown collection of knottings. Also, it is helpful to visualize results of calculations as we develop our code as a debugging aid.

We use GeomView [14] and Hew [34] as our principal visualization tools. Geomview has a good viewer for $\mathbb{R}^4$ and can monitor energy flow and indicate ways to improve computation in Surface Evolver. Hew has been useful for more detailed rapid analysis with its ability to do real-time slicing of surfaces in $\mathbb{R}^4$ thus revealing detail hidden by occlusion in viewing projections of the surface.

7 Ordering Knots

Here are some questions that provide motivation for our project. What is a reasonable scheme for constructing a table of knotted spheres? What is a good small collection of simplest examples of knotted spheres? Tables of knotted circles have been constructed since the first days of classical knot theory. For knotted spheres not only is there no standard table of simple examples, there is not even a standard accepted scheme for constructing such a table.

Here is a possible approach. For each knot type k we define a pair of numbers (n_k, p_k), where n_k is a non-negative integer and $0 < p_k \leq 1$. Let $\mathcal{J}_n$ denote the collection of all knots types of spheres in $\mathbb{R}^4$ that can be realized geometrically as the n-times refined icosahedron. Then n_k is the smallest integer such that $k \in \mathcal{J}_{n_k}$. If $k \in \mathcal{J}_{n_k}$ let $\mathcal{P}(k)$ denote the probability that a random n_k-times refined icosahedron in Q^4 has knot type k. It seems reasonable to assume that if we have distinct knot types k, k', with $n_k = n_{k'}$, that $\mathcal{P}(k) \neq \mathcal{P}(k')$. Assuming this, we can define an ordering of all knots by: $k < k'$ if either $n_k < n'_k$, or if $n_k = n'_k$ and $\mathcal{P}(k) > \mathcal{P}(k')$. The idea here is a high probability knot is "simpler".

One appeal of such a classification is that it seems more natural than other proposed schemes which depend special positioning of the knotting.

References

1. Artin, E. (1925) Zur Isotopie zweidimensionalen Flachen im $\mathbb{R}^4$, Abh. Math. Sem. Univ. Hamburg, 174—177.

2. Auckly, D. Sadun, L. (1997) A Family of Möbius Invariant 2-Knot energies. In: Geometric Topology: The Proceedings of the 1993 Georgia International Topology Conference, W. Kazez, ed. AMS/IP Studies in Advanced Mathematics 2, Part 1 235—258.

3. Brakke, K. A., The Surface Evolver, http://www.susqu.edu/brakke/evolver/evolver.html

4. Buck, G. R. (1994) J. Knot Theory and Its Ramifications **3** 355–363

5. Burde, G., Zieschang, H., Knots, (1985) de Gruyter Studies in Math. **5** , Walter de Gruyter

6. Carter, J. S., Saito M., (1993) Reidemeister Moves for Surface Isotopies and Their Interpretation as Moves to Movies, Jour. Knot Theory and its Ram., (2), 251-284, .

7. Crowell, P. R., Fox, R. H., (1992) Introduction to knot theory, (reissue 1997) Grad. Texts Math. **57**, Springer-Verlag

8. Deguchi, T., Tsurusaki, K., (1994) A Statistical Study of Random Knotting Using the Vassiliev Invariants, J. Knot Theory and Its Ramifications **3** 321—353.

9. Diao, Y., Pippenger, N., Sumners, D. W., (1994) On Random Knots, J. Knot Theory and Its Ramifications **3** 419—429.

10. Freedman, M. H., (1982) The topology of four-dimensional manifolds, J. Diff. Geom **17** 357—453.

11. Freedman, M. H., He, Zheng-Xu, and Wang, Zhenghan, (1994) On the "energy" of Knots and Unknots, Ann. Math. **139** 1—50.

12. Fox, R. (1962) A quick trip through knot theory. In: Topology of 3-Manifolds and Related Topics (Proc. Univ. Georgia Inst.,1961), Prentice-Hall, 120—167.

13. Fukuhara, S. (1988) Energy of a Knot. In: (ed. Matsumoto, Mizutani, Morita) A Fete of Topology, Academic Press.

14. Geomview software, originally developed at the Geometry Center, Minneapolis, Minn. current version at http://www.geomview.org/

15. Hanson, A.J., videotape entitled "$knot^4$," exhibited in Small Animation Theater o SIGGRAPH 93, Anaheim, CA, August 1–8, 1993. Published in SIGGRAPH Video Review 93, Scene 1 (1993).

16. Kawauchi, A., (1996) A Survey of Knot Theory, Birkhäuser Verlag.

17. Kusner, R., Sullivan, J. M. (1996) Möbius Energies for Knots and Links, Surfaces and Submanifolds. In: Geometric Topology, International Press, 570—604.

18. Millett, K. C. (1994) Knotting of Regular Polygons in 3-Space, J. Knot Theory and Its Ramifications **3** 263—278

19. O'Hara, Jun (1992): Family of energy functionals of knots, Topology and its Appl. **48** 147–161.

20. O'Hara, Jun (1992) Energy Functionals of Knots. In: (ed. K. H. Doverman) Topology-Hawaii, World Scientific 201–214.

21. Hillman, J., (1989) 2-knots and their Groups, Aust. Math. Soc Lect. Ser, **5** Cambridge University Press,

22. Pippenger, N. (1989) Knots in Random Walks, Discrete Appl. Math **25** 273—278.

23. Randell, R, (1988) A molecular conformation space MATH/CHEM/COMP 1987 Studies in Physical and Theoretical Chemistry **54** (Ed Lacher, R. C.) Elsevier 125—156

24. Randell, R, (1994) An Elementary Invariant of Knots, J. Knot Theory and Its Ramifications **3** 279—286.

25. Randell, R, (1998) Invariants of Piecewise-Linear Knots Knot Theory Banach Center Publications **42**, Inst, Math. Pol. Acad. Sci. 307-319

26. van Rensburg, E. J. J.,and Whittington, S. G. (1989) Self-avoiding surfaces , J. Phys A: Math. Gen **25** 49839—4958.

27. van Rensburg, E. J. J. (1992) Surfaces in the hypercubic lattice, J. Phys A: Math. Gen **22**.

28. van Rensburg, E. J. J. (1994) Statistical Mechanics and Topology of Surfaces in $\mathcal{Z}^d$, J. Knot Theory and Its Ramifications **3** 321—378.

29. Rolfson, D., Knots and Links (1976) Publish or Perish Press.

30. Roseman, D. (1975) Projections of knots, Fund. **89** Math 99—110

31. Roseman, Dennis, (1992) Design of a Mathematicians' Drawing Program, "Computer Graphics Using Object-Oriented Programming", Ed. S. Cunningham, J. Brown, N. Craghill and M. Fong, John Wiley & Sons, 279-296.

32. Roseman, D., (1994) Unraveling in 4 Dimensions, video, produced at the Geometry Center.

33. Roseman, D., (1998), Reidemeister-type moves for surfaces in four dimensional space, Knot Theory, Banach Center Publications, , Inst. Math. Polish Acad. Sci. **42** 347—380.

34. Roseman, D., with Berdine, J, Hew, a higher dimensional viewing software.

35. Roseman, D. (1999): Elementary Topology, Prentice-Hall.

36. Roseman, D. (2002) Untangling some spheres in R^4 by energy minimizing flow, To appear: (Conference Proceedings of Las Vegas AMS special session on Physical Knot Theory)

37. Soteros, C. E., Sumners, D. W., Whittington, S. G. (1999) Linking of Random p-Spheres in $\mathcal{Z}^d$, J. Knot Theory and Its Ramifications **3** 49—70.

38. Sumners, D. W., Whittington, S. G. (1988) Knots in self-avoiding walks, J. Phy. A: Math. Gen **21** 1689-1694.

39. Yajima, T, (1963) On the fundamental groups of knotted 2-manifolds in the 4-sphere, J. Math. Osaka City Univ **13** 63—71.

3D Loop Detection and Visualization in Vector Fields

Thomas Wischgoll[1] and Gerik Scheuermann[1]

Computer Science Department, University of Kaiserslautern, Germany
{*wischgol,scheuer*}*@informatik.uni-kl.de*

Summary. Visualization has developed a tendency to use mathematical analysis to obtain and present important data properties. In three-dimensional fluid flows, engineers are interested in several important features. One type are recirculation zones where the fluid stays for a long time. This plays a key role in combustion problems since recirculation allows a completion of chemical reactions which usually have a smaller time scale than fluid dynamics. Strong indicators for such recirculation zones are looping streamlines in a steady vector field or in the time steps of unsteady data. The article presents a method for the detection of such loops by analyzing streamlines approaching them.

1 Introduction

Many problems in natural science and engineering involve vector fields. Fluid flows, electric and magnetic fields are nearly everywhere, so measurements and simulations of vector fields still increase dramatically. As with other data, too, analysis is much slower and needs improvement. Mathematical methods together with visualization can provide help in this situation. In most cases, the scientist or engineer is interested in integral curves of the vector field like streamlines in fluid flows or magnetic field lines. The qualitative nature of these curves can be studied with topological methods developed originally for dynamical systems. Especially in the area of fluid mechanics, topological analysis and visualization have been used with success [3], [6], [11], [15].

This article concentrates on a specific topological class of integral curves, namely loops, also called limit cycles. These integral curves are closed, so that a particle traveling along the curve loops around forever. Their importance stems from the fact that quite often neighboring integral curves either tend toward the loop or come from there (i. e. tend toward the loop after reversing the direction of time). This is a well-established result from dynamical systems theory [4], [8]. It may be noted that loops may also behave like saddles in this three-dimensional case, but then they do not indicate recirculation and therefore this case is less relevant in our context.

Several publications have dealt with related topics. Hepting et al. [7] study invariant tori in four-dimensional dynamical systems by using suitable projections into three dimensions to enable detailed visual analysis of the tori. Wegenkittel et al. [18] present visualization techniques for known features of dynamical systems. Bürkle et al. [1] use a numerical algorithm developed by some of the coauthors [2] to visualize the behavior of more complicated dynamical systems. In the numerical literature, we can find several algorithms for the calculation of closed curves in dynamical systems [10], [17], but these algorithms are tailored to deal with smooth dynamical systems where a closed form solution is given.

Since visualization deals in many cases with piecewise linear, bilinear or trilinear data, we present an algorithm tailored to this situation that can be integrated into standard integral curve computation algorithms. While computing an integral curve, we track the visited cells checking for repetition. If we find a revisited cell, we check if the integral curve stays in the same cell cycle forever. For this, we look at the boundary of the cell cycle in question and check if the integral curve can cross the boundary. We have proposed a similar algorithm for the two dimensional case earlier [19]. In contrast to this two dimensional case where it is sufficient to find integral curves bounding the region where the integral curve can go, we have to work with integral surfaces in this article. We use a simplified version of Hultquist's algorithm [9] to construct these surfaces.

2 Mathematical Background

This section gives the necessary theoretical background and the terms for our algorithm. We restrict our consideration in this article to steady, linearly interpolated vector fields defined on a tetrahedral grid

$$v : \mathbb{R}^3 \supset D \to \mathbb{R}^3, \quad (x, y, z) \mapsto v(x, y, z).$$

D is assumed to be bounded. This is the situation for many experimental or simulated vector fields that have to be visualized. We are interested in the behavior of integral curves

$$c_a : \mathbb{R} \to \mathbb{R}^3, \qquad t \mapsto c_a(t)$$

with the properties

$$c_a(0) = a$$

$$\frac{\partial c_a}{\partial t}(t) = v(c_a(t)).$$

For Lipschitz continuous vector fields, one can prove the existence and uniqueness of integral curves c_a through any point $a \in D$, see [8], [12]. The actual computation of integral curves is usually done by numerical algorithms like Euler methods, Runge-Kutta-Fehlberg methods or Predictor/Corrector methods [16].

The topological analysis of vector fields considers the asymptotic behavior of integral curves. The α-limit set of an integral curve c is defined by $\{p \in$

$\mathbb{R}^3 | \exists (t_n)_{n=0}^\infty \subset \mathbb{R}, t_n \to -\infty, \lim_{n\to\infty} c(t_n) \to p\}$. The ω-limit set of an integral curve c is defined by $\{p \in \mathbb{R}^3 | \exists (t_n)_{n=0}^\infty \subset \mathbb{R}, t_n \to \infty, \lim_{n\to\infty} c(t_n) \to p\}$. If the α- or ω-limit set of an integral curve consists of only one point, this point is a critical point or a point at the boundary ∂D. (It is usually assumed that the integral curve stays at the boundary point forever.)

The most common case of a α- or ω-limit set in a vector field containing more than one inner point of the domain is a loop or limit cycle [8]. This is an integral curve c_a, so that there is a $t_0 \in \mathbb{R}$ with

$$c_a(t + nt_0) = c_a(t) \quad \forall n \in \mathbb{N}.$$

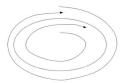

Fig. 1. A loop may attract integral curves in its neighborhood.

Figure 1 shows a typical example. Such a loop is called structurally stable if, after small changes, the vector field still contains a loop.

3 Loop Detection

The principal to detect loops in a three dimensional vector field is similar to the two dimensional case [19]. But there are some significant differences. To avoid confusion we start with a short notation:

Notation 3.1 (Actually investigated streamline) *We use the term actually investigated streamline to describe the streamline we check if it runs into a loop.*

To reduce computational cost we first only integrate the streamline using a Runge-Kutta-method of fourth order with an adaptive stepsize control. Every cell that is crossed by the streamline is stored during the computation. If a streamline approaches a loop it has to reenter the same cell again. This results in a *cell cycle*:

Definition 3.2 (Cell cycle) *Let s be a streamline in a given vector field V. Further, let G be a set of cells representing an arbitrary tetrahedral grid without any holes. Let $C \subset G$ be a finite sequence $c_0, \ldots, c_n$ of neighboring cells where each cell is crossed by the streamline s in order and $c_0 = c_n$. If s crosses every cell in C in this order again while continuing, C is called a cell cycle.*

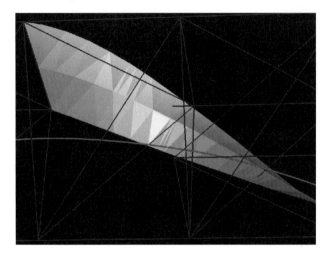

Fig. 2. Backward integrated surface.

This cell cycle identifies a region where we want to see if it can be left by
the streamline. To check this, we have to consider every backward integrated
streamline starting at an arbitrary point on a face of the boundary of the
cell cycle. Looking at the edges of a face we can see directly that it is not
sufficient to just integrate streamlines backward. Figure 2 shows an example.
We integrated backward a streamsurface starting at an edge of the cell cycle. The streamlines starting at the vertices of that edge leave the cell cycle
earlier than the complete surface. So it may be possible that a part of the
streamsurface stays inside the cell cycle although the backward integrated
streamlines starting at the vertices leave it. Consequently, we have to find
another definition for exits than in the two dimensional case.

Definition 3.3 (Potential Exit Edges) *Let C be a cell cycle in a given
tetrahedral grid G as in Def. 3.2. Then we call every edge at the boundary
of the cell cycle a* potential exit edge. *Analogue to the two dimensional case
we define a line on a boundary face where the vector field is tangential to the
face as a* potential exit edge.

Due to the fact that we use linear interpolation inside the tetrahedrons
we can show that there will be at least a straight line on the face where the
vector field is tangential to the face or the whole face is tangential to the
vector field. An isolated point on the face where the vector field is tangential
to the face cannot occur.

When dealing with edges as exits we have to compute a streamsurface
instead of streamlines to consider every point on an exit edge. This leads us
to the following notation.

Notation 3.4 (Backward integrated streamsurface) *We use the term* backward integrated streamsurface *to describe the streamsurface we integrate by taking the negative vectors of the vector field starting at a potential exit edge in order to validate this exit edge.*

Analogue to the definition in the 2D case we define *real exit edges*.

Definition 3.5 (Real exit edge) *Let E be a potential exit edge of a given cell cycle C as in definition 3.3. If the backward integrated streamsurface does not* **completely** *leave the cell cycle after one full turn through C then this edge is called a* real exit edge.

For the backward integrated streamsurface we use a simplified version of the streamsurface algorithm introduced by Hultquist [9]. Since we do not need a triangulation of the surface we only have to process the integration step of that algorithm. Initially we start the backward integration at the vertices of the edge. If the distance between these two backward integrations is greater than a special error limit we start a new backward integration in between. This continues with the two neighboring integration processes until we created an approximation of the streamsurface that respects the given error limit.

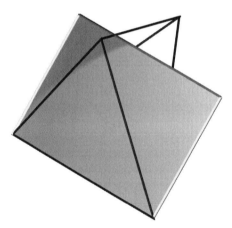

Fig. 3. Backward integration in one cell.

The integration stops if the whole streamsurface leaves the cell cycle or if we completed one full turn through the cell cycle. But to construct the surface properly we may have to continue a backward integration process across the boundary of the cell cycle. This is due to the fact that some part of the streamsurface is still inside the cell but the backward integrated streamline already left it. Figure 3 shows a simplified example. Both streamlines - shown as yellow lines - leave the cell, in fact they leave right after they started. But the integration process must be continued until the whole surface created by these two streamlines leaves the cell. This is marked by the red line at the end of the streamsurface.

With these definitions and motivations we can formulate the main theorem for our algorithm:

Theorem 3.6 *Let C be a cell cycle as in definition 3.3 with no singularity inside and E the set of potential exit edges. If there is no real exit edge among the potential exit edges E or there are no potential exit edges at all then there exists a loop inside the cell cycle.*

Proof:
Let C be a cell cycle with no real exit edges. Every backward integrated streamsurface leaves the cell cycle C completely. It is obvious that we cannot leave the cell cycle if every backward integration starting at an arbitrary point on a face of the boundary of the cell cycle C leaves the cell cycle. So we have to prove that the actually integrated streamline cannot leave the cell cycle C.

We look at each face of the boundary of the cell cycle C. Let Q be an arbitrary point on a face F of the boundary of the cell cycle C. Let us assume that the backward integrated streamline starting at Q converges to the actually investigated streamline. We have to show that this is a contradiction. We have two different cases:

1.: The edges of face F are exit edges and there is no point on F where the vector field is tangential to F.
   From a topological point of view the streamsurfaces starting at all edges of F build a tube that leaves the cell cycle. Since the backward integrated streamline starting at Q converges to the actually investigated streamline it does not leave the cell cycle. Consequently, it has to cross the tube built by the streamsurfaces. But streamlines cannot cross each other and therefore a streamline cannot cross a streamsurface.

2.: There is a potential exit edge e on the face F that is not a part of the boundary of F.
   Obviously, the potential exit edge e divides the face F into two parts. In one part there is outflow out of the cell cycle C while at the other part there is inflow into C. We do not need to consider the part with outflow any further because every backward integrated streamline starting at a point of that part immediately leaves the cell cycle C.
   The backward integrated surface starting at the potential exit edge e and parts of the backward integrated streamsurfaces starting at the boundary edges of the face F build a tube again from a topological point of view. Consequently, the backward integrated streamline starting at Q has to leave the cell cycle C.

We have shown that the backward integrated streamline starting at the point Q has to leave the cell cycle also. Since there is no backward integrated streamline converging to the actually investigated streamline at all the streamline will never leave the cell cycle.   ❏

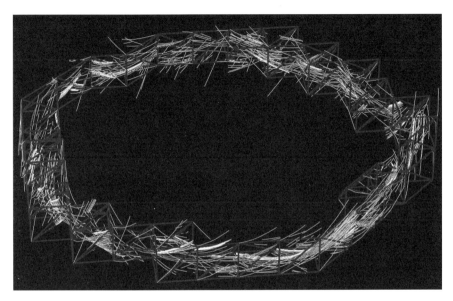

Fig. 4. Loop including cell cycle and backward integrations.

With theorem 3.6 we are able to describe our algorithm in detail. It is quite similar to the two dimensional case and mainly consists of three different states:

1. streamline integration: identifying one cell change after the other, check at each cell if we reached a cell cycle.
2. checking for exits: going backwards through the crossed cells and looking for potential exit edges.
3. validating exit: integrating backwards a curve from potential exit through the whole cell cycle.

Figure 4 (Color Plate 46 on page 451) shows an example of our backward integration step. There, also the loop is drawn in red and the cell cycle is shown in blue. Every backward integrated streamsurface leaves the cell cycle. According to Theorem 3.6, there exists a loop inside this cell cycle. Then we can find the exact location by continuing the integration process of the streamline that we actually investigate until the difference between two successive turns is small enough. This numerical criterion is sufficient in this case since we have shown that the streamline will never leave the cell cycle.

4 Results

Fig. 5. Loop including some streamlines.

To test our implementation we created a synthetic dataset which includes one loop. We first produced a two dimensional vector field which is symmetrical with respect to the y-axis. Additionally, all vectors residing at the y-axis where zero. Then we rotated it around the y-axis and distorted it a little bit to get a three dimensional flow. Figure 5 shows the result. The loop is colored red. To visualize a little bit of the surrounding flow several streamlines are drawn. Obviously, every streamline is attracted by the loop. After a short time, while the streamline spirals around the loop, it completely merges into it. We can see in this example that the loop in this three dimensional flow acts like a sink.

Figure 6 (Color Plate 47 on page 451) shows the same loop with two streamsurfaces. The streamsurfaces are – just like the streamlines – attracted by the loop. The streamsurface gets smaller and smaller while it spirals around the loop. After a few turns around the closed streamline it is only slightly wider than a streamline and finally it totally merges with the loop.

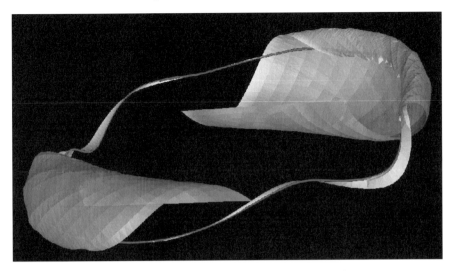

Fig. 6. Loop in a 3D vector field with streamsurfaces.

We used a rather arbitrary color scheme for the surface to enhance the three dimensional impression.

Acknowledgments

This research was supported by the DFG project "Visualisierung nicht-linearer Vektorfeldtopologie". Further, we like to thank Tom Bobach, Holger Burbach, Stefan Clauss, Jan Frey, Christoph Garth, Martin Öhler, Max Langbein, Aragorn Rockstroh, René Schätzl and Xavier Tricoche for their programming efforts and Inga Scheler for helping with some of the figures. The continuous support of all members of the computer graphics and visualization team in Kaiserslautern gives us a productive working environment.

References

1. D. Bürkle, M. Dellnitz, O. Junge, M. Rumpf, and M. Spielberg. Visualizing complicated dynamics. In A. Varshney, C. M. Wittenbrink, and H. Hagen, editors, *IEEE Visualization '99 Late Breaking Hot Topics*, pp. 33 – 36, San Francisco, 1999.
2. M. Dellnitz and O. Junge. On the Approximation of Complicated Dynamical Behavior. *SIAM Journal on Numerical Analysis*, 36(2), pp. 491 – 515, 1999.
3. A. Globus, C. Levit, and T. Lasinski. A Tool for Visualizing the Topology of Three-Dimensional Vector Fields. In G. M. Nielson and L. Rosenblum, editors, *IEEE Visualization '91*, pp. 33 – 40, San Diego, 1991.

4. J. Guckenheimer and P. Holmes. *Dynamical Systems and Bifurcation of Vector Fields*. Springer, New York, 1983.
5. R. Haimes. Using residence time for the extraction of recirculation regions. *AIAA Paper 99-3291*, 1999.
6. J. L. Helman and L. Hesselink. Visualizing Vector Field Topology in Fluid Flows. *IEEE Computer Graphics and Applications*, 11(3), pp. 36–46, May 1991.
7. D. H. Hepting, G. Derks, D. Edoh, and R. R. D. Qualitative analysis of invariant tori in a dynamical system. In G. M. Nielson and D. Silver, editors, *IEEE Visualization '95*, pp. 342 – 345, Atlanta, GA, 1995.
8. M. W. Hirsch and S. Smale. *Differential Equations, Dynamical Systems and Linear Algebra*. Academic Press, New York, 1974.
9. J. P. M. Hultquist. Constructing stream surface in steady 3d vector fields. In *Proceedings IEEE Visualization 1992*, pp. 171–177. IEEE Computer Society Press, Los Alamitos CA, 1992.
10. M. Jean. Sur la méthode des sections pour la recherche de certaines solutions presque périodiques de syst'emes forces periodiquement. *International Journal on Non-Linear Mechanics*, 15, pp. 367 – 376, 1980.
11. D. N. Kenwright. Automatic Detection of Open and Closed Separation and Attachment Lines. In D. Ebert, H. Rushmeier, and H. Hagen, editors, *IEEE Visualization '98*, pp. 151–158, Research Triangle Park, NC, 1998.
12. S. Lang. *Differential and Riemannian Manifolds*. Springer, New York, third edition, 1995.
13. K. Museth, A. Barr, and M. W. Lo. Semi-immersive space mission design and visualization: Case study of the "terrestrial planet finder" mission. In *Proceedings IEEE Visualization 2001*, pp. 501–504. IEEE Computer Society Press, Los Alamitos CA, 2001.
14. G. M. Nielson, H. Hagen, and H. Müller, editors. *Scientific Visualization, Overviews, Methodologies, and Techniques*. IEEE Computer Society, Los Alamitos, CA, USA, 1997.
15. G. Scheuermann, B. Hamann, K. I. Joy, and W. Kollmann. Visualizing local Vetor Field Topology. *Journal of Electronic Imaging*, 9(4), 2000.
16. J. Stoer and R. Bulirsch. *Numerische Mathematik 2*. Springer, Berlin, 3 edition, 1990.
17. M. van Veldhuizen. A New Algorithm for the Numerical Approximation of an Invariant Curve. *SIAM Journal on Scientific and Statistical Computing*, 8(6), pp. 951 – 962, 1987.
18. R. Wegenkittel, H. Löffelmann, and E. Gröller. Visualizing the Behavior of Higher Dimensional Dynamical Systems. In R. Yagel and H. Hagen, editors, *IEEE Visualization '97*, pp. 119 – 125, Phoenix, AZ, 1997.
19. T. Wischgoll and G. Scheuermann. Detection and Visualization of Closed Streamlines in Planar Flows. *IEEE Transactions on Visualization and Computer Graphics*, 7(2), 2001.
20. P. C. Wong, H. Foote, R. Leung, E. Jurrus, D. Adams, and J. Thomas. Vector fields simplification – a case study of visualizing climate modeling and simulation data sets. In *Proceedings IEEE Visualization 2000*, pp. 485–488. IEEE Computer Society Press, Los Alamitos CA, 2000.

Part III

Geometric Modelling

Minkowski Geometric Algebra and the Stability of Characteristic Polynomials

Rida T. Farouki and Hwan Pyo Moon

Department of Mechanical and Aeronautical Engineering, University of California, Davis, CA 95616, USA. *farouki@ucdavis.edu*

Summary. A polynomial **p** is said to be Γ–stable if all its roots lie within a given domain Γ in the complex plane. The Γ–stability of an entire family of polynomials, defined by selecting the coefficients of **p** from specified complex sets, can be verified by (i) testing the Γ–stability of a single member, and (ii) checking that the "total value set" $\mathcal{V}_*$ for **p** along the domain boundary $\partial\Gamma$ does not contain 0 ($\mathcal{V}_*$ is defined as the set of all values of **p** for each point on $\partial\Gamma$ and every possible choice of the coefficients). The methods of *Minkowski geometric algebra* — the algebra of point sets in the complex plane — offer a natural language for the stability analysis of families of complex polynomials. These methods are introduced, and applied to analyzing the stability of *disk polynomials* with coefficients selected from circular disks in the complex plane. In this context, $\mathcal{V}_*$ may be characterized as the union of a one–parameter family of disks, and we show that the Γ–stability of a disk polynomial can be verified by a finite algorithm (a counterpart to the Kharitonov conditions for rectangular coefficient sets) that entails checking that at most two real polynomials remain positive for all t, when the domain boundary $\partial\Gamma$ is a given polynomial curve $\boldsymbol{\gamma}(t)$. Furthermore, the "robustness margin" can be determined by computing the real roots of a real polynomial.

Keywords. Minkowski geometric algebra, robust control, stability analysis, complex polynomials

1 Introduction

In "robust control" problems one wishes to ensure certain stability properties for dynamic systems with prescribed uncertainty in their defining parameters [1, 3]. The impetus for the recent growth of interest in such problems can be traced to the papers [35, 36] of Kharitonov, which revealed remarkably simple extensions of the classical Routh–Hurwitz stability criterion [25, 42] to the case of real or complex "interval polynomials" (i.e., sets of polynomials defined by independently choosing coefficients from prescribed real intervals or rectangular complex domains).

The stability analysis of characteristic polynomials whose coefficients are drawn from given "uncertainty sets" often incurs complicated manipulations of sets in the complex plane. Although such manipulations arise naturally in a variety of contexts, and often have valuable geometric as well as algebraic interpretations, the enunciation of a formal algebra for complex sets is absent from standard texts [17, 48, 54] on the "geometry of complex numbers." In some recent papers [19, 20, 21, 22] we have attempted to redress this neglect, by sketching basic concepts and procedures for a *Minkowski geometric algebra* of complex sets, and its applications in areas such as geometrical optics, image processing and mathematical morphology, and the extension of real interval arithmetic to sets of complex numbers.

Since complex number addition is equivalent to vector summation in $\mathbb{R}^2$, adding two complex sets yields exactly the same result as the *Minkowski sum* of two point sets in the plane. Minkowski sums have been extensively studied [27, 30, 33, 43, 55] in both their theoretical and computational aspects. The notion of a *Minkowski product* of two complex sets, however, does not appear to have attracted much attention. In [20] we gave a comprehensive treatment of such products. We showed, in particular, that the theory and algorithms for Minkowski *sums* can be readily adapted to analyze and compute Minkowski *products*, by virtue of the mapping $\mathbf{z} \to \log \mathbf{z}$. We also introduced in [20] the *implicitly–defined complex sets*, a natural generalization of Minkowski sums and products, which offer exact descriptions of complex sets that cannot be characterized as "simple" Minkowski combinations.

Our goal in this paper is to introduce these new methods into the stability analysis of characteristic polynomials, and to illustrate their use by applying them to the stability of *disk polynomials* (i.e., polynomials whose coefficients are drawn from prescribed circular disks in the complex plane). We begin in §2 by reviewing some key concepts from the Minkowski geometric algebra of complex sets. The construction of Minkowski sums, products, and implicitly–defined sets relies extensively upon the theory of families of plane curves and their envelopes, which we briefly outline in §3 (with particular emphasis on one–parameter families of circles). After setting the stability of complex disk polynomials in its proper context in §4 and §5, we address the problem of nominal (Hurwitz) stability in §6, which is shown to be reducible to checking the positivity of two real polynomials. This result is extended to Γ–stability, for domains Γ with polynomial boundary curves, in §7. We also investigate the *robustness margin* of disk polynomials (the factor by which the coefficient disks can be magnified without compromising stability) in §8. Finally, in §9 we summarize our results and suggest directions for further research.

2 Minkowski Geometric Algebra

Minkowski geometric algebra [19] is concerned with the complex sets that are generated by algebraic operations on values drawn independently from given complex sets. The *Minkowski sum* and *Minkowski product*

$$\mathcal{A} \oplus \mathcal{B} = \{\, \mathbf{a} + \mathbf{b} \mid \mathbf{a} \in \mathcal{A} \text{ and } \mathbf{b} \in \mathcal{B} \,\},$$
$$\mathcal{A} \otimes \mathcal{B} = \{\, \mathbf{a} \times \mathbf{b} \mid \mathbf{a} \in \mathcal{A} \text{ and } \mathbf{b} \in \mathcal{B} \,\}, \tag{1}$$

for complex–set operands[1] $\mathcal{A}$ and $\mathcal{B}$ are the simplest elements of this algebra — they yield the sets populated by sums and products of all pairs of complex numbers chosen freely from $\mathcal{A}$ and $\mathcal{B}$. Minkowski difference $\ominus$ and division $\oslash$ operations may be analogously defined, but their introduction is not essential. We can always write $\mathcal{A} \ominus \mathcal{B} = \mathcal{A} \oplus (-\mathcal{B})$ and $\mathcal{A} \oslash \mathcal{B} = \mathcal{A} \otimes \mathcal{B}^{-1}$ instead, the negation and reciprocal of set $\mathcal{B}$ being defined by

$$-\mathcal{B} = \{ -\mathbf{b} \mid \mathbf{b} \in \mathcal{B} \} \qquad \text{and} \qquad \mathcal{B}^{-1} = \{ \mathbf{b}^{-1} \mid \mathbf{b} \in \mathcal{B} \} \,.$$

From the definitions (1), one can readily see that the Minkowski sum and product operations are commutative and associative. Unlike "ordinary" real or complex arithmetic, however, the Minkowski product does *not* distribute over Minkowski sums — in general, we have

$$(\mathcal{A} \oplus \mathcal{B}) \otimes \mathcal{C} \subseteq (\mathcal{A} \otimes \mathcal{C}) \oplus (\mathcal{B} \otimes \mathcal{C}) \,.$$

Whereas the Minkowski sum (with "+" interpreted as vector addition) has been extensively studied [27, 30, 33, 34, 43, 55], interest in the Minkowski product (with "×" regarded as complex number multiplication) developed only recently [20]. A basic interpretation of Minkowski geometric algebra is that it defines the natural generalization of (real) *interval arithmetic* [46, 47] to sets of complex numbers. However, the two–dimensional nature of the Minkowski geometric algebra endows it with much richer geometrical content than interval arithmetic — for example, it offers an intuitive language for the generation, description, and analysis of planar geometries, and for wavefront reflection or refraction problems in geometric optics [18, 19].

Minkowski sums and products on "simple" operands (e.g., circular disks) admit closed–form evaluation as the regions bounded by certain higher–order algebraic curves [19] — the Minkowski product of two disks, for example, is bounded by (the outer loop of) a classical quartic curve, known [38, 41] as the *Cartesian oval* [28]; see Figure 1. This result can be generalized to provide an exact closed–form description [22] for the boundary of a Minkowski product of $N \geq 3$ circular disks. For more general operands, bounded by (piecewise) analytic curves, methods are available to approximate the Minkowski sum or product boundary to any prescribed geometrical tolerance.

For domains $\mathcal{A}$ and $\mathcal{B}$ as operands, these algorithms employ Gauss maps to systematically subdivide their boundaries $\partial\mathcal{A}$ and $\partial\mathcal{B}$ into corresponding segments. The sums and products of corresponding points on such segments, as they are simultaneously traced, yield points on (potential) segments of the Minkowski sum and product boundaries $\partial(\mathcal{A} \oplus \mathcal{B})$ and $\partial(\mathcal{A} \otimes \mathcal{B})$. An elegant feature of this theory is that Minkowski products can be transformed into Minkowski sums, by invoking the map $\mathbf{z} \to \log \mathbf{z}$. This leads to consideration of novel geometrical concepts, such as the logarithmic curvature, logarithmic Gauss map, and logarithmic convexity of plane curves [20].

---

[1] We denote real and complex variables by italic and bold characters, respectively. Sets of complex numbers are denoted by upper–case calligraphic characters.

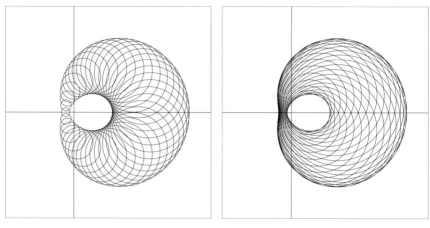

Fig. 1. The two realizations of a Cartesian oval as the boundary of the Minkowski product of two circles in the complex plane, corresponding to the multiplication — i.e., scaling/rotation — of one circle by each point of the other circle, and vice–versa.

As a generalization of Minkowski sums and products, we have introduced [19, 20] "implicitly–defined sets" of the form

$$\mathcal{A} \, \circledf \, \mathcal{B} = \{ \, \mathbf{f}(\mathbf{a}, \mathbf{b}) \mid \mathbf{a} \in \mathcal{A} \text{ and } \mathbf{b} \in \mathcal{B} \, \}, \tag{2}$$

i.e., sets of values generated by a function $\mathbf{f}(\cdot, \cdot)$ — typically, a polynomial — with arguments $\mathbf{a}$ and $\mathbf{b}$ drawn from given complex sets $\mathcal{A}$ and $\mathcal{B}$. Whereas Minkowski sums and products can be interpreted as unions of translated and scaled/rotated copies of one set, expression (2) is even more versatile: it can be regarded as the union of a family of *conformal mappings* of one set.

Expression (2) clearly subsumes Minkowski sums and products as special instances, and under appropriate conditions on $\mathbf{f}$ the Minkowski sum/product algorithms can be extended [20] to perform boundary evaluations for sets of the form (2). It is not easy, however, to formulate such algorithms to accept arbitrary (polynomial) functions $\mathbf{f}$; each case incurs specific implementation details. Alternately, one can formulate *bounding sets* in terms of Minkowski combinations, by replacing $\mathbf{a}, \mathbf{b}, \ldots$ with $\mathcal{A}, \mathcal{B}, \ldots$ and $+, \times$ with $\oplus, \otimes$ in the expression[2] for $\mathbf{f}$. If $\mathbf{f}(\mathbf{a}, \mathbf{b}) = \mathbf{a}\mathbf{b} + \mathbf{b}^2$ (see Figure 2), for example, we have

$$\mathcal{A} \, \circledf \, \mathcal{B} \subseteq (\mathcal{A} \oplus \mathcal{B}) \otimes \mathcal{B} \subseteq (\mathcal{A} \otimes \mathcal{B}) \oplus (\mathcal{B} \otimes \mathcal{B}). \tag{3}$$

Now since Minkowski products are commutative and associative, we can uniquely define the nth *Minkowski power* $\bigotimes^n \mathcal{A}$ of a given complex set $\mathcal{A}$ by

$$\bigotimes^n \mathcal{A} = \underbrace{\mathcal{A} \otimes \mathcal{A} \otimes \cdots \otimes \mathcal{A}}_{n \text{ times}} = \{ \, \mathbf{z}_1 \mathbf{z}_2 \cdots \mathbf{z}_n \mid \mathbf{z}_i \in \mathcal{A} \text{ for } i = 1, \ldots, n \, \}.$$

---

[2] Note that, since Minkowski sums and products do not obey the distributive law, there may be several (non–equivalent) ways to do this.

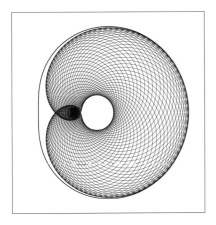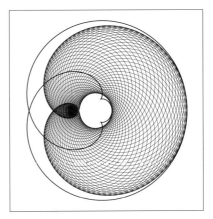

Fig. 2. The implicitly–defined complex set specified by (2), where $\mathcal{A}$ and $\mathcal{B}$ are the circles $|\mathbf{z}| = 1$ and $|\mathbf{z} - 1| = 1$, and we take $\mathbf{f}(\mathbf{a}, \mathbf{b}) = \mathbf{ab} + \mathbf{b}^2$. Also shown are the boundaries of the simpler sets defined by the two Minkowski combinations in (3).

Note that each value $\mathbf{z} \in \bigotimes^n \mathcal{A}$ arises from a product of n values $\mathbf{z}_1, \mathbf{z}_2, \ldots, \mathbf{z}_n$ *independently chosen* from $\mathcal{A}$. Conversely, an nth *Minkowski root* $\bigotimes^{1/n} \mathcal{A}$ of a set $\mathcal{A}$ is defined by the property that

$$\bigotimes^n (\bigotimes^{1/n} \mathcal{A}) = \mathcal{A}, \tag{4}$$

i.e., it is a set of complex values whose n^{th} Minkowski power coincides with the given complex set $\mathcal{A}$. Equivalently, we may write

$$\{\, \mathbf{z}_1 \mathbf{z}_2 \cdots \mathbf{z}_n \mid \mathbf{z}_i \in \bigotimes^{1/n} \mathcal{A} \text{ for } i = 1, \ldots, n \,\} = \mathcal{A}. \tag{5}$$

Again, note that $\mathbf{z}_1, \mathbf{z}_2, \ldots, \mathbf{z}_n$ are *independently chosen* values from $\bigotimes^{1/n} \mathcal{A}$. We refer to "a" — rather than "the" — Minkowski root since, if it exists, it is not ordinarily unique. Minkowski powers and roots should not be confused with the simpler "ordinary" powers and roots,

$$\mathcal{A}^n = \{\, \mathbf{z}^n \mid \mathbf{z} \in \mathcal{A} \,\} \qquad \text{and} \qquad \mathcal{A}^{1/n} = \{\, \mathbf{z} \mid \mathbf{z}^n \in \mathcal{A} \,\},$$

which are just the image and pre–image of $\mathcal{A}$ under the map $\mathbf{z} \to \mathbf{z}^n$. One can easily verify the inclusion relations

$$\bigotimes^{1/n} \mathcal{A} \subseteq \mathcal{A}^{1/n} \qquad \text{and} \qquad \mathcal{A}^n \subseteq \bigotimes^n \mathcal{A}.$$

Perhaps the simplest non–trivial example is the Minkowski square root of a circular disk $\mathcal{D}$. Provided that the origin does not lie within the disk, $\bigotimes^{1/2} \mathcal{D}$ admits a closed–form representation as (one loop of) a classical quartic curve, the *ovals of Cassini* (which degenerates to a *lemniscate of Bernoulli* when the origin lies on the perimeter of the disk); see Figure 3. This generalizes readily

[21] to the Minkowski n^{th} root, $\bigotimes^{1/n}\mathcal{D}$. The situation is more complicated, however, if $0 \in \mathcal{D}$ — a *composite* curve (portions of the Cassini ovals and of a higher–order curve) is then required to describe a Minkowski root [21].

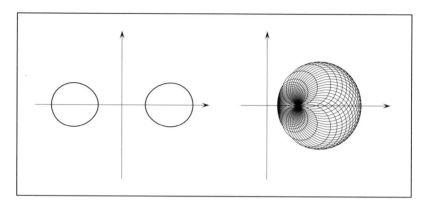

Fig. 3. An example of the ovals of Cassini (left). Each loop is the Minkowski square root of a circular disk: by performing a complex multiplication (scaling/rotation) of either loop by each of its points, we obtain the circular disk shown on the right.

Our interest here is in applying the theory and computational methods of Minkowski geometric algebra to stability analysis of dynamic systems with varying or uncertain parameters. Despite its fundamental nature and diverse potential applications, the systematic development of this geometric algebra of complex sets has surprisingly failed to attract much attention. Although Minkowski operations that admit tractable closed–form boundary evaluations are limited to simple operands, the availability of algorithms for more general complex sets [20] argues for a wider utilization of these methods.

3 Families of Curves and Envelopes

Since envelopes of families of planar curves play a central role in Minkowski geometric algebra and the stability analysis of complex disk polynomials, it is useful to review some of their basic properties. If $C(t)$ is a one–parameter family of curves, continuously dependent on a (real) parameter t, there are several approaches to defining its envelope — for example:

1. the envelope $\mathcal{E}$ is a plane curve that is tangent, at each of its points, to *some* curve in the family $C(t)$;
2. the envelope $\mathcal{E}$ is the locus, as t varies, of the intersection points of "neighboring" curves $C(t)$ and $C(t + \Delta t)$, in the limit $\Delta t \to 0$;

3. if S is the surface obtained by "stacking" each curve $C(t)$ at height $z = t$ above the (x, y) plane, the envelope $\mathcal{E}$ is the projection of the *silhouette* of S (as viewed along the z–axis) onto the (x, y) plane.

These definitions are not always precisely equivalent, and are subject to some technical qualifications under exceptional circumstances (which we shall not delve into here; see [8, 12, 13, 23] for a more detailed treatment).

Now a curve family may be defined by an implicit equation $f(t, x, y) = 0$ or by a parameterization $\mathbf{r}(t, s)$, where each t identifies a different curve and s specifies position on it. In the former case, the envelope equation is given (when f is a polynomial) by

$$e(x, y) = \mathrm{Resultant}_t(f, f_t) = 0,$$

i.e., by eliminating t between the two equations $f = 0$ and $\partial f / \partial t = 0$. In the latter case, the points of each curve t that lie on the envelope are identified by the condition that $\mathbf{r}_t = \partial \mathbf{r}/\partial t$ and $\mathbf{r}_s = \partial \mathbf{r}/\partial s$ are parallel vectors. This can, in principle, be solved for a function $s(t)$ that gives s in terms of t, and by substitution we have a parameterization for the envelope,

$$\mathbf{e}(t) = \mathbf{r}(t, s(t)).$$

Consider now the case of a one–parameter family of circles, which can be characterized by a *center curve* $\mathbf{c}(t)$ and *radius function* $R(t)$, so that

$$\mathbf{r}(t, \theta) = \mathbf{c}(t) + R(t)(\cos \theta, \sin \theta). \tag{6}$$

The points that each circle contributes to the envelope are identified by the condition that $\partial \mathbf{r}/\partial t$ and $\partial \mathbf{r}/\partial \theta$ are parallel, which yields the equation

$$c'_x(t) \cos \theta + c'_y(t) \sin \theta + R'(t) = 0, \tag{7}$$

with the two solutions

$$\cos \theta = \frac{-R'(t)c'_x(t) \pm c'_y(t)\sqrt{c'^2_x(t) + c'^2_y(t) - R'^2(t)}}{c'^2_x(t) + c'^2_y(t)},$$

$$\sin \theta = \frac{-R'(t)c'_y(t) \mp c'_x(t)\sqrt{c'^2_x(t) + c'^2_y(t) - R'^2(t)}}{c'^2_x(t) + c'^2_y(t)}. \tag{8}$$

By substituting these expressions into (6), we obtain a parameterization $\mathbf{e}(t)$ of the envelope. Because of the radical in (8) this is not, in general, a *rational* parameterization — even if both $\mathbf{c}(t)$ and $R(t)$ are rational.[3]

Now suppose the family of circles (6) occupy, for all t, a planar domain $\mathcal{D}$. For arbitrary choices of $\mathbf{c}(t)$ and $R(t)$, the boundary $\partial \mathcal{D}$ of this domain is, in general, only a *subset* of the envelope. Two circumstances may disqualify the envelope points at a given t from belonging to the true domain boundary:

---

[3] A rational envelope is guaranteed by incorporating a special algebraic structure — the *Minkowski Pythagorean hodograph* [45] — into the description $\{\mathbf{c}(t), R(t)\}$.

(a) equations (8) yield complex values, because the argument of the radical is negative — this will occur when "neighboring" circles lie completely *inside* or *outside* of each other, instead of intersecting;

(b) the values (8) are real, but the envelope points nevertheless lie *inside* the domain $\mathcal{D}$ — this occurs if the envelope exhibits *self–intersections*.

Computing the boundary $\partial \mathcal{D}$ of the domain $\mathcal{D}$ occupied by the family of circles (6) thus entails a two–step process: (i) compute the complete envelope $\mathbf{e}(t)$, excluding the intervals in t that yield complex points; and (ii) apply a *trimming* process to this envelope, whereby segments lying in the interior of $\mathcal{D}$, delineated by self–intersections, are identified and discarded.

The *medial axis transform* (MAT) of a planar domain $\mathcal{D}$ defines a family of circles of a special type. The "medial axis" or "skeleton" of $\mathcal{D}$ is the locus of centers of all maximal circles, touching the domain boundary $\partial \mathcal{D}$ in at least two points, that can be inscribed within $\mathcal{D}$. In general, the medial axis is a locus that exhibits a finite number of *bifurcations* (centers of inscribed circles with more than two boundary contact points) and *termination points* (centers of circles with two coincident boundary contact points).

Now if $\mathbf{c}(t)$ is a parameterization of (a segment of) the medial axis, we can define a *radius function* $R(t)$ on it, specifying the size of the inscribed circle centered at each point along it. The union $\{\mathbf{c}(t), R(t)\}$ of all the medial axis segments and their superposed radius functions then constitutes the *medial axis transform* of the domain $\mathcal{D}$. The MAT embodies complete information on the shape of $\mathcal{D}$, and the domain boundary $\partial \mathcal{D}$ is precisely identical[4] to the envelope of the family of circles with centers $\mathbf{c}(t)$ and radii $R(t)$. Medial axis transforms have proven useful in diverse application contexts, such as pattern analysis and shape recognition [7, 9]; NC tool path generation [16, 31, 49]; finite element meshing [29, 58, 59]; and font design [15].

4 Stability of Characteristic Polynomials

In the control of dynamic systems governed by ordinary differential equations, the issue of stability (i.e., precluding the possibility of exponentially–growing oscillations) is of paramount concern. The usual approach to this problem is to convert the differential equation into an algebraic equation, by taking the Laplace transform. The roots of the resulting *characteristic polynomial* must have negative real parts if the system is to be stable. If the system parameters are not specified precisely, or are subject to perturbations, one must ensure the stability of each member of a *family* of polynomials, whose coefficients are drawn from certain sets. Such systems are said to exhibit "robust" stability with respect to the uncertainties or perturbations in their parameters.

Consider the characteristic polynomial of a linear n^{th} order system

---

[4] If the center curve $\mathbf{c}(t)$ and radius function $R(t)$ represent the MAT of a planar domain, neither of the circumstances (a) or (b) above can arise.

$$\mathbf{p}(\mathbf{s}) = \mathbf{a}_n \mathbf{s}^n + \mathbf{a}_{n-1} \mathbf{s}^{n-1} + \cdots + \mathbf{a}_1 \mathbf{s} + \mathbf{a}_0 \qquad (9)$$

in the complex Laplace transform variable, $\mathbf{s} = \sigma + i\omega$. We do not assume here that $\mathbf{p}(\mathbf{s})$ is monic (i.e., the highest–order coefficient is unity). Although, for the case of fixed coefficients, such an assumption incurs no loss of generality, this is no longer true when the coefficients are imagined to come from complex sets of a particular form (e.g., rectangles or circular disks).

For fixed coefficients, the polynomial (9) has the factorization

$$\mathbf{p}(\mathbf{s}) - \mathbf{a}_n \prod_{k=1}^{n} (\mathbf{s} - \mathbf{z}_k),$$

$\mathbf{z}_1, \ldots, \mathbf{z}_n$ being its n (not necessarily distinct) roots. Writing $\mathbf{z}_k = \sigma_k + i\omega_k$ for $k = 1, \ldots, n$, the system characterized by the polynomial (9) is said to be *stable* or *Hurwitz* when $\sigma_k < 0$ for $k = 1, \ldots, n$.

However, "nominal" stability may not be an adequate guarantee of system performance. The system design must usually incorporate rapid damping of oscillations and fast response times, which entails placing the roots $\mathbf{z}_1, \ldots, \mathbf{z}_n$ in a specified domain Γ in the left–half of the complex plane. A system that satisfies such a root–placement condition is said to be Γ–*stable*.

In "robust" control problems, the coefficients of (9) are allowed to vary in a specified manner, and one must ensure Hurwitz or Γ–stability for all possible variations [3] — which could arise from measurement errors, environmental perturbations, uncertainty in physical quantities, or tuning of performance–optimization parameters. A hierarchy of situations may be enumerated, in which the characteristic polynomial coefficients are considered to be[5]

0a precisely–specified real values a_k;
0b precisely–specified complex values $\mathbf{a}_k$;
1a independent real values a_k varying within specified real intervals $[\underline{a}_k, \overline{a}_k]$;
1b independent complex values $\mathbf{a}_k$ varying within specified domains $\mathcal{A}_k \subset \mathbb{C}$;
2a real values $a_k = f_k(q_1, \ldots, q_r)$ given by functions of parameters q_i varying independently within specified real intervals $[\underline{q}_i, \overline{q}_i]$;
2b complex values $\mathbf{a}_k = \mathbf{f}_k(\mathbf{q}_1, \ldots, \mathbf{q}_r)$ given by functions of parameters $\mathbf{q}_i$ varying independently within specified domains $\mathcal{Q}_i \subset \mathbb{C}$.

Note that correlations exist among the coefficients in cases 2a,b when two or more coefficients depend on common parameters. Note also that the general complex case is inherently more complicated than the real case, since $\mathcal{A}_k$ and $\mathcal{Q}_i$ above can be two–dimensional domains of general shape — not just tensor products of one–dimensional intervals (i.e., rectangles).

A sufficient–and–necessary condition for stability in case 0a is given by the Routh–Hurwitz criterion [25, 32, 53], which requires the $n \times n$ determinant

---

[5] The motivation for studying characteristic polynomials with *complex* coefficients arises from transformations of the complex plane and problems in dynamics, vibrations, and the design of band–pass of band–rejection filters [3, 4, 11].

$$\Delta_n = \begin{vmatrix} a_1 & a_3 & a_5 & a_7 & a_9 & \cdot & \cdot & \cdot \\ a_0 & a_2 & a_4 & a_6 & a_8 & \cdot & \cdot & \cdot \\ 0 & a_1 & a_3 & a_5 & a_7 & \cdot & \cdot & \cdot \\ 0 & a_0 & a_2 & a_4 & a_6 & \cdot & \cdot & \cdot \\ 0 & 0 & a_1 & a_3 & a_5 & \cdot & \cdot & \cdot \\ 0 & 0 & a_0 & a_2 & a_4 & \cdot & \cdot & \cdot \\ & \cdot & \cdot & \cdot & \cdot & \cdot & & \\ & & \cdot & \cdot & \cdot & \cdot & \cdot & \end{vmatrix} \qquad (10)$$

and all minors $\Delta_{n-1}, \ldots, \Delta_1$, obtained by successively deleting the last row and column, to be positive (entries a_k with $k > n$ in (10) are zero).

The Routh–Hurwitz criterion can be extended to the case 0b of complex coefficients [6, 24, 42]. In this context, one forms a sequence of "partitioned" determinants, the real and imaginary parts of the coefficients $\mathbf{a}_k$ appearing separately in Hurwitz–like sequences with appropriate signs, in diametrically opposed quadrants of the array. Again, a sufficient–and–necessary condition for stability is that the n determinants $\Delta_n, \ldots, \Delta_1$ thus defined be positive. A comprehensive treatment of this case may be found in [42].

The first results concerning the stability of *families* of polynomials were obtained by Kharitonov [35]; see also [3, 44, 62]. He demonstrated that robust stability can be established in case 1a by verifying that just *four* specially–constructed members from the complete family of polynomials defined by interval coefficients, $a_k \in [\underline{a}_k, \overline{a}_k]$ for $k = 0, \ldots, n$, satisfy the Routh–Hurwitz condition. These "Kharitonov polynomials" are

$$\mathbf{p}_1(\mathbf{s}) = \underline{a}_0 + \underline{a}_1 \mathbf{s} + \overline{a}_2 \mathbf{s}^2 + \overline{a}_3 \mathbf{s}^3 + \underline{a}_4 \mathbf{s}^4 + \underline{a}_5 \mathbf{s}^5 + \cdots,$$
$$\mathbf{p}_2(\mathbf{s}) = \underline{a}_0 + \overline{a}_1 \mathbf{s} + \overline{a}_2 \mathbf{s}^2 + \underline{a}_3 \mathbf{s}^3 + \underline{a}_4 \mathbf{s}^4 + \overline{a}_5 \mathbf{s}^5 + \cdots,$$
$$\mathbf{p}_3(\mathbf{s}) = \overline{a}_0 + \underline{a}_1 \mathbf{s} + \underline{a}_2 \mathbf{s}^2 + \overline{a}_3 \mathbf{s}^3 + \overline{a}_4 \mathbf{s}^4 + \underline{a}_5 \mathbf{s}^5 + \cdots,$$
$$\mathbf{p}_4(\mathbf{s}) = \overline{a}_0 + \overline{a}_1 \mathbf{s} + \underline{a}_2 \mathbf{s}^2 + \underline{a}_3 \mathbf{s}^3 + \overline{a}_4 \mathbf{s}^4 + \overline{a}_5 \mathbf{s}^5 + \cdots. \qquad (11)$$

They can be interpreted as follows — as the coefficients a_k vary independently over their respective intervals $[\underline{a}_k, \overline{a}_k]$, the values $\mathbf{p}(i\omega)$ assumed by (9), for a fixed point $\mathbf{s} = i\omega$ on the imaginary axis, cover a rectangle in the complex plane: the corners of this rectangle are $\mathbf{p}_1(i\omega)$, $\mathbf{p}_2(i\omega)$, $\mathbf{p}_3(i\omega)$, $\mathbf{p}_4(i\omega)$.

This rectangle is called the *value set* at $\mathbf{s} = i\omega$ of the polynomial (9) with interval coefficients [3]. We denote it by $\mathcal{V}(\omega)$ and observe that, as ω varies, its dimensions and location change. The property that the polynomials (11) are Hurwitz for all ω is equivalent to the condition that $0 \notin \mathcal{V}(\omega)$ for all ω.

This Zero Exclusion Condition for the value set can be generalized from nominal (Hurwitz) stability to Γ–stability [3]. If the domain boundary $\partial\Gamma$ admits a regular parameterization, $\gamma(t)$ for $-\infty < t < +\infty$, we can define a value set $\mathcal{V}(t)$ for (9) as the set of all values of $\mathbf{p}(\gamma(t))$ when the coefficients vary over their prescribed real intervals (or complex domains) — the system is then guaranteed to be Γ–stable when $0 \notin \mathcal{V}(t)$ for all t.

Although it may be computation–intensive, the Zero Exclusion Condition based on value sets is sufficiently versatile to accommodate polynomials with correlated coefficients (arising from common parameter dependency) in cases 2a,b and domains $\mathcal{A}_k$ and $\mathcal{Q}_i$ of arbitrary shape in cases 1b and 2b.

A generalization [36] of Kharitonov's theorem treats the case of complex coefficients $\mathbf{a}_k = a_k + i\alpha_k$ varying independently within rectangular domains in the complex plane, $(a_k, \alpha_k) \in [\underline{a}_k, \overline{a}_k] \times [\underline{\alpha}_k, \overline{\alpha}_k]$. To establish stability of the entire family of polynomials, it suffices to verify that *eight* specially–constructed (complex) polynomials are Hurwitz — these polynomials reduce to the four polynomials (11) if $\underline{\alpha}_k = \overline{\alpha}_k = 0$ for each k; see also [11].

Although more germane to practical robust control problems, cases 2a and 2b are much more challenging than cases with independent coefficient uncertainties. Minkowski products and implicitly–defined sets play a key role in the robustness analysis of systems with parameter–dependent coefficients. We defer substantive treatment of such cases to a future study, however, and focus here on a more basic problem that entails use of Minkowski sums (with complex scalings only). Specifically, we are interested in the analysis of case 1b for non–rectangular coefficient sets $\mathcal{A}_k$ in the complex plane.

The choice of circular disks as the coefficient sets $\mathcal{A}_k$ offers a compelling illustration of the power of Minkowski geometric algebra: the *total value set* $\mathcal{V}_*$ — i.e., the union of all the value sets $\mathcal{V}(t)$ as t varies — admits a natural characterization as a one–parameter family of disks with centers on a known curve $\mathbf{c}(t)$ and radii given by a known function $R(t)$, from which the boundary $\partial\mathcal{V}_*$ can be explicitly constructed. In fact, this construction is not essential. The stability of a "complex disk polynomial" can be ascertained directly from $\mathbf{c}(t)$ and $R(t)$ — it suffices to check that the single *real* function

$$\rho(t) = |\mathbf{c}(t)|^2 - R^2(t)$$

is positive for all t (which can be accomplished by a finite algorithm).

The conversion of stability problems into tests for the positivity of certain functions has also been considered [56, 61, 63] for characteristic polynomials with parameter–dependent coefficients. In this context, the "test functions" are *multivariate* polynomials, and the positivity test is based upon expressing them in the (tensor–product) multivariate Bernstein basis.

5 Complex Disk Polynomials

Interval arithmetic [46, 47] is a formal algebra for intervals of real numbers, that allows (equally probable) errors or uncertainties in the input parameters to be propagated through a calculation. The result of an arithmetic operation $\star \in \{+, -, \times, \div\}$ on two intervals $[a, b]$ and $[c, d]$ is the set of real values obtained by applying $\star$ to all pairs of operands drawn from these intervals:

$$[a, b] \star [c, d] = \{x \star y \mid x \in [a, b] \text{ and } y \in [c, d]\}. \tag{12}$$

Specifically, one can easily verify that

$$[\,a,b\,] + [\,c,d\,] = [\,a+c,b+d\,]\,,$$
$$[\,a,b\,] - [\,c,d\,] = [\,a-d,b-c\,]\,,$$
$$[\,a,b\,] \times [\,c,d\,] = [\,\min(ac,ad,bc,bd), \max(ac,ad,bc,bd)\,]\,,$$
$$[\,a,b\,] \div [\,c,d\,] = [\,a,b\,] \times [\,1/d,1/c\,]\,, \tag{13}$$

where division is usually defined only for denominators such that $0 \notin [\,c,d\,]$.

The simplest generalization of interval arithmetic to the complex numbers involves "tensor products" of intervals in the real and imaginary parts of a complex variable, i.e., sets of complex numbers that correspond to *rectangular domains* in the complex plane. Such sets exhibit closure under addition, but not multiplication. Gargantini and Henrici [26] were the first to consider an arithmetic for sets of complex numbers defined by *circular disks*, motivated by the problem of confining the roots of a polynomial with precisely–known coefficients inside complex disks. As noted in §2 above, the Minkowski sum of two disks is also a disk, but the Minkowski product is more involved. To ensure closure of the "circular arithmetic," the usual practice [2, 50, 52] is to replace the true Minkowski product of two disks with centers c_1, c_2 and radii R_1, R_2 by a *bounding disk* with center and radius

$$\mathbf{c} = \mathbf{c}_1 \mathbf{c}_2\,, \qquad R = |\mathbf{c}_1| R_2 + |\mathbf{c}_2| R_1 + R_1 R_2\,.$$

Although this is not the *smallest* disk that contains the Minkowski product, it has the useful property of being centered on the product $\mathbf{c}_1 \mathbf{c}_2$ of the operand centers. Another application context for disk arithmetic is the generalization [40] of Bézier curves to "uncertainty disks" as control points.

The most common generalization of the Kharitonov theorem has been to sets of real coefficients $a_0, \ldots, a_n$ that lie within a polytope [5, 39] or a sphere [3, 57] in $\mathbb{R}^{n+1}$. For complex coefficients $\mathbf{a}_0, \ldots, \mathbf{a}_n$, the case where the real and imaginary parts lie individually within "$(n+1)$–dimensional diamonds" has also been addressed [10]. We are interested here in the stability of *disk polynomials* — i.e., complex polynomials with coefficients $\mathbf{a}_0, \ldots, \mathbf{a}_n$ that vary over prescribed circular disks $\mathcal{A}_0, \ldots, \mathcal{A}_n$ in the complex plane:

$$\mathbf{a}_k \in \mathcal{A}_k = \{\, \mathbf{c}_k + \rho\, e^{i\theta} \mid 0 \leq \rho \leq R_k\,,\ 0 \leq \theta < 2\pi \,\}\,. \tag{14}$$

Each coefficient disk $\mathcal{A}_k$ is described by its center $\mathbf{c}_k$ and radius R_k. The problem of Hurwitz stability of disk polynomials has previously been treated in [14, 51]. We re–visit this problem, from a more geometrical perspective, in §6, and give its extension to the case of Γ–stability in §7. We also treat the problem of finding the *robustness margin* (the greatest factor by which the disk radii can be enlarged without sacrificing robust stability) in §8.

A disk polynomial $\mathbf{p}(\mathbf{s})$ has the attractive property that, since it depends *linearly* on the coefficients (14), its value set $\mathcal{V}(t)$ at any point of a complex curve $\boldsymbol{\gamma}(t)$ is just a Minkowski sum of (scaled/rotated) disks, which is always simply a disk. Hence we can characterize the value sets as a one–parameter family of circular disks as t varies. In principle, this characterization allows us to explicitly construct the boundary of the total value set

$$\mathcal{V}_* = \bigcup_t \mathcal{V}(t) \tag{15}$$

by the approach described in §3 above.

An alternative approach to characterizing the disk polynomial specified by the coefficients (14) is to regard each instance $(\mathbf{a}_0, \dots, \mathbf{a}_n)$ as a point $\mathbf{a}$ in $(n+1)$–dimensional complex space, $\mathbb{C}^{n+1}$. We can then identify the disk polynomial with the "uncertainty volume" in $\mathbb{C}^{n+1}$ defined by

$$\mathcal{A} = \{\, \mathbf{a} = (\mathbf{a}_0, \dots, \mathbf{a}_n) \in \mathbb{C}^{n+1} \;:\; |\mathbf{a}_k - \mathbf{c}_k| < R_k \text{ for } k = 0, \dots, n \,\}.$$

This volume is a *tensor product of $n+1$ circular disks* (not easy to visualize, since the simplest case is a four–dimensional entity). When all the disks have the same radius, $R_0 = \cdots = R_n \, (= R, \text{ say})$ this volume can be written as

$$\mathcal{A} = \{\, \mathbf{a} \in \mathbb{C}^{n+1} \;:\; \| \mathbf{a} - \mathbf{c} \|_{\infty, \mathbb{C}^{n+1}} < R \,\},$$

where $\mathbf{c} = (\mathbf{c}_0, \dots, \mathbf{c}_n)$ and $\| \cdot \|_{\infty, \mathbb{C}^{n+1}}$ denotes the ℓ_∞ norm in $\mathbb{C}^{n+1}$, namely

$$\| \mathbf{a} \|_{\infty, \mathbb{C}^{n+1}} = \max_{0 \le k \le n} |\mathbf{a}_k| \qquad \text{for} \quad \mathbf{a} = (\mathbf{a}_0, \dots, \mathbf{a}_n) \in \mathbb{C}^{n+1}.$$

This uncertainty volume $\mathcal{A}$ is the ℓ_∞ ball with center $\mathbf{c}$ and radius R.

6 Hurwitz Stability of Disk Polynomials

Consider first the Hurwitz stability of the disk polynomial (9) whose complex coefficients $\mathbf{a}_0, \dots, \mathbf{a}_n$ vary independently over uncertainty disks $\mathcal{A}_0, \dots, \mathcal{A}_n$ given by (14). To avoid the possibility of a drop in degree, we assume that the highest–order coefficient disk $\mathcal{A}_n$ does not contain the origin, i.e., $|\mathbf{c}_n| > R_n$. We also assume the disk polynomial has at least one member that is Hurwitz stable. The Zero Exclusion Condition [3] can then be applied to the value set $\mathcal{V}(t)$ for $\mathbf{p}$ along the imaginary axis. Since the polynomial (9) has *complex* rather than real coefficients, we must apply the Zero Exclusion Condition to the *whole* imaginary axis [3], not just the positive half.

Because the coefficients vary independently, the value set $\mathcal{V}(t) = \mathbf{p}(\mathrm{i}t)$ on the imaginary axis can be expressed as the Minkowski combination

$$\mathbf{p}(\mathrm{i}t) = (\mathcal{A}_n \otimes \{\mathrm{i}^n t^n\}) \oplus \cdots \oplus (\mathcal{A}_1 \otimes \{\mathrm{i}t\}) \oplus \mathcal{A}_0.$$

The generic term $\mathcal{A}_k \otimes \{\mathrm{i}^k t^k\}$ above is the Minkowski product of a disk and a singleton set — which yields a scaling and rotation of the disk $\mathcal{A}_k$ by the complex value $\mathrm{i}^k t^k$. The value set $\mathcal{V}(t)$ can thus be regarded as a Minkowski sum of $n+1$ disks, whose centers and radii depend on the "frequency" t and on the centers and radii of the coefficient disks $\mathcal{A}_0, \ldots, \mathcal{A}_n$.

Now the Minkowski summation of any number of disks is a trivial task — the result is a disk whose center is given by the vector sum of the individual centers, and whose radius equals the sum of the individual radii. Hence, the value set $\mathcal{V}(t)$ defines a one-parameter family of disks,

$$\mathcal{V}(t) \;=\; \{\, \mathbf{c}(t) + \rho\,e^{i\theta} \mid 0 \leq \rho \leq R(t),\; 0 \leq \theta < 2\pi \,\}$$

whose centers and radii are specified for each t by

$$\mathbf{c}(t) \;=\; \sum_{k=0}^{n} (i\,t)^k \mathbf{c}_k\,, \qquad R(t) \;=\; \sum_{k=0}^{n} R_k |t|^k\,.$$

The following example illustrates the behavior of a representative value set.

Example 5. Consider the complex–coefficient characteristic polynomial

$$\mathbf{s}^4 + 6\mathbf{s}^3 + (14 - i)\mathbf{s}^2 + (15 - 3i)\mathbf{s} + 6 - 2i \tag{16}$$

whose zeros are at $-1, -2, -1+i$ and $-2-i$. Since all roots have negative real parts, this polynomial is Hurwitz stable. We now define a disk polynomial by allowing each of the coefficients to suffer random perturbations within disks of radius $R_0 = \cdots = R_4 = 0.8$ about their nominal values — note that, since $R_4 < |\mathbf{c}_4| = 1$, this disk polynomial is of invariant degree.

The value set $\mathcal{V}(t)$ along the imaginary axis is a one–parameter family of disks, specified by the center curve and radius function

$$\mathbf{c}(t) = t^4 - 6i\,t^3 - (14 - i)t^2 + (3 - 15i)t + 6 - 2i\,,$$
$$R(t) = 0.8\,(t^4 + |t|^3 + t^2 + |t| + 1)\,.$$

Figure 4 illustrates the behavior of $\mathcal{V}(t)$ as t varies. Since the total value set (15) does not contain the origin, the disk polynomial is Hurwitz stable.

Now since the value set $\mathcal{V}(t)$ is always a disk, the Zero Exclusion Condition is satisfied if we can verify for each t that the distance $|\mathbf{c}(t)|$ of its center from the origin exceeds its radius $R(t)$. This is evidently a much simpler task than explicitly constructing the total value set (15). Consider, therefore, the real–valued function

$$\rho(t) = |\mathbf{c}(t)|^2 - R^2(t)$$
$$= \left[\sum_{k=0}^{n} \mathrm{Re}\,(i^k \mathbf{c}_k) t^k \right]^2 + \left[\sum_{k=0}^{n} \mathrm{Im}\,(i^k \mathbf{c}_k) t^k \right]^2 - \left[\sum_{k=0}^{n} R_k |t|^k \right]^2, \tag{17}$$

which we call the *test function* for the Hurwitz stability of the disk polynomial defined by (9) and (14). Clearly, it suffices to show that $\rho(t) > 0$ for all $t \in \mathbb{R}$ to guarantee Hurwitz stability of this disk polynomial.

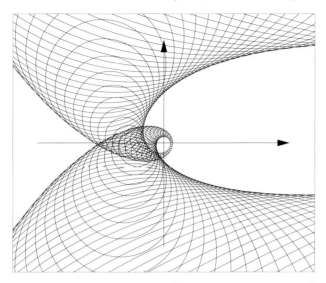

Fig. 4. Value set of the disk polynomial defined by allowing each coefficient of (16) to vary in a "perturbation disk" of radius 0.8 about its nominal value.

For the case of Hurwitz stability, the test function $\rho(t)$ is not a polynomial in t, since it depends on the absolute value $|t|$. However, we may write

$$\rho(t) = \begin{cases} \rho_+(t) & \text{if } t \geq 0, \\ \rho_-(t) & \text{if } t \leq 0, \end{cases} \tag{18}$$

where $\rho_+(t)$ and $\rho_-(t)$ are the polynomials defined, respectively, by replacing the terms $R_k|t|^k$ in (17) with $R_k t^k$ and $(-1)^k R_k t^k$. The positivity of $\rho(t)$ for all t can then be established by verifying that these two polynomials satisfy $\rho_+(t) > 0$ for $t \geq 0$ and $\rho_-(t) > 0$ for $t \leq 0$. We shall see in §7 that the test for general Γ–stability can also be reduced to checking the positivity of two polynomials (for all t values), provided that the stability domain boundary $\partial\Gamma$ is described by a polynomial or rational curve, $\boldsymbol{\gamma}(t)$.

Now the test function (17) has highest–order term $(|\mathbf{c}_n|^2 - R_n^2)t^{2n}$, whose coefficient is positive by the invariant–degree assumption. Hence, $\rho(t) \to +\infty$ as $t \to -\infty$ or $t \to +\infty$, and this observation leads to the following result:

Proposition 1. *Suppose the disk polynomial* $\mathbf{p}(\mathbf{s})$ *is of invariant degree, and contains at least one Hurwitz–stable member. Then* $\mathbf{p}(\mathbf{s})$ *is robustly Hurwitz if and only if the corresponding test function (17) has no real roots.*

By virtue of this result, we reduce the problem of checking if all complex roots of a family of complex polynomials lie to the left of the imaginary axis, into the problem of existence of real roots for a single real–valued function. The existence of real roots for $\rho(t)$ can be determined by Sturm's theorem [60] in combination with lower and upper bounds on its real roots.

Consider first the polynomial $\rho_+(t)$ for $t \geq 0$, which we write as

$$\rho_+(t) = a_{2n}t^{2n} + a_{2n-1}t^{2n-1} + \cdots + a_1t + a_0$$

where $a_{2n} > 0$, as noted above, and we assume that $a_k < 0$ for at least one of $k = 0, \ldots, 2n - 1$ (otherwise, we would know by Descartes Law of Signs [60] that $\rho_+(t)$ has no positive real roots). There are several ways [37] to express an upper bound on the positive roots of $\rho_+(t)$ — the simplest,

$$r_{max} = 1 + \max_{0 \leq i \leq 2n-1} \frac{|a_i|}{a_{2n}}, \tag{19}$$

gives a bound on the absolute value of all (real and complex) roots. A tighter bound for just the *real* roots is given by

$$r_{max} = 1 + \left[\max_{0 \leq i \leq 2n-1} - \frac{a_i}{a_{2n}} \right]^{1/k}, \tag{20}$$

where k is the largest index such that $a_k < 0$ (note that, by supposition, the argument of the kth root above is a *positive* number). If desired, a lower bound r_{min} on the positive roots of $\rho_+(t)$ can also be found [37]: if r_{max} is the upper bound (19) or (20) for the polynomial $t^{2n}\rho_+(1/t)$, then $r_{min} = 1/r_{max}$. Alternately, we may prefer to simply choose $r_{min} = 0$.

Sturm's theorem [37, 60] provides a means to determine the exact number of real roots of a polynomial inside a given interval $[a, b]$. We shall apply it here to $\rho_+(t)$ on $t \in [r_{min}, r_{max}]$. A Sturm sequence $\rho_0(t), \rho_1(t), \ldots, \rho_m(t)$ is constructed by taking $\rho_0(t) = \rho_+(t)$, $\rho_1(t) = \rho'_+(t)$, and defining subsequent polynomials by the equation

$$\rho_{k-1}(t) = \rho_k(t)q_k(t) - \rho_{k+1}(t) \qquad \text{for } k = 1, 2, \ldots,$$

where $\deg(\rho_{k+1}) < \deg(\rho_k)$. This defines $q_k(t)$ as the *quotient* and $-\rho_{k+1}(t)$ as the *remainder* in the division $\rho_{k-1}(t)/\rho_k(t)$ of successive polynomials, and the degree reduction in successive steps ensures that the Sturm sequence will comprise a finite number $m \ (\leq 2n)$ of polynomials beyond $\rho_+(t)$.

Now let $V(r_{min})$ and $V(r_{max})$ be the number of *sign changes* in the values $\rho_0(r_{min}), \ldots, \rho_m(r_{min})$ and $\rho_0(r_{max}), \ldots, \rho_m(r_{max})$ of the Sturm polynomials at $t = r_{min}$ and $t = r_{max}$. Then Sturm's theorem[6] states that

$$\text{# of real roots of } \rho_+(t) \text{ on } t \in (r_{min}, r_{max}) = V(r_{min}) - V(r_{max}).$$

In particular, $V(r_{min}) = V(r_{max}) \Longleftrightarrow \rho_+(t)$ has no real roots in $[r_{min}, r_{max}]$.

To complete the verification that the test function (17) has no real roots, we must also check for the existence of negative real roots of $\rho_-(t)$. This is accomplished in a manner analogous to the test for positive roots of $\rho_+(t)$.

---

[6] Application of Sturm's theorem requires that: (i) r_{min}, r_{max} should not be roots of $\rho_+(t)$, which is easily checked; and (ii) $\rho_+(t)$ should have no *multiple* roots, which we may ensure by dividing it by $\gcd(\rho_+(t), \rho'_+(t))$.

7 Γ–Stability of Disk Polynomials

In §6 we derived a simple method to test whether a disk polynomial $\mathbf{p(s)}$ is robustly Hurwitz–stable, i.e., all its roots lie in the left half of the complex plane as its coefficients vary over the disks (14). We now extend this result to general stability domains Γ, bounded by polynomial or rational curves.

Let $\boldsymbol{\gamma}(t)$ be a regular parameterization of the domain boundary $\partial\Gamma$, and let $\mathcal{V}(t)$ be the value–set for $\mathbf{p}(\boldsymbol{\gamma}(t))$, i.e., the set of all values assumed by $\mathbf{p(s)}$ at $\mathbf{s} = \boldsymbol{\gamma}(t)$ when the coefficients $\mathbf{a}_k$ vary independently over their respective disks $\mathcal{A}_k$. Because of the independence of the coefficient variations, $\mathcal{V}(t)$ can be expressed as the Minkowski combination

$$\mathcal{V}(t) \;=\; (\mathcal{A}_n \otimes \{\boldsymbol{\gamma}^n(t)\}) \;\oplus\; \cdots \;\oplus\; (\mathcal{A}_1 \otimes \{\boldsymbol{\gamma}(t)\}) \;\oplus\; \mathcal{A}_0 \,.$$

Note that the generic term $\mathcal{A}_k \otimes \{\boldsymbol{\gamma}^k(t)\}$ above is the Minkowski product of a disk and a singleton set, which yields a scaling and rotation of the disk $\mathcal{A}_k$ by the complex value $\boldsymbol{\gamma}^k(t)$. Introducing the polar form

$$\boldsymbol{\gamma}(t) \;=\; r(t)\,\mathrm{e}^{\mathrm{i}\,\phi(t)} \tag{21}$$

for the boundary curve, the scaled and rotated disk is described by

$$\mathcal{A}_k \otimes \{\boldsymbol{\gamma}^k(t)\} \;=\; \{\, r^k(t)\mathrm{e}^{\mathrm{i}k\phi(t)}\mathbf{c}_k + \rho\,\mathrm{e}^{\mathrm{i}\theta} \mid 0 \le \rho \le R_k r^k(t)\,,\; 0 \le \theta < 2\pi \,\}\,.$$

Hence, the value set $\mathcal{V}(t)$ can be regarded as the Minkowski sum of $n+1$ disks, whose centers and radii depend on the parameter t and on the centers and radii of the coefficient disks $\mathcal{A}_0, \ldots, \mathcal{A}_n$. In the present context, one can readily see that $\mathcal{V}(t)$ corresponds to a one–parameter family of circular disks

$$\mathcal{V}(t) \;=\; \{\, \mathbf{c}(t) + \rho\,\mathrm{e}^{\mathrm{i}\theta} \mid 0 \le \rho \le R(t)\,,\; 0 \le \theta < 2\pi \,\}$$

whose centers and radii are specified for each t by

$$\mathbf{c}(t) \;=\; \sum_{k=0}^{n} \mathbf{c}_k \boldsymbol{\gamma}^k(t)\,, \quad R(t) \;=\; \sum_{k=0}^{n} R_k r^k(t)\,. \tag{22}$$

If the coefficient disks all have the same radius, $R_0 = \cdots = R_n$, we have

$$R(t) \;=\; R_0 \,\frac{1 - r^{n+1}(t)}{1 - r(t)}\,.$$

Now as t ranges over all values required to trace the boundary $\boldsymbol{\gamma}(t)$ of Γ, the stability of $\mathbf{p(s)}$ is guaranteed when the the region (15) — the union of the value sets for each t does not contain the origin. This "total value set" $\mathcal{V}_*$ is the union of the family of disks with centers and radii given by (22).

The boundary of the set $\mathcal{V}_*$ corresponds to (a subset[7] of) the *envelope* of the family of circles described by (6) and (22). The points on each circle that

---

[7] Portions of the envelope may lie in the interior of $\mathcal{V}_*$ — hence the qualification that the boundary $\partial\mathcal{V}_*$ is, in general, a *subset* of the envelope (see §3).

contribute to the envelope are identified by equations (8) — substituting the latter into (6) yields an explicit parameterization $\mathbf{e}(t)$ of the envelope.

Example 6. Consider the stability region $\Gamma = \{\, x + iy \mid x^2 - y^2 \geq 1,\ x < 0 \,\}$ whose boundary $\partial\Gamma$ is the branch of a hyperbola defined by

$$\boldsymbol{\gamma}(t) = (-\cosh t, \sinh t), \quad -\infty < t < +\infty.$$

The polar representation (21) of $\boldsymbol{\gamma}(t)$ is characterized by the functions

$$r(t) = \sqrt{\cosh 2t}, \quad \phi(t) = \pi - \tan^{-1}(\tanh t),$$

where we take $-\pi/2 \leq \tan^{-1}\theta \leq +\pi/2$. We are interested in the Γ–stability of the quadratic

$$\mathbf{p}(\mathbf{s}) = \mathbf{a}_2\mathbf{s}^2 + \mathbf{a}_1\mathbf{s} + \mathbf{a}_0$$

whose coefficients are independently selected from disks of the form (14) with $\mathbf{c}_2 = 1$, $\mathbf{c}_1 = p + q$, $\mathbf{c}_0 = pq$ and $R_0 = R_1 = R_2$ — i.e., $\mathbf{p}(\mathbf{s})$ has nominal real roots $\mathbf{z}_1 = -p$, $\mathbf{z}_2 = -q$ and coefficients satisfying $|\mathbf{a}_k - \mathbf{c}_k| \leq R_0$, $k = 0, 1, 2$. The total value set (15) is then the union of a one–parameter family of disks described by the radius function and center curve

$$R(t) = R_0(1 + \sqrt{\cosh 2t} + \cosh 2t), \tag{23}$$
$$\mathbf{c}(t) = 1 + pq - (p + q)\cosh t + i\,[(p + q) - 2\cosh t]\,\sinh t. \tag{24}$$

This is actually a *rational* curve of degree 4, with parameterization

$$\frac{(1 + pq)(1 - \tau^2)^2 - (p + q)(1 - \tau^4) + i\,2\tau\,[(p + q)(1 - \tau^2) - 2(1 + \tau^2)]}{(1 - \tau^2)^2}$$

— obtained by writing $\tau = \tanh\frac{1}{2}t$ and making use of the identities $\cosh t = (1 + \tau^2)/(1 - \tau^2)$ and $\sinh t = 2\tau/(1 - \tau^2)$. In fact, $\mathbf{c}(t)$ is always a rational curve of degree mn whenever $\boldsymbol{\gamma}(t)$ is rational and of degree m, and $\mathbf{p}(\mathbf{s})$ is a polynomial of degree n. Finally, substituting

$$R'(t) = R_0 \sinh 2t \left[\frac{1}{\sqrt{\cosh 2t}} + 2 \right],$$
$$\mathbf{c}'(t) = -(p + q)\sinh t + i\,[(p + q)\cosh t - 2\cosh 2t],$$
$${c'_x}^2(t) + {c'_y}^2(t) = \cosh 2t\,[(p + q)^2 - 4(p + q)\cosh t + 4\cosh 2t],$$

in (8) yields an explicit parameterization (which we refrain from writing out) for the envelope of the family of disks defined by (23) and (24).

In Figure 5 we illustrate the family of disks and its envelope for the cases specified by $R_0 = 0.25$, $p = 3$, and the three values $q = 0.2$, 1, and 2. It is interesting to note that, even in these "special" examples of Γ–stability of a quadratic disk polynomial, the envelope exhibits a rather intricate structure that depends sensitively on the single parameter q we vary. It is evident in all these cases that the boundary $\partial\mathcal{V}_*$ of the total value set is a *subset* of the envelope. Clearly, the disk polynomial is Γ–stable when $q = 0.2$ since $0 \notin \mathcal{V}_*$; it becomes unstable for $q = 1$, and stable again when q increase to 2.

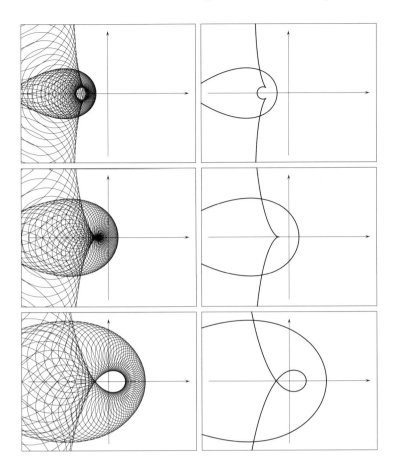

Fig. 5. The family of disks (left) and their envelopes (right) defined by the radius function (23) and center curve (24), for values $R_0 = 0.25$, $p = 3$, and (from top to bottom) $q = 0.2$, 1, and 2. Only the $q = 0.2$ and 2 cases are Γ–stable.

In §6 we showed that establishing Hurwitz stability for a disk polynomial does not require an explicit evaluation of the total value set (15) — it suffices to verify that a single (real) function is always positive. For a more general stability region Γ, bounded by a given polynomial curve $\gamma(t)$, the problem is more difficult. Nevertheless, we shall see that Γ–stability in this case can be tested by verifying the positivity of (at most) two real polynomials.

The difficulty stems from the fact that, for an arbitrary polynomial curve $\gamma(t) = x(t) + \mathrm{i}\,y(t)$, the center locus $\mathbf{c}(t)$ in (22) is always a polynomial curve, but the radius $R(t)$ is *not* a polynomial function, since we have

$$r(t) \;=\; \sqrt{x^2(t) + y^2(t)} \tag{25}$$

in the polar representation (21). Hence, $|\mathbf{c}(t)|^2 - R^2(t) \geq 0$ is not, in general, a polynomial inequality. To remedy this, we separate $R(t)$ into components involving even powers and odd powers of $r(t)$ only, as follows

$$R(t) = P(t) + Q(t)r(t),$$

where

$$P(t) = \sum_{k=0}^{\lfloor \frac{1}{2}n \rfloor} R_{2k} r^{2k}(t) \quad \text{and} \quad Q(t) = \sum_{k=0}^{\lfloor \frac{1}{2}(n-1) \rfloor} R_{2k+1} r^{2k}(t).$$

Note that, when $\boldsymbol{\gamma}(t)$ does not go through the origin and $R_0, \ldots, R_n > 0$, the functions $R(t)$, $P(t)$, $Q(t)$, $r(t)$ are positive for all t. The stability condition

$$|\mathbf{c}(t)|^2 - R^2(t) \geq 0$$

can then be formulated as

$$F(t) = |\mathbf{c}(t)|^2 - P^2(t) - Q^2(t)r^2(t) \geq 2P(t)Q(t)r(t). \tag{26}$$

Now since the right–hand side is always positive, this can only be satisfied if the polynomial $F(t)$ on the left is positive for all t. The first step is thus to form $F(t)$, and check its positivity. If it fails this test, we stop since $\mathbf{p}(\mathbf{s})$ cannot be Γ–stable. If it passes, we proceed to form the polynomial

$$G(t) = F^2(t) - 4P^2(t)Q^2(t)r^2(t). \tag{27}$$

Since we know at this stage that $F(t)$ and $2P(t)Q(t)r(t)$ are both positive for all t, the inequality (26) is exactly equivalent to the requirement that $G(t)$ be positive for all t. Thus, Γ–stability of a disk polynomial can be established, for any domain Γ bounded by a polynomial curve $\boldsymbol{\gamma}(t)$, by showing that the two real polynomials $F(t)$ and $G(t)$ remain positive for all t.

From expressions (22) one can easily see that, if $\mathbf{p}(\mathbf{s})$ is of degree n and $\boldsymbol{\gamma}(t)$ is of degree m, we have $\deg(F) = 2mn$ and $\deg(G) = 4mn$. Furthermore, under the assumption that $\mathbf{p}(\mathbf{s})$ is of invariant degree (i.e., $|\mathbf{c}_n|^2 > R_n^2$), both of these polynomials have positive highest–order coefficients, so $F(t) \to +\infty$ and $G(t) \to +\infty$ as $t \to -\infty$ or $t \to +\infty$. We may summarize as follows:

Proposition 2. *Let Γ be a domain in the complex plane, with the polynomial curve $\boldsymbol{\gamma}(t) = x(t) + \mathrm{i}\,y(t)$ as boundary. Then, if the disk polynomial $\mathbf{p}(\mathbf{s})$ is of invariant degree and includes at least one Γ–stable member, it is robustly Γ–stable if and only if the test functions (26) and (27) have no real roots.*

To verify the absence of real roots for (26) and (27), we use an adaptation of the method described in §6, based on root bounds and Sturm sequences. Finally, we note that it is not difficult to generalize the above arguments to domains Γ bounded by *rational* curves of the form

$$\boldsymbol{\gamma}(t) = \frac{X(t) + \mathrm{i}\,Y(t)}{W(t)},$$

where $W(t)$, $X(t)$, $Y(t)$ are polynomials with $\gcd(W, X, Y) = 1$.

8 Robustness Margin of Disk Polynomials

Suppose we have verified the Γ–stability of a disk polynomial $\mathbf{p}(\mathbf{s})$. It is then natural to ask — what is the greatest factor σ_m by which we can magnify its coefficient disks (14) without sacrificing robust Γ–stability? We call this factor σ_m the *robustness margin* of the disk polynomial defined by (14).

Let $\mathbf{p}(\mathbf{s})$ be the disk polynomial whose coefficient disks $\mathcal{A}_k$ have centers $\mathbf{c}_k$ and radius R_k for $k = 0, \ldots, n$, and let $\boldsymbol{\gamma}(t)$ be a parameterization of the boundary of the stability region Γ. The total value set (15) is then the union of a one–parameter family $\mathcal{V}(t)$ of disks with centers $\mathbf{c}(t)$ and radii $R(t)$ given by expressions (22), and the Zero Exclusion Condition is satisfied if and only if the test function $\rho(t) = |\mathbf{c}(t)|^2 - R^2(t)$ is positive for all t.

We now introduce a new disk polynomial $\mathbf{p}_\sigma(\mathbf{s})$, whose coefficient disks have radii σR_k (where $\sigma > 0$), and we denote by $\mathcal{V}_\sigma(t)$ the value set for $\mathbf{p}_\sigma(\mathbf{s})$ corresponding to the point $\boldsymbol{\gamma}(t)$ on $\partial\Gamma$. Note that this satisfies the inclusion condition $\mathcal{V}_{\sigma_1}(t) \subset \mathcal{V}_{\sigma_2}(t)$ if $\sigma_1 < \sigma_2$. The stability margin σ_m is the smallest value of σ such that, for some t, the value set $\mathcal{V}_\sigma(t)$ just contains the origin. Alternately, we can define σ_m in terms of the test function

$$\rho_\sigma(t) = |\mathbf{c}(t)|^2 - \sigma^2 R^2(t)$$

for $\mathbf{p}_\sigma(\mathbf{s})$. Namely, σ_m is the smallest σ value for which $\rho_\sigma(t)$ has real roots.

For Hurwitz stability, computing the robustness margin is relatively easy, since the test function can be decomposed into the polynomials (18). Suppose that $\rho_\sigma(t)$ is likewise decomposed into $\rho_{\sigma+}(t)$ for $t \geq 0$ and $\rho_{\sigma-}(t)$ for $t \leq 0$. Since the stability of the disk polynomial $\mathbf{p}(\mathbf{s})$ ensures the positivity of $\rho_\sigma(t)$ for $\sigma = 1$, the first real root of $\rho_\sigma(t)$ will appear as a multiple root when σ is increased above unity. The σ values which cause $\rho_\sigma(t)$ to have multiple roots can be computed by solving the equation

$$\Delta(\sigma) = \text{Resultant}_t(\rho_\sigma(t), \rho'_\sigma(t)) = 0 \tag{28}$$

Here $\Delta(\sigma)$, the *discriminant* of $\rho_\sigma(t)$, is a polynomial in σ. In fact, we must solve equation (28) twice, once with $\rho_{\sigma+}(t)$ and once with $\rho_{\sigma-}(t)$.

Now suppose $\sigma_1, \ldots, \sigma_k$ are the roots of (28) with $\rho_{\sigma+}(t)$. One or more multiple roots t of $\rho_{\sigma+}(t)$ is associated with each of these values — we keep the smallest value among $\sigma_1, \ldots, \sigma_k$ that yields a *positive* multiple root, and we call this value σ_+. Analogously, among the roots of (28) with $\rho_{\sigma-}(t)$, we keep the smallest value that yields a *negative* multiple root, and call it σ_-. The robustness margin for the disk polynomial is then $\sigma_m = \min(\sigma_+, \sigma_-)$.

A somewhat more subtle approach to the robustness margin is required for Γ–stability. If the domain boundary $\partial\Gamma$ has the parameterization $\boldsymbol{\gamma}(t) = x(t) + \mathrm{i}y(t)$, the test function $\rho_\sigma(t)$ for the disk polynomial $\mathbf{p}_\sigma(\mathbf{s})$ is

$$\rho_\sigma(t) = |\mathbf{c}(t)|^2 - \sigma^2 R^2(t). \tag{29}$$

Here $|\mathbf{c}(t)|^2$ is a polynomial in t, but $R(t)$ contains the radical term (25). To identify multiple roots of $\rho_\sigma(t)$, we must solve the simultaneous equations

$$\rho_\sigma(t) = \rho'_\sigma(t) = 0$$

in t and σ. Now the presence of the radical (25) prevents us from taking the resultant with respect to t, but we can easily eliminate σ instead, since σ^2 appears linearly in both equations. This yields a single equation[8]

$$|\mathbf{c}(t)|^2 R'(t) - \mathbf{c}(t) \cdot \mathbf{c}'(t) R(t) = 0$$

in t (each root of this equation is a multiple root of $\rho_\sigma(t)$ for some σ value). Multiplying by $r(t)$ and re–arranging, we can transform this into an equation of the form

$$A(t) + B(t)\, r(t) = 0 \,,$$

where $A(t)$ and $B(t)$ are both polynomials. To solve this, we take squares to form the polynomial equation

$$A^2(t) - B^2(t)\, r^2(t) = 0 \,.$$

Among the real roots of this polynomial, we have to discard those that are roots of $A(t) - B(t)\, r(t) = 0$ rather than $A(t) + B(t)\, r(t) = 0$; this is easily accomplished by comparing the signs of $A(t)$ and $B(t)$ at each root.

For each remaining root t, the corresponding σ value satisfies $\rho_\sigma(t) = 0$, and hence is given by $\sigma = |\mathbf{c}(t)|/R(t)$. Finally, the smallest of the σ values thus determined defines the robustness margin σ_m for Γ–stability of $\mathbf{p}(\mathbf{s})$. The following example illustrates the above method of computing the robustness margin, using the discriminant of the test function.

Example 7. Consider the characteristic polynomial (16) in Example 5. We construct a disk polynomial by allowing its coefficients to vary within complex disks of radii $R_0 = \cdots = R_4 = 1$ about the nominal values (we choose disks of equal radii for simplicity; the method is by no means limited to this case). The test function (29) is then given by

$$\rho_\sigma(t) = (t^4 - 14t^2 + 3t + 6)^2 + (-6t^3 + t^2 - 15t - 2)^2 \\ - \sigma^2(t^4 + |t|^3 + t^2 + |t| + 1)^2 \,.$$

For $t > 0$, this can be re–written as a polynomial of degree 8, namely

$$\rho_{\sigma+}(t) = (1 - \sigma^2)t^8 - 2\sigma^2 t^7 + (8 - 3\sigma^2)t^6 - (6 + 4\sigma^2)t^5 + (389 - 5\sigma^2)t^4 \\ - (90 + 4\sigma^2)t^3 + (62 - 3\sigma^2)t^2 + (96 - 2\sigma^2)t + 40 - \sigma^2 \,,$$

and the discriminant (28) for $\rho_{\sigma+}(t)$ yields the following polynomial[9] in σ:

---

[8] Note that $\mathbf{c}(t) \cdot \mathbf{c}'(t)$ is a *real* polynomial, since it is half the derivative of $|\mathbf{c}(t)|^2$.
[9] For brevity, the coefficients are shown with only 3 significant digits, although the calculation was performed exactly using a computer algebra system. Such large coefficients commonly arise in resultants, even with "simple" input polynomials.

$$
\begin{aligned}
&- 7.23 \times 10^{16}\, \sigma^{22} + 8.64 \times 10^{19}\, \sigma^{20} - 3.74 \times 10^{22}\, \sigma^{18} + 7.23 \times 10^{24}\, \sigma^{16} \\
&- 6.02 \times 10^{26}\, \sigma^{14} + 2.47 \times 10^{28}\, \sigma^{12} - 5.12 \times 10^{29}\, \sigma^{10} + 4.88 \times 10^{30}\, \sigma^{8} \\
&- 1.46 \times 10^{31}\, \sigma^{6} + 1.92 \times 10^{31}\, \sigma^{4} - 1.19 \times 10^{31}\, \sigma^{2} + 2.85 \times 10^{30} \ .
\end{aligned}
$$

At any root σ of this polynomial, the test function $\rho_{\sigma+}(t)$ possesses a multiple root t. The following table lists all the positive roots σ, and the corresponding multiple roots of $\rho_{\sigma+}(t)$. Among these solutions, we must choose the smallest σ that generates a positive multiple root t, namely, $\sigma_+ = 0.9481$.

positive root σ of $\Delta(\sigma)$	0.9481	5.9430	6.3771	22.3676
multiple root t of $\rho_{\sigma+}(t)$	11.16	−0.22	0.08	−0.99

By similar means, we find the value σ_- associated with the test function $\rho_{\sigma-}(t)$ to be $\sigma_- = 0.9518$, with $t = -12.10$ as the corresponding negative multiple root. The stability margin $\sigma_m = \min(\sigma_-, \sigma_+)$ for the characteristic polynomial (16) with coefficient disk radii $R_0 = \cdots = R_4 = 1$ is thus equal to $\sigma_+ = 0.9481$, and the critical frequency is $t = 11.16$. The fact that $\sigma_m < 1$ indicates that a uniform coefficient radius $R = 1$ defines an unstable disk polynomial: the total value set encroaches on the origin as R increases from 0.8 (the case shown in Figure 4) through the critical value 0.9841 to unity.

9 Closure

The methods of Minkowski geometric algebra offer a systematic approach to problems of robust stability analysis. Although Minkowski operations that admit closed–form evaluations are limited to "simple" operands, algorithms are available to compute general Minkowski sums, products, and implicitly–defined sets to any desired precision. Whereas prior work on the manipulation of complex sets has been more algebraic in emphasis, these algorithms rely extensively on the use of sophisticated *geometrical* methods.

As a preliminary illustration of the application of Minkowski geometric algebra, we have analyzed the Hurwitz and Γ stability of polynomials whose coefficients vary independently over circular disks in the complex plane. We have shown that stability for such "disk polynomials" can be established by verifying that just *two* (real) polynomials are free of real roots. This compares favorably with the Kharitonov result [36] requiring the Hurwitz condition to be satisfied by *eight* (complex) polynomials to establish robust stability for a family of polynomials whose coefficients vary over complex rectangles.

In fact, verifying the robust stability of disk polynomials entails using only a limited subset of the full repertoire of Minkowski operations on complex sets — namely, (scaled) Minkowski sums. In future studies, we hope to investigate the role that Minkowski products and implicitly–defined complex sets play in analyzing the stability of polynomials with parameter–dependent coefficients.

Acknowledgment

This work was supported in part by the National Science Foundation under grant CCR–9902669.

References

1. J. Ackermann (1993), *Robust Control: Systems with Uncertain Physical Parameters*, Springer, London.
2. G. Alefeld and J. Herzberger (1983), *Introduction to Interval Computations* (translated by J. Rokne), Academic Press, New York.
3. B. R. Barmish (1994), *New Tools for Robustness of Linear Systems*, Macmillan, New York.
4. S. Barnett (1983), *Polynomials and Linear Control Systems*, Marcel Dekker, New York.
5. A. C. Bartlett, C. V. Hollot, and H. Lin (1988), Root locations of an entire polytope of polynomials: It suffices to check the edges, *Mathematics of Control, Signals, and Systems* **1**, 61–71.
6. H. Bilharz (1944), Bemerkung zu einem Satze von Hurwitz, *Zeitschrift für Angewandte Mathematik und Mechanik* **24**, 77–82.
7. H. Blum and R. N. Nagel (1978), Shape description using weighted symmetric axis features, *Pattern Recognition* **10**, 167–180.
8. V. G. Boltyanskii (1964), *Envelopes*, Macmillan, New York.
9. F. L. Bookstein (1979), The line–skeleton, *Computer Graphics and Image Processing* **11**, 123–137.
10. N. K. Bose and K. D. Kim (1989), Stability of a complex polynomial set with coefficients in a diamond and generalizations, *IEEE Transactions on Circuits and Systems* **36**, 1168–1174.
11. N. K. Bose and Y. Q. Shi (1987), A simple general proof of Kharitonov's generalized stability criterion, *IEEE Transactions on Circuits and Systems* **34**, 1233–1237.
12. J. W. Bruce and P. J. Giblin (1981), What is an envelope?, *Math. Gazette* **65**, 186–192.
13. J. W. Bruce and P. J. Giblin (1984), *Curves and Singularities*, Cambridge Univ. Press.
14. H. Chapellat, S. P. Bhattacharyya, and M. Dahleh (1990), Robust stability of a family of disc polynomials, *International Journal of Control* **51**, 1353–1362.
15. H. I. Choi, S. W. Choi, H. P. Moon, and N. S. Wee (1997), New algorithm for medial axis transform of plane domain, *Graphical Models and Image Processing* **59**, 463–483.
16. J. J. Chou (1989), *Numerical Control Milling Machine Toolpath Generation for Regions Bounded by Free Form Curves*, PhD thesis, University of Utah.
17. R. Deaux (1956), *Introduction to the Geometry of Complex Numbers* (translated from the French by H. Eves), F. Ungar, New York.
18. R. T. Farouki and J–C. A. Chastang (1992), Curves and surfaces in geometrical optics, *Mathematical Methods in Computer Aided Geometric Design II*, (T. Lyche & L. L. Schumaker, eds.), Academic Press, pp. 239–260.

19. R. T. Farouki, H. P. Moon, and B. Ravani (2001), Minkowski geometric algebra of complex sets, *Geometriae Dedicata* **85**, 283–315.

20. R. T. Farouki, H. P. Moon, and B. Ravani (2000), Algorithms for Minkowski products and implicitly–defined complex sets, *Advances in Computational Mathematics* **13**, 199–229.

21. R. T. Farouki, W. Gu, and H. P. Moon, (2000), Minkowski roots of complex sets, in *Geometric Modeling and Processing 2000*, IEEE Computer Society Press, Los Alamitos, CA, pp. 287–300.

22. R. T. Farouki and H. Pottmann, (2002), Exact Minkowski products of *N* complex disks, *Reliable Computing* **8**, to appear.

23. R. H. Fowler (1929), *The Elementary Differential Geometry of Plane Curves*, Cambridge Univ. Press.

24. E. Frank (1946), On the zeros of polynomials with complex coefficients, *Bulletin of the American Mathematical Society* **52**, 144–157 & 890–898.

25. F. R. Gantmacher (1960), *The Theory of Matrices*, Vol. 2, Chelsea, New York.

26. I. Gargantini and P. Henrici (1972), Circular arithmetic and the determination of polynomial zeros, *Numerische Mathematik* **18**, 305–320.

27. P. K. Ghosh (1988), A mathematical model for shape description using Minkowski operators, *Comput. Vision, Graphics, Image Process.* **44**, 239–269.

28. F. Gomes Teixeira (1971), *Traité des Courbes Spéciales Remarquables Planes et Gauches*, Tome I, Chelsea (reprint), New York.

29. H. N. Gursoy and N. M. Patrikalakis (1992), An automatic coarse and fine surface mesh generation scheme based on the medial axis transform, I: Algorithms *Engineering with Computers* **8**, 121–137.

30. H. Hadwiger (1957), *Vorlesungen über Inhalt, Oberfläche, und Isoperimetrie*, Springer, Berlin.

31. M. Held (1991), *On the Computational Geometry of Pocket Machining*, Springer–Verlag, Berlin.

32. A. Hurwitz (1895), On the conditions under which an equation has only roots with negative real parts, in *Selected Papers on Mathematical Trends in Control Theory*, Dover, New York, 1964, pp. 72–82 (translated from *Mathematische Annalen* **46**, 273–284).

33. A. Kaul (1993), Computing Minkowski sums, PhD Thesis, Columbia University.

34. A. Kaul and R. T. Farouki (1995), Computing Minkowski sums of plane curves, *Int. J. Comput. Geom. Applic.* **5**, 413–432.

35. V. L. Kharitonov (1978), Asymptotic stability of an equilibrium position of a family of systems of linear differential equations, *Differential'nye Uraveniya* **14**, 1483–1485.

36. V. L. Kharitonov (1978), On a generalization of a stability criterion, *Izvestiia Akademii nauk Kazakhskoi SSR, Seria fiziko–matematicheskaia* **1**, 53–57.

37. A. Kurosh (1980), *Higher Algbera* (translated by G. Yankovsky), Mir Publishers, Moscow.

38. J. D. Lawrence (1972), *A Catalog of Special Plane Curves*, Dover, New York.

39. H. Lin, C. V. Hollot, and A. C. Bartlett (1987), Stability of families of polynomials: geometric considerations in the coefficient space, *International Journal of Control* **45**, 649–660.

40. Q. Lin and J. G. Rokne (1998), Disk Bézier curves, *Computer Aided Geometric Design* **15**, 721–737.

41. E. H. Lockwood (1967), *A Book of Curves*, Cambridge Univ. Press.
42. M. Marden (1966), *Geometry of Polynomials* (2nd edition), American Mathematical Society, Providence, RI.
43. H. Minkowski (1903), Volumen und Oberfläche, *Math. Ann.* **57**, 447–495.
44. R. J. Minnechelli, J. J. Anagnost, and C. A. Desoer (1989), An elementary proof of Kharitonov's stability theorem with extensions, *IEEE Transactions on Automatic Control* **34**, 995–998.
45. H. P. Moon (1999), Minkowski Pythagorean hodographs, *Computer Aided Geometric Design* **16**, 739–753.
46. R. E. Moore (1966), *Interval Analysis*, Prentice–Hall, Englewood Cliffs, NJ.
47. R. E. Moore (1979), *Methods and Applications of Interval Analysis*, SIAM, Philadelphia.
48. T. Needham (1997), *Visual Complex Analysis*, Oxford Univ. Press.
49. H. Persson (1978), NC machining of arbitrarily shaped pockets, *Computer Aided Design* **10**, 169–174.
50. M. S. Petković and L. D. Petković (1998), *Complex Interval Arithmetic and Its Applications*, Wiley–VCH, Berlin.
51. B. T. Polyak, P. S. Scherbakov, and S. B. Shmulyian (1994), Construction of value set for robustness analysis via circular arithmetic, *International Journal of Robust and Nonlinear Control* **4**, 371–385.
52. H. Ratschek and J. Rokne (1984), *Computer Methods for the Range of Functions*, Ellis Horwood, Chichester.
53. E. J. Routh (1892), *Dynamics of a System of Rigid Bodies*, Macmillan, New York.
54. H. Schwerdtfeger (1979), *Geometry of Complex Numbers*, Dover, New York.
55. J. Serra (1982), *Image Analysis and Mathematical Morphology*, Academic Press, London.
56. D. D. Šiljak and D. M. Stipanović (1999), Robust D–stability via positivity, *Automatica* **35**, 1477–1484.
57. C. B. Soh, C. S. Berger, and K. P. Dabke (1985), On the stability properties of polynomials with perturbed coefficients, *IEEE Transactions on Automatic Control* **30**, 1033–1036.
58. V. Srinivasan, L. R. Nackman, J. M. Tang, and S. N. Meshkat (1992), Automatic mesh generation using the symmetric axis transform of polygonal domains, *IEEE Proceedings* **80**, 1485–1501.
59. T. K. H. Tam and C. G. Armstrong (1991), 2D finite element mesh generation by medial axis subdivision, *Advances in Engineering Software* **13**, 313–324.
60. J. V. Uspensky (1948), *Theory of Equations*, McGraw–Hill, New York.
61. V. Vicino and M. Milanese (1990), Robust stability of linear state space models via Bernstein polynomials, in *Control of Uncertain Systems* (D. Hinrichsen and B. Martensson, eds.), Birkhauser, Boston.
62. J. C. Willems and R. Tempo (1999), The Kharitonov theorem with degree drop, *IEEE Transactions on Automatic Control* **44**, 2218–2220.
63. M. Zettler and J. Garloff (1989), Robustness analysis of polynomial parameter dependence using Bernstein expansion, *IEEE Transactions on Automatic Control* **43**, 425–431.

Subdivision Invariant Polynomial Interpolation

Stefanie Hahmann[1], Georges-Pierre Bonneau[2], and Alex Yvart[1]

[1] Laboratoire LMC-IMAG, BP.53, F-38041 Grenoble cedex 9, France.
Stefanie.Hahmann@imag.fr
[2] Laboratoire iMAGIS-GRAVIR, INRIA Rhône-Alpes, 655, avenue de l'Europe,
F-38330 Montbonnot, France. *Georges-Pierre.Bonneau@imag.fr*

Summary. In previous works a polynomial interpolation method for triangular meshes has been introduced. This interpolant can be used to design smooth surfaces of arbitrary topological type. In a design process, it is very useful to be able to locate the deformation made on a geometric model. The previously introduced interpolant has the so-called strict locality property: when a mesh vertex is changed, only the surface patches containing this vertex are changed. This enables to locate the deformation at the size of the input triangles. Unfortunately this is not sufficient if the designer wants to add some detail at a smaller size than that of the input triangles. In this paper, we propose a modification of our interpolant, that enables to arbitrary refine the input triangulation, without changing the resulting surface. We call this property the subdivision invariance. After refinement of the input triangulation, the modification of one of the vertices will change the shape of the interpolant at the scale of the refined triangulation. In this way, it is possible to add details at an arbitrary fine scale.

Keywords. polynomial interpolation, subdivision invariance, geometric design

1 Introduction

Designing smooth surfaces of arbitrary topological type has attracted considerable attention in the last 20 years. Standard NURBS surfaces [2], [7], [10] are restricted to model surfaces homeomorphous to a disc. Basically two research directions have been followed in order to overcome this restriction. On the one hand, previous works have tried to interpolate or approximate arbitrary topological type polyhedron, using parametric polynomial surfaces. More recently, subdivision surfaces, already discovered in the mid 70's, have become very popular in the computer graphic community. These surfaces are defined as the limit of a refinement process starting on an initial control polyhedron. We will come in more detail on these related works in the next section. The present paper follows the first research direction: we define a smooth (tangent plane continuous) polynomial surface that interpolates a 2D-manifold triangulation of arbitrary topological type.

In comparison to our previous works on smooth polynomial interpolation [5], [6], this paper adds the following novelties and improvements:

- the concept of subdivision invariance is introduced,
- this concept is illustrated in the curve case,
- the interpolant of [6] is modified in order to fulfill the subdivision invariance property,
- examples of smooth surfaces interpolating locally refined triangulations of non tensor product topology are given.

The subdivision invariance property states in essence that, for some refinement (subdivision) of the input mesh, the interpolant doesn't change. More precisely, if the same interpolation method is applied on the original mesh, or on the refined mesh, the resulting smooth polynomial surfaces are identical. This property enables to refine an input mesh before editing it. In this way the designer can add details at arbitrary positions, and at arbitrary scale. If she/he wants to modify very locally the model, she/he only has to refine the input control mesh sufficiently until the size of the triangles is smaller than the size of the part of the surface that she/he wants to modify. However, the subdivision invariance property would not be useful in practice, if it would be necessary to entirely refine the input mesh before applying the interpolation method. Fortunately, since the interpolation method we are talking about in this paper is strictly local, meaning that a modification of one vertex of the input mesh modifies only the patches having this vertex in common, it is possible to refine locally the triangulation instead of refining it globally. Altogether, the subdivision invariance and the strict locality enable to compactly model smooth parametric polynomial surfaces with arbitrary fine details.

The remaining of the paper is structured as follows. In Section 2, we discuss related works. In Section 3 the concept of subdivision invariance is introduced and illustrated in the simple case of curve interpolation. Section 4 introduces the subdivision invariant interpolant of arbitrary triangulations. Results are given in Section 5. We conclude in Section 6, and give some possible future works.

2 Related Works

The literature on geometric design of arbitrary topological type smooth surfaces can be divided into two groups: those that deal with polynomial surfaces and those dedicated to subdivision surfaces.

Although subdivision surfaces were discovered already in the 70's, they became popular only recently. In comparison active research on polynomial surfaces of arbitrary topological type has been conducted earlier. Much of those works are concerned with polynomial interpolation of 2D-manifold triangulations. Piper [11], Shirman-Sequin [13] and Jensen [8] have used Clough-Tocher like splitting of the input triangles in order to develop a smooth G^1 continuous interpolant of low degree. The splitting process consists in inserting a new vertex in the interior of each input triangle, and inserting new edges between this vertex and the triangle vertices. Thus each input triangle is divided into three sub-triangles. In comparison, the present interpolant is based on a regular 4-split of the input triangles. This is clearly fully appropriate for successive refinements. Recursively splitting a triangle Clough-Tocher like very quickly leads to degenerate triangles, while the regular 4-split can be repeated any given number of times without flattening the triangle. Loop [9] has proposed a G^1 continuous surface that approximates a 2D-manifold triangulation. In comparison, the present method is dedicated to interpolation, and not approximation, and it is of lower degree.

Subdivision surfaces nowadays tend to become a de facto standard for the modeling of smooth surfaces of arbitrary topological type. Works on subdivision surfaces are too numerous to be referred here. A complete survey on subdivision surfaces can be found in [12]. Clearly subdivision surfaces - by construction - are perfectly adapted to the design of surfaces with details at any scale. They are naturally subdivision invariant. The butterfly scheme [1], [14] is the most closest subdivision scheme to our interpolation scheme, because it is interpolatory and based on a triangular input mesh. However, when the triangular mesh presents irregularities (long and thin triangles, long and short edges, etc) the resulting subdivision surface is generally not of sufficient fairness. A comparison is given in [6]. Furthermore, subdivision surfaces do not have an explicit parameterization, while our interpolant has an explicit, low degree polynomial parameterization. In some application areas, for example in the CAD/CAM industry, it is very important to know an explicit parameterization of a surface. This is probably the reason why subdivision surfaces have not yet been integrated in CAD/CAM software.

Even though the present work concerns the more general class of surfaces of arbitrary topological type, some related works about tensor product surfaces can be cited as well. Hierarchical B-splines [3] allow for local refinement of B-spline surface patches without modification of the shape. Both schemes, hierarchical B-spline and our triangular interpolation scheme [6] are subdivision invariant and therefore allow for hierarchical applications. But hierarchical B-splines only work for tensor product surfaces and cannot represent surfaces of arbitrary topological type.

3 Subdivision Invariance

An interpolating scheme can be viewed as a map that takes as input a discrete set of data, and parameters associated to these data. It outputs a continuously defined function that takes the prescribed data values at the corresponding parameters. This can be stated more mathematically using the following notations:

$$I : \quad (d_i \in E, \, t_i \in \Omega)_{i \in \mathcal{D}} \quad \longrightarrow \quad \begin{array}{l} f : \Omega \to E \\ f(t_i) = d_i \quad \forall i \in \mathcal{D} \end{array}$$

Ω is the parameter domain (e.g. an interval of $\mathbb{R}$ for curve interpolation, and the input mesh for our triangular interpolation scheme) containing all parameter values t_i. E is a set containing the data to be interpolated d_i. $\mathcal{D}$ is a discrete index set.

A subdivision scheme can be viewed as a map that takes as input a discrete set of data, and that outputs another discrete set of data that is typically more dense than the input set. This can be stated using the following notations:

$$S : (d_i \in E,\ t_i \in \Omega)_{i \in \mathcal{D}} \quad \longrightarrow \quad (e_j \in E, s_j \in \Omega)_{j \in \mathcal{E}},$$

where $\mathcal{D}$ and $\mathcal{E}$ are two discrete index sets, and $\mathcal{D}$ is included in $\mathcal{E}$. The subdivision scheme is said to be interpolating if $s_i = t_i$, $e_i = d_i$ for all $i \in \mathcal{D}$.

Definition : An interpolation scheme I is *invariant with respect to a subdivision scheme S* iff

$$I \circ S = I.$$

In other words, I is invariant relative to S iff it yields the same result before and after applying subdivision S.

If the interpolation mapping is continuous, and the subdivision scheme is converging towards a continuous limit, then the following identity holds

$$I = S^{\infty}.$$

Thus given any subdivision scheme, there is a unique interpolation mapping that is invariant relative to it. The converse is not true: given an interpolation mapping, there may be different subdivision schemes relative to which the interpolation mapping is invariant.

Hermite Curve Interpolation

We illustrate the subdivision invariance property on a simple curve interpolation example. The Hermite interpolation scheme H takes as input a set of points (position and derivatives) and parameters associated to these data, and outputs the unique piecewise cubic C^1 continuous polynomial function that interpolates these data.

Let $(d_i, d_i')_{i=0,\ldots,n}$ denote the position and derivatives to interpolate, and $(t_i)_{i=0,\ldots,n}$ the corresponding parameter values. Let f denote the Hermite interpolant. f is the unique piecewise cubic C^1 continuous polynomial such that $f(t_i) = d_i$ and $f'(t_i) = d_i'$. Using f, we can easily define a subdivision scheme S - in that case a Hermite subdivision scheme - relative to which the Hermite interpolant is invariant: Simply define

$$S_H(d_i, d_i', t_i) = (e_j, e_j', s_j),$$

where $(e_{2i}, e'_{2i}, s_{2i}) = (d_i, d'_i, t_i)$, and $(e_{2i+1}, e'_{2i+1}, s_{2i+1}) = \left(f\left(\frac{t_i+t_{i+1}}{2}\right), \right.$
$\left. f'\left(\frac{t_i+t_{i+1}}{2}\right), \frac{t_i+t_{i+1}}{2} \right)$. In other words S is an interpolatory subdivision scheme,
and the newly inserted data values are computed using the Hermite inter-
polant. It is almost trivial to prove that H is invariant relative to S_H. Figure
1 shows two pieces of the interpolant to the left, and at the middle the resulting three pieces after subdivision of the first segment. Since each of the
two first pieces after subdivision has the same position and derivative at their
end point than the original polynomial curve, and since all these polynomials
are cubic, they must be identical.

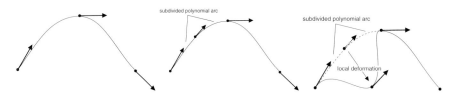

Fig. 1. Hermite interpolation - subdivision - local deformation.

Because H is invariant relative to S_H, and since H is a strictly local interpolation scheme, it is possible to add local details in the following way: use
S_H in order to locally refine the input data around the region that must be
modified, edit the subdivided data in this region, interpolate this new data
using H. The subdivision invariance ensures that the resulting smooth curve
differs from the original one only in the region of interest. This process is
illustrated in Figure 1.

4 Subdivision Invariant $\mathbf{G^1}$ Polynomial Triangular Interpolant

This section begins by a brief recall on a previous work on G^1 polynomial
triangular interpolation (Section 4.1). Then Section 4.2 will give the improvements made on this interpolant in order to satisfy the subdivision invariance
property.

4.1 $\mathbf{G^1}$ Polynomial Triangular Interpolation

In [6], a G^1 polynomial interpolation method for 2D-manifold triangulations
has been introduced. In this section, we briefly recall the main points of this
interpolant, useful for the remaining of the paper.

The interpolant maps each input triangle onto four Bézier patches of degree 5, each of them is parameterized over one of the 4 triangles obtained by the regular 4-split of the input triangle. Figure 2 illustrates the parameterization of the interpolant. Each group of 4 Bézier patches corresponding to one input triangle is called a macro-patch.

The interpolant is build in three steps

- the boundary curves of the macro-patch are determined,
- the first rows of control points on both sides of the boundary curve's control points are constructed. These are the control points from which the tangent planes along the boundary curves are computed. Thus they entirely control the G^1 continuity of the interpolant across the boundary curves, and therefore between the macro-patches.
- the remaining interior control points for each macro-patch are computed so that the 4 Bézier patches connect with C^1 continuity.

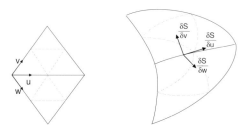

Fig. 2. Parameterization of the interpolant. left, parameter domain; right, macro-patches with tangent directions.

There are many degrees of freedom in this interpolant. The main degrees of freedom are

· the derivatives at the boundary curve's end points, and
· the twists at the mesh vertices.

Once the derivatives and the twists are chosen, the second derivatives of the boundary curves are directly fixed by the twist compatibility condition.

The main equation that governs the surface construction is the G^1 continuity condition between two patches:

$$\Phi(u)\frac{\partial S}{\partial u} = \mu(u)\frac{\partial S}{\partial v} + \nu(u)\frac{\partial S}{\partial w}. \tag{1}$$

Φ, μ and ν are piecewise linear functions defined along the common boundary curve. S is the interpolating surface. $\frac{\partial \cdot}{\partial u}$, $\frac{\partial \cdot}{\partial v}$, $\frac{\partial \cdot}{\partial w}$ are the three derivative operators in the parametric directions shown in Figure 2. G^1 condition (1) states that the three tangent vectors $\frac{\partial S}{\partial u}$, $\frac{\partial S}{\partial v}$, and $\frac{\partial S}{\partial w}$ are coplanar, thus they define a common tangent plane along the boundary curve. In order to find a polynomial solution to the G^1 condition (1), the boundary curve $S(u)$ is chosen that

$$\frac{\partial S}{\partial u} = \mu(u)\,\nu(u)\,H(u) \quad \text{along the boundary curve,} \tag{2}$$

where H is a piecewise continuous degree 2 polynomial. H (and S along the boundary) is uniquely defined by the choice of the first and second derivative of the boundary curves at the end points.

Once S is fixed along the boundary curves, the cross-derivatives $\frac{\partial S}{\partial v}$ and $\frac{\partial S}{\partial w}$ are computed by:

$$\begin{cases} \frac{\partial S}{\partial v} = \frac{1}{2}\big(\Phi(u)\nu(u)H(u)\big) + \nu(u)W(u) \\ \frac{\partial S}{\partial w} = \frac{1}{2}\big(\Phi(u)\mu(u)H(u)\big) - \mu(u)W(u) \end{cases} \tag{3}$$

It is trivial to see that (3) implies (1): simply multiply the first equation in (3) by $\mu(u)$, the second equation in (3) by $\nu(u)$, and add them up.

In [6] W has been chosen as a piecewise C^0 degree 2 polynomial. But it could have been chosen as well as a cubic, without increasing the degree of the interpolant.

After having computed the boundary curves and the cross boundary derivatives using (2) and (3), 15 control points remain to choose inside each macro-patch (see fig. 3). 6 out of these 15 can be freely chosen, the other 9 control points are fixed by the C^1 continuity conditions between the 4 Bézier patches.

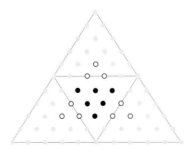

Fig. 3. control points of a macro-patch which os composed of four quintic Bézier triangles. The 15 inner control points are highlighted. The 6 free control points are shown as black dots.

4.2 Subdivision invariance

This section first gives the subdivision operator relative to which the previously introduced interpolation scheme will be made invariant. Then it is shown how the subdivision invariance may be fulfilled.

Subdivision operator

The subdivision operator takes as input the position, the n first derivatives along the boundary curves, and the n twists at each mesh vertex of valence n. Recall (Section 4.1) that these quantities are among the degrees of freedom of the interpolation operator. Figure 4 shows to the left the input of the subdivision operator. The subdivision operator outputs the same data (position, first derivatives and twists) at the input mesh vertices and at the surface points corresponding to the midpoints of each edge (fig. 4-right). In order to compute this subdivision operator, the interpolating surface has first to be computed, and then it must be evaluated at the midpoints of each edge.

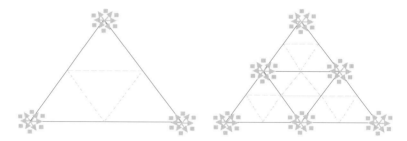

Fig. 4. subdivision operator.

Subdivision invariant interpolation

Beside the first derivatives and twists at the mesh vertices, the other degrees of freedom of the interpolation operator are:

- the values of the (piecewise linear C^0) functions Φ, μ, and ν at the edge midpoints
- the piecewise cubic polynomial function W along the edges. The constraints on W are its position and derivative at the end points, and the C^0 continuity at the edge midpoint.
- the 6 free inner control points.

All these degrees of freedom were previously chosen using heuristics. For example the 6 free inner control points were computed by a least squares energy minimization. The key idea to define a subdivision invariant interpolant is to

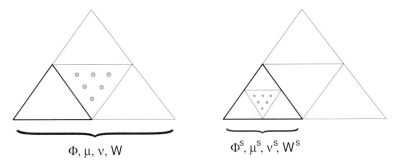

Φ, μ, ν, W $\Phi^s, \mu^s, \nu^s, W^s$

Fig. 5. degrees of freedom of the interpolant before (left) and after(right) subdivision.

choose these degrees of freedom such that they are not modified by the subdivision operator. Figure 5 shows these degrees of freedom for one macro-patch before (left) and after (right) the subdivision operator. We will concentrate on the bottom left Bézier patch of the original macro-patch (highlighted in fig. 5-left). Let Φ, ν, μ, W denote the functions controlling the G^1 continuity along the bottom edge of the macro-patch before subdivision. Let Φ^s, μ^s, ν^s, W^s denote the functions controlling the G^1 continuity along the bottom edge of the bottom left Bézier patch after subdivision. We have to choose the degrees of freedom so that $\Phi = \Phi^s$, $\mu = \mu^s$, $\nu = \nu^s$, $W = W^s$ along their common interval of definition. This property is fulfilled if and only if all the functions are chosen as *one* polynomial piece, e.g. the piecewise linear functions must be chosen linear: $[\Phi, \mu, \nu](\frac{1}{2}) = \frac{1}{2}\Big([\Phi, \mu, \nu](0) + [\Phi, \mu, \nu](1)\Big)$ and W must be the cubic Hermite interpolant of $W(0)$, $W'(0)$, $W(1)$, $W'(1)$ (instead of piecewise quadratic as in the previous method). It remains to fix the 6 free inner control points. In order to do this, we use the Bézier subdivision algorithm of Goldman [4] to 4-split the original Bézier patches. Then we simply copy the 6 inner control points of this Bézier patch in the corresponding Bézier patch of the subdivided macro-patch.

5 Results

Figure 6 (Color Plate 9 on page 430) shows three local refinements of an input icosahedron mesh (left column), the smooth polynomial surface interpolating this triangulation (middle column), and the Bézier control mesh (right column). Instead of globally refining the input mesh, only one edge is splitted at each subdivision step, and the two neighbouring triangles are 4-splitted. The reader may notice discontinuities in the pictures in the left and right columns. This doesn't mean that the interpolating surface is not tangent plane continuous: indeed the middle column clearly illustrates the G^1 continuity of this surface. The gaps in the left and right columns are due to the fact that the same boundary curve is represented by control polygons at different subdivision levels. Thus although both control polygons define the same curve, they are not identical.

Figure 7 illustrates the edition of a vertex in the original model (top) and at the finest level of subdivision (bottom). The locality of the deformations can be clearly seen in this example.

6 Conclusion

In this paper the concept of subdivision invariant interpolation was introduced. This concept is very useful in practice, since it tells exactly when an interpolation scheme can be used to add details at any scale on a given geometric model. The subdivision invariance has been illustrated on a simple curve interpolation scheme. Then the previously introduced triangular polynomial interpolant [6] has been modified in order to fulfill the subdivision invariance property. Examples were given that illustrate the ability to add detail on a smooth surface of arbitrary topological type. In future works, the subdivision invariant interpolation scheme will serve as a basis for the definition of a multiresolution smooth surface representation based on polynomial patches of low degree.

Acknowledgements

This work was partially supported by the European Community 5-th framework program, with the Research Training Network MINGLE (Multiresolution IN Geometric modELing, HPRN-1999-00117).

References

1. Dyn N., Levin D., and Gregory J. (1990): A butterfly subdivision scheme for surface interpolation with tension control. ACM Transactions on Graphics **9** (2), 160–169
2. Farin G. (1996): Curves and Surfaces for Computer Aided Geometric Design. Academic Press, New York, 4th edition
3. Forsey D., Bartels R. (1988): Hierarchical B-spline refinement. In Proceedings of SIGGRAPH 88 , ACM SIGGRAPH, 1205–212
4. Goldman R.N. (1983): Subdivision algorithms for Bézier triangles. CAD **15**, 159–166
5. Hahmann S., Bonneau G-P. (2000): Triangular G^1 interpolation by 4-splitting domain triangles. Computer Aided Geometric Design **17**, 731–757
6. Hahmann S., Bonneau G-P. (2002): Parametric surfaces over arbitrary triangulations. IEEE Transactions on Visualization and Computer Graphics, to appear

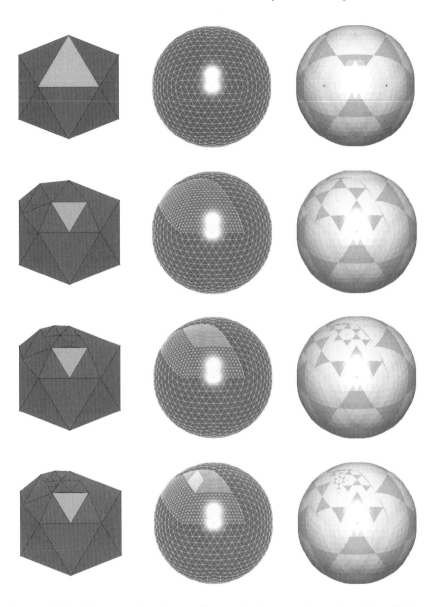

Fig. 6. Subdivision invariant local refinement of a smooth surface. From left to right are shown the input triangulation, the smooth surface together with its Bézier control mesh and the color shaded Bézier control mesh: to each input triangle correspond 4 Bézier patches, the central one is shaded in red. From top to bottom the successive refinement of the icosahedron input mesh is shown.

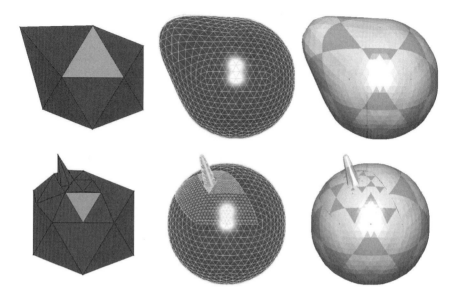

Fig. 7. Deformation of input mesh at coarsest and finest level of subdivision.

7. Hoschek J. and Lasser D. (1993): Fundamentals of Computer Aided Geometric Design. A.K. Peters

8. Jensen (1987): Assembling triangular and rectangular patches and multivariate splines. In: Farin, G. (ed) Geometric Modeling: Algorithms and new Trends. SIAM, 203–220

9. Loop C. (1994): A G^1 triangular spline surface of arbitrary topological type. Computer Aided Geometric Design **11**, 303–330

10. Piegl L. (1987): On NURBS: a Survey. Computer Graphics and Applications **11** (1), 55–71

11. Piper B.R. (1987): Visually smooth interpolation with triangular Bézier patches. In: Farin, G. (ed) Geometric Modeling: Algorithms and new Trends. SIAM, 221–233

12. Schröder P., Zorin D. (1999): Subdivision for Modeling and Animation. SIGGRAPH course notes.

13. Shirman L.A., Séquin C.H. (1987): Local surface interpolation with Bézier patches. Computer Aided Geometric Design **4**, 279–295

14. Zorin D., Schröder P., and Sweldens W. (1996): Interpolating subdivision for meshes with arbitrary topology. In Proceedings of SIGGRAPH 96 , ACM SIGGRAPH, 189–192

Another Metascheme of Subdivision Surfaces

Heinrich Müller and Markus Rips

Informatik VII, University of Dortmund, D-44221 Dortmund, Germany.
mueller@ls7.informatik.uni-dortmund.de

Summary. A subdivision surface is defined by a polygonal mesh which is iteratively refined into an infinite sequence of meshes converging to the desired smooth surface. A framework of systematic classification and construction of subdivision schemes is presented which is based on a single operation, the calculation of the average of the vertices incident to a vertex, edge, or face. More complex subdivision schemes are constructed by concatenation of a collection of elementary subdivision schemes. Properties of these schemes are discussed. In particular it is shown how known subdivision schemes fit into this framework.

Keywords. subdivision metascheme, sudivision surfaces, geometric modelling

1 Introduction

A subdivision surface is defined by a polygonal mesh which is iteratively refined into an infinite sequence of meshes converging to the desired smooth surface. Classical subdivision schemes are those of Catmull/Clark [2], Doo/Sabin [4], and Loop [11]. More recent developments are the butterfly scheme by Dyn et al. [5], midedge subdivision by Peters and Reif [15], $\sqrt{3}$-subdivision by Kobbelt [8] with modifications and extensions by Labsik et al. [10] and Oswald et al. [14], and $4-8$-subdivision by Velho [21]. Schemes for surfaces of higher degree are in the focus of work of Prautzsch [16], Zorin et al. [22], Stam [17], and Oswald et al. [14]. This list certainly is not complete, and basically there are many more possibilities of defining subdivision schemes. The question arises how they can be systematically constructed and classified.

A first attempt of systematic classification and construction of subdivision schemes was undertaken by Kohler [9]. He has found twelve basic types of meshes, by considering assignments of cells between the original and the subdivided mesh. For example, if a vertex of the new mesh is calculated from the vertices of a face of the old one, this means an assignment of type F→V. If the new vertices are connected according to the adjacency of the corresponding faces, we get a mesh corresponding to the well known dual graph of planar graph theory. In this mesh, the faces can be assigned one-to-one to the vertices of the original mesh (V→F), and its edges correspond one-to-one to an edge "perpendicular" to the original one (E→E). In Kohler's classification, shown in Fig. 1, this mesh is of type 4. From the view of construction, Kohler

relation type	vertex to	edge to	face to
1	vertex	vertex	vertex
2	face	vertex	vertex
3	vertex	edge	vertex
4	face	edge	vertex
5	vertex	face	vertex
6	face	face	vertex
7	vertex	vertex	face
8	face	vertex	face
9	vertex	edge	face
10	face	edge	face
11	vertex	face	face
12	face	face	face

Fig. 1. Classification of subdivision schemes according to Kohler. The table expresses assignments of cell types of the old mesh to cell types of the new mesh.

has suggested an algebraic scheme based on two operators named *doubling operator* and *averaging operator*. The doubling (or refinement) and the averaging operators have been used before as a tool to describe the structure of concrete subdivision schemes [16, 19].

Another framework for subdivision scheme is described by Zorin et al. [22]. They distinguish between two components, a *topological split rule* describing how the connectivity of the control mesh is refined, and a *geometric rule* which determines the new control point positions from the old positions. A split can be either *primal*, that is a face split, or *dual*, that is a vertex split, also cf. [7]. A typical geometric rule is averaging of neighboring control points.

Oswald et al. [14] also distinguish between topology and geometry. The topology is changed by an *upsampling operator* which usually refines the mesh. The geometry is changed by iterative transition to the dual mesh in the sense of the "dual graph" of graph theory, calculating the location of the dual vertices as average of the vertices of the corresponding face (and some more, in a modified version). Taubin [18] also uses duality in the sense of graph theory. An even number of iterative applications of dual graph calculation yields a graph which is combinatorially the same as the initial one, and thus is denoted as "primal".

The "metascheme", that is a general principle of construction and classification, presented in the following is a slight revision of an earlier scheme developed by the authors [13], mainly with respect to terminology. The new metascheme is as general as Kohler's, and more general than the other approaches just mentioned. Like those other approaches it distinguishes more clearly between the combinatorical structure of the mesh and its geometry. The doubling operator is eliminated. The fundamental operation of the new metascheme is *vertex assignment* which may concern vertices, edges, or faces of the given mesh. For the different types of vertex assignment to a vertex, an edge, or a face, a topological *filter support* or *topological (filter) mask* is defined. The topological filter support of a new vertex is a set consisting of original vertices which may have influence on the location of the new vertex. The way the location is calculated is expressed in a *geometric filter function* or *geometric (filter) mask* which is a function using the vertices of the filter support. A typical geometric dependency is the possibly weighted average of the locations of the vertices of the topological mask.

The main contribution of this paper is the development of subdivision schemes based on these ideas, and the discussion of structural properties of the resulting meshes. A subdivision scheme is defined as the concatenation of *elementary subdivision schemes* which are introduced in section 2. The focus lies on the combinatorical structure of the generated meshes and the structural dependency of the new vertices on the original ones. Geometric dependency is excluded from detailed discussion.

An interesting question is whether the original mesh can be recalculated from the resulting mesh. This question is illuminated from a combinatorical and a computational point of view in section 3, and further considered in the subsequent section 4. In section 4, a selection of composed subdivision schemes of our metascheme is presented which in particular includes the classical schemes already mentioned. This selection is restricted to schemes up to order 3, that is at most three elementary schemes are concatenated. Section 5 outlines schemes of heigher order which are related to classical concepts of smooth surfaces, like B-spline surfaces [6] and box spline surfaces [1].

2 Elementary Schemes Based on Vertex Assignment

A surface mesh consists of three types of cells of different dimension: vertices (V), edges (E), and faces (F). An elementary subdivision scheme is obtained in three steps.

1. Selection of a combination of cell types.
2. Assignment of vertices to the cells of selected type and definition of topological and geometric filter masks of the new vertices.
3. Definition of the interconnecting mesh.

Step 1 includes seven different combinations: just vertices (V), just edges (E), just faces (F), vertices and edges (VE), vertices and faces (VF), edges and faces (EF), and vertices, faces, and edges together (VEF).

The assignment of vertices in step 2 is straightforward. Intuitively, the vertex assigned to a vertex, called V-vertex, can be understood as the vertex itself, the vertex assigned to an egde, called E-vertex, as some sort of midpoint of the edge, and the vertex assigned to a face, called F-vertex, as some sort of center of the face. This intuitive interpretation leads to the definition of *canonical topological masks* (*CTM*). The CTM of a V-vertex consists of the vertex to which it is assigned. The CTM of an E-vertex has two elements, the vertices of the edge inducing the E-vertex. Analogously, the CTM of an F-vertex consists of the vertices of the face to which the F-vertex has been assigned.

A *canonical geometric mask* related to the canonical masks is averaging. In the case of a V-vertex, filtering by averaging yields the original vertex, whereas the midpoint of the original edge and the center of the original face are delivered for an E-vertex and an F-vertex, respectively. Averaging is a special case of a *linear geometric mask*, that is the new point results from a linear combination of the original points.

For a V-vertex v, a second type of masks looks quite natural, too. The topological mask consists of the original vertex and the vertices adjacent to the original vertex of v, that is those sharing a common edge with v. As geometric mask, averaging the locations of the vertices in the topological mask is again a canonical choice. We call this mask *adjacency mask of a V-vertex*.

The interconnecting mesh calculated in step 3 is obtained by connecting the new vertices resulting from step 2. For this purpose, the incidences of the cells in the original mesh are taken into consideration. A variety of meshes results if not all types of incidences are considered. For example, in the case of combination EF, the possible incidences are neighboring E-vertices ("ee"), and neighboring E- and F-vertices ("ef"). "ff"-incidence has to be excluded because it either gives rise to a conflict with the other types, or it, as the only type, excludes the E-vertices from the new mesh. Similarly, "ee" as only type is impossible because it prevents the occurrence of F-vertices in the new mesh. "ef" as only type is possible, but the simultaneous consideration of "ee" and "ef" is impossible.

The number of possibilities of step 3 can be somewhat restricted by the reasonable constraint that a combination of incidences again yields a planar mesh with convex cells if applied to a planar mesh with convex cells.

Combination V and V_{av}

The combination V is trivial in that it yields the topology of the original mesh. As already mentioned, the canonical topological mask of a vertex consists of the vertex itself. Here and in the following the construction is graphically visualized by showing an original planar mesh with convex faces and the resulting mesh, cf. Fig. 2. In the figures, the original mesh is dashed and slightly moved to the upper left.

Combination V_{av} also yields a mesh with the original topology. The difference to V is in the definition of the topological mask which is now the adjacency mask.

According to Kohler's classification, both constructions are of type 9.

Combination E

For the combination E, only one reasonable interconnection of the E-vertices is possible. Two E-vertices are connected if they share a common vertex and a

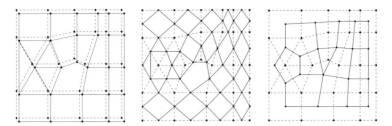

Fig. 2. Combination V (left), E (middle), and F (right). The original mesh is drawn dashed and shifted slightly to the upper left.

common face (Fig. 2). In Kohler's classification, the construction has type 8.

Combination F

For combination F, the only reasonable interconnection is the dual graph construction, that is, to link two F-vertices by an edge if their corresponding faces share an edge (Fig. 2).

A special property of this construction is that its application on the resulting mesh yields a mesh with the same combinatorical structure as the original one. According to Kohler, the construction is of type 4.

Combination VE

In the case of combination VE, we have two intuitive versions of interconnecting meshes. The first version (Fig. 3) only uses edges between a V-vertex and an E-vertex. A V-vertex is connected with an E-vertex if and only if its corresponding vertex in the original mesh is a vertex of the edge from which the E-vertex is derived. In the classification of Kohler this construction is of type 7.

The construction of version 1 is not very useful. The reason is that the basic form of the original mesh is maintained, only its edges are subdivided by vertices. This means that two incident faces share more than one edge.

Version 2 links E-vertices to E-vertices and V-vertices to E-vertices. Two E-vertices get connected if their edges share a common vertex and a common face. A V-vertex is linked to an E-vertex if its corresponding vertex in the original mesh is a vertex of the edge related to the E-vertex. An example is depicted in Fig. 3. The classification according to Kohler is again 7.

A particular property of this construction is that a triangular mesh is again transferred into a triangular mesh. From the view of mesh topology, the subdivision scheme is the same as for Loop surfaces [11] and butterfly surfaces [5] which however have different masks.

Combination VF

In the case of combination VF, we have three versions. The first version (Fig. 3) has only edges between V-vertices and F-vertices. A V-vertex is

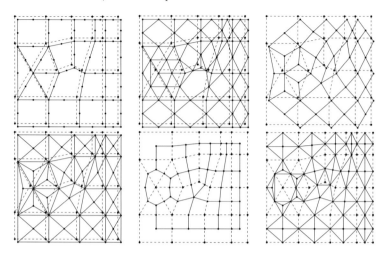

Fig. 3. From left to right, and top to bottom: combination VE/version 1, VE/version 2, VF/version 1, VF/version 2, EF/version 1, and EF/version 2. The edges of the original mesh are dashed.

connected to an F-vertex if the corresponding old vertex of the V-vertex is incident to the face of the F-vertex. This construction is of type 5 in Kohler's classification.

The construction has the interesting property that the resulting mesh has only four-sided faces. This means in particular that a quadrilateral initial mesh is subdivided into a quadrilateral mesh.

Somewhat unfavorable, but acceptable because of that property, is that the resulting faces are not convex in general. A non-convex face may arise for two triangular faces which have a very short edge in common, and for which the perpendicular projection of the vertex opposite to the edge onto the line through the edge lies strictly outside of the edge.

The second version (Fig. 3) connects V-vertices to F-vertices, and V-vertices to V-vertices. Like in version 1, a V-vertex is connected to an F-vertex if the corresponding vertex of the V-vertex is incident to the face of the F-vertex. Additionally, two V-vertices get connected by an edge if they are the two vertices of a common edge.

The third version is analogous to the second version, but connects F-vertices to F-vertices instead of V-vertices to V-vertices. Two F-vertices are connected if their faces are incident to a common edge.

The construction is of type 3 in Kohler's classification. It has the favorable property that it transfers any initial mesh into a triangular mesh. Hence, in particular, an initial triangular mesh is again subdivided into a triangular mesh.

Velho et al. [21] have invented their $4 - 8$-scheme in order to generalize four-directional box splines to arbitrary meshes. It is basically of type $V_{av}F1$. However, every iterative application of this operation has to be terminated by $V_{av}F2$ in order to achieve the $4 - 8$-connectivity. Furthermore, a given arbitrary mesh has to be transferred into a quadrilateral mesh first. This task can in principle be achieved by applying VF1 or any other scheme which yields four-sided faces, but a more efficient, irregular approach has been proposed in [21].

Combination EF

We have two versions for the combination EF. The first version (Fig. 3) only has edges between E-vertices and F-vertices. An E-vertex gets an edge to an F-vertex if and only if its corresponding edge belongs to the boundary of the face corresponding to the F-vertex. The resulting faces need not necessarily be convex, as Fig. 3 shows. In Kohler's classification, this construction is of type 3.

In the second version (Fig. 3), edges between E-vertices, and edges between E-vertices and F-vertices occur. Two E-vertices are connected if their edges share a common vertex and a common face. An E-vertex is linked to an F-vertex if and only if its corresponding edge belongs to the boundary of the face corresponding to the F-center. In Kohler's classification we have type 2 for this construction.

Combination VEF

For combination VEF, we have identified four versions of construction:

Version 1: Edges from V-vertices to E-vertices and from E-vertices to F-vertices (Fig. 4, top left),

Version 2: Edges from V-vertices to F-vertices and from E-vertices to F-vertices (Fig. 4, top right),

Version 3: Edges from V-vertices to E-vertices and from V-vertices to F-vertices (Fig. 4, bottom left),

Version 4: Edges from V-vertices to E-vertices, from E-vertices to F-vertices, and from V-vertices to F-vertices (Fig. 4, bottom right),

with the following rules of connection:

V-vertex to E-vertex: A V-vertex is connected to an E-vertex if its corresponding vertex belongs to the corresponding edge of the E-vertex.

V-vertex to F-vertex: A V-vertex is connected to an F-vertex if its corresponding vertex belongs to the boundary of the corresponding face of the F-vertex.

E-vertex to F-vertex: An E-vertex is connected to an F-vertex if its corresponding edge lies on the boundary of the face corresponding to the F-vertex.

They all yield meshes of type 1 in Kohler's classification.

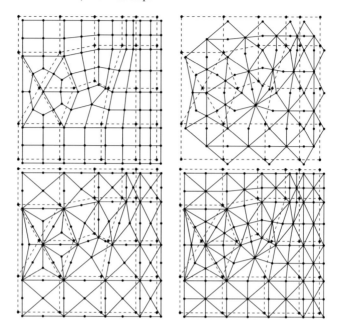

Fig. 4. Combination VEF, versions 1, 2, 3, and 4, from left to right and top to bottom. The edges of the original mesh are dashed.

Version 1 has the special property that it always yields quadrilateral meshes, independent of the initial mesh. Its faces are convex if the faces of the given mesh have been convex. The combinatorical structure of the resulting mesh is the same as that of the Catmull-Clark-scheme [2] which, however, has different masks.

Version 2 also yields quadrilateral meshes. In general, however, the resulting faces need not to be convex.

Version 3 has the same properties. However, it is unfavorable because it generates more than two common edges between adjacent faces.

Version 4 yields triangular meshes independent from the initial mesh.

Summary

The table in Fig. 5 summarizes the discussion of the previous sections. The first column shows the combination of vertex types. In the second column, the pairs of vertex types are listed for which edges are introduced in the new mesh. The last column notes special properties of the construction scheme.

combination	edge types	Kohler type	specialties
V	vv	9	
E	ee	8	
F	ff	4	
VE	ve	7	
	ve, ee	7	triangular meshes; Loop mesh topology;
VF	vf	5	quadrilateral meshes; does not preserve convex faces;
	vf, vv	3	
EF	ef	2	
	ef, ee	2	
VEF	ve, ef	1	quadrilateral meshes; preserves convex faces; Catmull-Clark mesh topology;
	vf, ef	1	quadrilateral meshes; does not preserve convex faces;
	ve, vf	1	quadrilateral meshes; does not preserve convex faces; unfavorable face adjacency;
	ve, ef, vf	1	triangular meshes;

Fig. 5. Summary of the elementary subdivision schemes based on vertex assignment. The first column shows the combination of vertex types. In the second column, the pairs of vertex types are listed for which edges are introduced in the new mesh. The last column shows special properties of the construction scheme.

3 Inversion

Inversion means to recalculate the initial mesh from the mesh resulting by applying a subdivision scheme. An invertible scheme maintains the information contained in the original mesh, whereas a non-invertible scheme looses information. Inversion has been used by Müller and Jaeschke [12] to calculate level-of-detail representations of special types of subdivision meshes which allow dynamic refinement and coarsening without explicitly storing a hierarchy of resolution. The issue of loss of information has also been discussed by Taubin [18] from the different view of signal analysis.

The geometric masks of a subdivision scheme induce a set of formula

$$\mathbf{q}_j = \mathbf{f}_j(\mathbf{p}_0, \ldots, \mathbf{p}_n), \ j = 0, \ldots, m.$$

The formulas describe how the vertices $\mathbf{q}_j$, $j = 0, \ldots, m$, of the new mesh are calculated from the vertices $\mathbf{p}_i$, $i = 0, \ldots, n$, of the initial mesh.

A necessary condition of invertibility is *combinatorical invertibility*. Combinatorical invertibility means $m \geq n$. If $m < n$, a trick to get a related mesh with $m > n$ is to apply the elementary subdivision scheme F. Under this scheme, vertices of the initial mesh become faces in the new one, and faces just become the vertices.

In our focus on linear subdivision schemes, the functions $\mathbf{f}_j$ are linear, so that the complete calculation is described by a $(m+1) \times (n+1)$ matrix $\mathbf{f}$. If $\mathbf{f}$ has a regular $(n+1) \times (n+1)$-submatrix, then complete invertibility is given.

Of particular interest is *local invertibility*. Local invertibility means that the set J of row indices can be represented as a union of subsets J_k, $k = 0, \ldots, p$, so that

$$\mathbf{q}_j = \mathbf{f}_j(\mathbf{p}_0, \ldots, \mathbf{p}_n), \ j \in J_k,$$

is invertible for $k = 0, \ldots, p$. The necessity of combinatorical invertibility for invertibility implies that only up to $|J_k|$ different points $\mathbf{p}_i$ have nonzero coefficients in the matrix induced by the $\mathbf{f}_j$, $j \in J_k$, for linear geometric masks. Let I_k be the set of indices of these points.

I_k has to have the property that the points with indices in I_k induce a connected submesh of the initial mesh which is local in some sense. An example is the existence of a constant D so that the diameter of the submesh, seen as a graph, is less than D, for all $k = 0, \ldots, p$. Another example is that the subgraphs only contain a number of faces bounded by D. In this sense, the Doo/Sabin scheme and the Catmul/Clark scheme, the latter restricted to meshes with quadrilateral faces, are examples of locally invertible subdivision schemes.

For the analysis of local combinatorical invertibility, the concrete geometric masks are not required. Of interest is the information about which input vertices influence the result. These vertices are contained in the masks. For the following discussion we take into consideration all vertices of the mask. An interesting question of further research is whether invertibility can be achieved in some cases if not all vertices of the topological mask influence the geometric mask.

Inversion from V-Vertices

Concerning the elementary schemes of section 2, those schemes which contain the V-vertices are trivially locally invertible because the inversion of a V-vertex is a vertex of the original mesh.

Inversion from E-Vertices

The schemes which contain the E-vertices are invertible if all faces have an odd number of vertices. Triangular meshes, for example, have this property. The reason is that in this case the vertices of a face can be recalculated from the centers of its edges. For even faces this observation does not hold. If at least one odd face exists in a mesh, recalculation is efficiently possible by first inverting this face, and then iteratively continuing on neighboring faces which then become invertible also in the even case because at least one of its vertices is known.

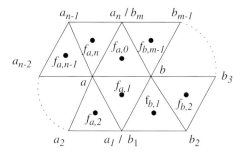

Fig. 6. A triangular subgraph which is combinatorically invertible from F-vertices. However, computational invertibility is in general impossible.

Inversion from F-Vertices

Just from F-vertices, an inversion is not possible in general. The reason is that the number of F-vertices resulting from a given mesh may be less than the number of its vertices. Hence the necessary condition of combinatorical invertibility, $m \geq n$, cannot be satisfied. A simple example is a cube. The cube has eight vertices, but only six faces. Hence the subdivided mesh has only six vertices from which the original eight ones cannot be determined.

Triangular meshes are combinatorically locally invertible if they can be covered by subgraphs of the type depicted in Fig. 6. It is induced by the triangles incident with two adjacent vertices a and b. If the degrees of a and b are $n+1$ and $m+1$, respectively, the number of vertices of the subgraph is $n+m$. The number of faces is $n+m$, too, so that combinatorical invertibility from the F-vertices is given.

Experimental investigations with n and m up to 5 have shown, however, that the corresponding filter matrices obtained by averaging are not regular and hence not (computationally) invertible.

If the elementary schemes are combined, it may happen that the resulting composed subdivision scheme may become invertible. For example, if an E-subdivision is followed by an F-subdivision, a one-two-one correspondence between the vertices of the original mesh and those of the resulting mesh which are the F-vertices of those faces of the intermediate mesh which correspond to the vertices of the original one. Similarly, if an F-subdivision is followed by an F-subdivision, there is a one-two-one assignment between the vertices of the resulting mesh and the vertices of the original mesh. However, if the number of vertices of the mesh resulting from the first F scheme application is less than that of the original mesh, combinatorical invertibility cannot be expected because information has been lost by intermediately reducing the number of vertices.

Relaxation of the Center Rule

Up to now we have chosen the vertices of the subdivision mesh as the average of the vertices of cells of the given mesh. A generalization is achieved by replacing the average by any other (linear) formula operating on the same points. This principle is particularly interesting for composed subdivision schemes which emerge from the concatenation of a constant number of elementary schemes. For each vertex of the new mesh those vertices of the original mesh are determined which are involved in its calculation. By replacing the formula resulting from the concatenation with any other formula operating on these vertices, further subdivision schemes can be derived. Examples will be given in section 4.

With respect to invertibility, an alternative choice of the subdivision formula may convert a subdivision scheme which is combinatorically invertible to a scheme which is also computationally invertible.

4 Selected Composed Subdivision Schemes

From the elementary subdivision schemes, further subdivision schemes can be derived by concatenating a finite number of elementary subdivision steps. The number of elementary subdivision steps that is concatenated is called the *order* of the subdivision scheme. The topological mask $M_k(v)$ of a vertex **v** of a scheme composed of k elementary schemes describes the dependency of a new vertex on vertices of the initial mesh. It is inductively defined by

$$M_k(v) := \bigcup_{v' \in M(v)} M_{k-1}(v')$$

where $M(v)$ is the mask of v of the last applied scheme, and $M_{k-1}(v')$ is the mask of the composed scheme before applying the last scheme. M_1 denotes the mask of the first scheme of the sequence of composition. A canonical geometric mask is analogously obtained by concatenating the filter functions defining the composed elementary scheme.

In the following, we present a number of subdivision schemes up to order 3. In the first three subsections we show how several well-known subdivision schemes fit into this framework. These schemes cover the types 1, 7, and 12 of Kohler's classification. The remainder sections provide a composed subdivision scheme for each further type of Kohler's classification, in particular for type 6 which was neither covered by the elementary nor by the composed subdivision schemes discussed up to then.

Subdivision Scheme VEF1/F

The subdivision scheme obtained by executing VEF1 followed by F yields a mesh which is combinatorically equivalent to Doo/Sabin scheme [4], with the same topological mask. However, the geometric mask of the Doo/Sabin scheme differs from the canonical geometric mask obtained by averaging.

The VEF1/F-scheme is combinatorially invertible. This can be seen from the observation that the new mesh consists of three types of faces, those corresponding to a vertex of the original mesh, those corresponding to an edge, and those corresponding to a face. In the last case, the new face has the same number of vertices as the corresponding old one. The vertices of the new face just depend on the vertices of the corresponding old one which means that this configuration is combinatorially invertible. In fact, this means that the VEF1/F scheme is locally combinatorially invertible.

It can be easily seen that, for a closed initial mesh, the degree of all vertices of the resulting mesh is 4.

Subdivision Scheme VEF1/F/F

The subdivision scheme VEF1/F/F is obtained from the scheme VEF1/F of the previous subsection by appending another elementary subdivision step of type F. The scheme VEF1/F/F is combinatorially equivalent to the Catmull/Clark scheme [2] and has the same topological mask between the new and old vertices. However, the concrete formulas are again different.

If the original mesh is closed and all its faces have exactly four vertices, the VEF1/F/F-scheme is combinatorially invertible. This can be seen by considering the subgraph induced by a vertex and its incident faces of the original mesh. In the new mesh, we choose the V-vertex of that vertex, the E-vertices of the incident edges, and the F-vertices of the incident faces. These vertices just depend on those of our subgraph, and their number is the same as the number of vertices of the subgraph.

An interesting point is that, independent from the original mesh, the derived mesh has the property that all faces have exactly four vertices. This means that even for arbitrary initial meshes the iterations of the VEF1/F/F-scheme are combinatorially invertible, possibly except the first iteration.

Subdivision Scheme VE2/F/F

The subdivision scheme is obtained from the scheme VEF1/F/F in the previous section by replacing VEF1 with the elementary subdivision step VE2 (Fig. 7). It turns out that the resulting subdivision scheme is combinatorically equivalent to the Loop scheme and has the same topological masks, but different geometric masks.

Another scheme which yields Loop connectivity is $F/V_{av}/EF2$. This scheme yields topological masks of the resulting vertices which are sufficient large for the butterfly scheme [5]. While the topological masks are tight for the resulting vertices related to the edges of the original mesh, they are too large for the vertices corresponding to the vertices of the original mesh. The latter vertices are in fact the vertices of the original mesh because the butterfly scheme is interpolatory.

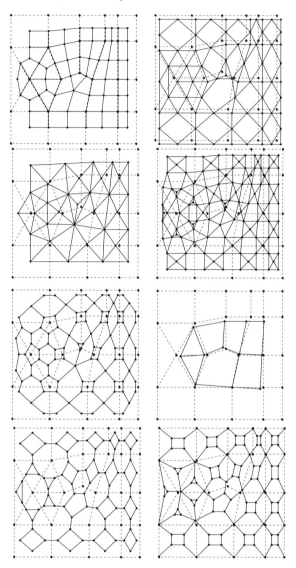

Fig. 7. From left to right and top to bottom: subdivision schemes VEF1/F/F, VE2/F/F, F/VF2, E/EF2, VE2/F, F/F, VF2/F, and EF2/F. The edges of the original mesh are dashed.

Subdivision Scheme EF2/F/F

The combinatorical structure of the subdivision scheme EF2/F/F is the same as for EF2. Its Kohler type is 2. The difference between both schemes lies in the dependency structure of the new vertices on the old one. The extension of EF2 by double application of an F-step expands the dependency region for the new vertices.

Subdivision Scheme F/VF2

The F/VF2-scheme generates two new triangular faces for each edge of the given mesh (Fig. 7). This implies that the resulting mesh is triangular, independent from the initial mesh. The type of the mesh is 3 in Kohler's classification.

Subdivision Scheme F/EF2

If the given mesh is triangular, then the subdivision scheme F/EF2 yields a mesh with the same topology as Loop subdivision. Furthermore, the topological masks of the vertices are the same as those of Loop subdivision [11]. The weights of the Loop scheme are different from the canonical geometric mask of the F/EF2 scheme. The scheme has Kohler type 7.

Subdivision Scheme E/EF2

Under the subdivision scheme E/EF2, the initial mesh induces triangles around its vertices and for its faces, whereas the edges induce quadrilaterals (Fig. 7). This means that only these two types of faces will occur in any further iteration. The type of this scheme according to Kohler is 5.

Subdivision Scheme VE2/F

For each edge of the old face, the subdivision scheme VE2/F generates a polygonal face with 6 vertices. For each vertex, a face is generated with the same number of edges as the degree of the vertex. The faces induce vertices with degree equal to the number of vertices of the face. An interesting observation is that all other vertices of the new edge have degree 3. These observations only hold for inner vertices, edges, and faces, cf. Fig. 7. In Kohler's classification, the mesh has type 6.

Subdivision Scheme F/F

The combinatorical structure of the new mesh derived according to the scheme F/F is identical to that of the initial mesh, at least for inner cells or closed meshes, respectively (Fig. 7). Hence these scheme cannot be seen as a real subdivision scheme. For those it is expected that the number of cells is monotonously increasing during iteration. However, the new geometric location of the vertices is different. The location of a new vertex is a weighted average of the locations of its surrounding vertices. This equals low-pass filtering in image processing. Hence the F/F-scheme is an example of a low-pass filter for meshes.

As already mentioned in section 3, the F/F-scheme is combinatorically invertible, but in general not computationally. The type in Kohler's classification is 9.

Subdivision Scheme VF2/F

The subdivision scheme VF2/F generates a new face for each vertex and each face of the old mesh, for inner cells of an open mesh or for a closed mesh (Fig. 7). The degree of all new inner vertices is 4. The scheme is of type 10 in Kohler's classification.

Subdivision Scheme VF3/VF3

If the given mesh is triangular, then the subdivision scheme VF3/VF3 yields a mesh with the same topology as $\sqrt{3}$-subdivision [8]. Furthermore, the topological masks of the vertices are the same. The weights of the $\sqrt{3}$-scheme are different from the canonical geometric mask of the VF3/VF3 scheme.

Subdivision Scheme EF2/F

The new mesh generated by the scheme EF2/F consists of faces induced by edges and faces of the old mesh. The faces induced by edges have six vertices, the vertex number of the faces is inherited by the old faces (Fig. 7). The vertices of the new mesh correspond to vertices of the old one, and they are of the same degree. All other new vertices have degree 3, as long as they are inner vertices of the new mesh. The type according to Kohler is 11.

An Example

For illustration, all the subdivision schemes presented in this section have been applied to the mesh of a mushroom depicted in the top left of Fig. 8 (Color Plate 25 on page 440). The rest of Fig. 8 (Color Plate 25 on page 440) shows the meshes generated in one step.

5 Subdivision Schemes of Higher Order

For regular quadrilateral meshes, that is quadrilateral meshes with inner vertices of degree 4, it is known that the VEF1/F-scheme converges towards surfaces which can be described as quadratic B-spline surfaces. Analogously, the VEF1/F/F-scheme lets converge the iteration to cubic B-spline surfaces. In general, it is known that the VEF1/F^k-scheme converges to B-spline surfaces of degree $k + 1$.

The VEF1/F^k-schemes can also be applied to meshes of arbitrary topology and thus yield a generalization of the B-spline concept to arbitrary control meshes. In the case of $k = 1$, the Doo/Sabin-scheme consists in a slight variation of the formula, but not the combinatorics of subdivision, which matches with the original scheme for regular quadrilateral meshes, and hence offers another generalization of quadratic B-spline surfaces. Analogously, the Catmull/Clark-scheme emerged by modifying the subdivision formula for $k = 2$, and thus is a generalization of cubic B-spline surfaces.

From the view of mesh combinatorics it turns out that combinatorics of the resulting meshes for an odd k always is the same as that of the VEF1/F-

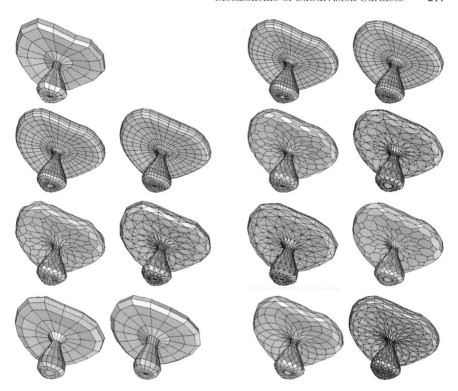

Fig. 8. First row, left: The mesh of a mushroom which is used for illustration of the different subdivision schemes. First row, right pair: The Doo/Sabin and the Catmull/Clark scheme. Second row, left pair: The VEF1/F-scheme and the VFE1/F/F-scheme. Second row, right pair: The EF2/F-scheme and the EF2/F/F-scheme. Third row, left pair: The VE2/F-scheme and the VE2/F/F-scheme. Third row, right pair: The F/VF2-scheme and the VF2/F-scheme. Fourth row, left pair: The F-scheme and the F/F-scheme. Fourth row, right: The E-scheme and the E/EF2-scheme.

scheme. Analogously, for even k we always get a mesh combinatorically equivalent to the mesh obtained by the VEF1/F/F-scheme. The difference is the region of dependency of a new vertex on vertices of the old mesh which increases with k.

The subdivision schemes for regular quadrilateral surfaces can be derived by using the tensor product construction of this sort of surfaces from curves. For curves an analogous framework as for surfaces can be developed. In that case we have only two types of cells: vertices and edges. The new vertex assigned to a vertex is the vertex itself, the vertex assigned to an edge corresponds to the center of the edge. There are three elementary schemes obtained from combination of these types of vertices: V, E, and VE. The involved new vertices are connected to a new curve according to their neighborhood on the

scheme	Kohler type	specialties
VEF1/F/F	1	combinatorically Catmull/Clark; resulting mesh of vertex degree 4; locally combinatorically invertible;
VEF1/F	12	combinatorically Doo/Sabin; resulting mesh of face degree 4; locally combinatorically invertible for meshes with quadrilateral faces;
EF2/F/F	2	triangular mesh;
EF2/F	11	
F/VF2	3	triangular mesh;
VF2/F	10	resulting mesh of vertex degree 4;
F/F	9	low-pass filter; globally combinatorically invertible;
E/EF2	5	mesh of triangles and quadrilaterals;
VE2/F	6	
VE2/F/F	7	combinatorically Loop; triangular resulting mesh for a triangular initial mesh;

Fig. 9. Survey of properties of the composed subdivision schemes. The properties listed in the column "specialties" usually concern closed meshes.

old one, which is unique. It is known, that iteration of the composed scheme VE/E yields a curve which can be represented as a quadratic B-spline curve. This scheme is also known as Chaikin-scheme [3]. The scheme VE/E/E leads to a cubic B-spline curve, and in general VE/E^k results in a B-spline curve of degree $k + 1$.

The curve scheme VE/E^k is combinatorically invertible. This can be seen by taking $k + 1$ consecutive new vertices, starting with the first one. It turns out that these vertices depend on exactly $k + 1$ consecutive vertices of the original polygonal chain. It can be shown that a grouping in sequences of $k + 1$ points can be found which cover all new vertices. Beyond that, it can be shown that the schemes are also computationally invertible.

For general meshes it is not clear whether the schemes VEF1/F^k are invertible for all k. We know that the schemes for $k = 1$ and $k = 2$ are combinatorically invertible. One idea might be to demonstrate invertibility is to show this property for subsequences of elementary schemes in the chain. For example, F/F is combinatorically invertible, and hence from

$$\text{VEF1/F}^k = \text{VEF1/F}^{k \bmod 2}(\text{F}^2)^{k \text{ div } 2}$$

we might conclude that VEF1/F^k is too. However, we already know that F/F is not invertible if the number of faces after application of the first F is less than the number of vertices. We also know that for even k all meshes, except possibly the first one, are quadrilateral. From Euler's formula for closed meshes of genus γ, $v - e + f = 2 - 2\gamma$, where v, e, and f are the number of vertices, faces, and edges, respectively, and the observation that $4f = 2e$, we get $f = v + 2\gamma - 2$ for this type of surfaces. Hence, for $\gamma = 0$, the number of faces is less than the number of vertices which prevents invertibility. Nevertheless, it cannot be concluded from this discussion that VEF1/F^k cannot be inverted because the vertices of the initial mesh have to be related to the vertices of the derived mesh.

In the past there have been several attempts to transfer the B-spline concept to other types than regular quadrilateral meshes. One example are the box spline surfaces [1]. Box spline surfaces are defined for regular triangular meshes, that is the degree of every inner vertex is 6. Box spline surfaces can also be represented as subdivision surfaces, cf. e.g. [19]. From that representation the scheme VE2/(FF)^k can be derived for surfaces.

The schemes VEF1/F^k and VE2/(FF)^k induce B-spline and box spline surfaces which have the same spline degree in every direction. However, in general these surfaces may have different spline degrees in every direction. In that case, subdivision schemes can be derived from the corresponding curve schemes. However, because of the unsymmetry, these schemes do not seem to fit immediately in our metascheme. Fig. 9 summarizes the discussion.

6 Concluding Remarks

We have proposed a metascheme which allows to derive subdivision operators for arbitrary meshes. These operators are uniform in that they are applied everywhere on the mesh in one step of iteration in order to get the next level of refinement. An alternative approach is to decide on the number of iterations dependent on local requirements of, and on, the mesh related to an approximative representation of the limit surface. This approach is one possibility of adaptive subdivision schemes. A problem to be solved is to join the faces of the different levels [13, 8].

An alternative approach to the regular subdivision schemes of this paper and the related approach of adaptivity just mentioned is irregular subdivision which takes into account the local geometry of the mesh immediately. An example of this kind of irregular subdivision scheme is quasi $4 - 8$-subdivision of Velho [20]. A problem of irregular subdivision is that convergence analysis seems to be more difficult than for the regular schemes. An interesting question related to the metascheme is to find further suitable irregular subdivision schemes based on regular subdivision schemes.

The development of geometric masks yielding smooth surfaces for concrete schemes derivable with the metascheme is an interesting task for the future. Of particular interest are interpolatory schemes, and approximative schemes which avoid shrinkage during iteration, as addressed by Taubin [18].

References

1. C. de Boor, K. Höllig, S.D. Riemenschneider, *Box Splines*, Springer-Verlag, New-York, 1993
2. E. Catmull, J. Clark, *Recursively generated B-spline surfaces of arbitrary topological meshes*, Computer Aided Design 10(6) (1978) 350–355
3. G.M. Chaikin, *An algorithm for high speed curve generation*, Computer Graphics and Image Processing 3 (1974) 346–349
4. D. Doo, M. Sabin, *Behaviour of recursive division surfaces near extraordinary points*, Computer Aided Design 10(6) (1978) 356–360
5. N. Dyn, D. Levin, J.A. Gregory, *A butterfly subdivision scheme for surface interpolation with tension control*, ACM Trans. on Graphics 9(2) (1990) 160–169
6. G. Farin, *Curves and Surfaces for CAGD*, 3rd edition, Academic Press, 1993
7. A. Habib, J. Warren, *Edge and vertex insertion for a class of C^1 subdivision surfaces*, Comp. Aided Geom. Des. 16(4) (1999) 223–247
8. L. Kobbelt, $\sqrt{3}$ *subdivision*, Proc. SIGGRAPH 2000, 103–112, 2000
9. M. Kohler, *A Meta Scheme for Interactive Refinement of Meshes*, Visualization and Mathematics (Ch. Hege, K. Polthier, eds.), Springer-Verlag, 1998
10. U. Labsik, G. Greiner, *Interpolatory $\sqrt{3}$-subdivision*, Proc. Eurographics 2000, Computer Graphics Forum 19(3) (2000) 131–138
11. C. Loop, *Smooth Subdivision Surfaces Based on Triangles*, Master's Thesis, Department of Mathematics, University of Utah, 1987
12. H. Müller, R. Jaeschke, *Adaptive Subdivision Curves and Surfaces*, Proc. Computer Graphics International 1998 (CGI'98), IEEE Computer Society Press, 1998, 48–58
13. H. Müller, M. Rips, *Another Metascheme of Subdivision Surfaces*, Research Report 713, Department of Computer Science, Univ. of Dortmund, Germany, 1999
14. P. Oswald, P. Schröder, *Composite primal/dual $\sqrt{3}$-subdivision schemes*, submitted, http://cm.bell-labs.com/who/poswald/
15. J. Peters, U. Reif, *The simplest subdivision scheme for smoothing polyhedra*, ACM Trans. on Graphics 16(4), 1997, 420–431
16. H. Prautzsch, *Smoothness of subdivision surfaces at extraordinary points*, Advances in Computational Mathematics 9 (1998) 377–389
17. J. Stam, *On subdivision schemes generalizing uniform B-spline surfaces of arbitrary degree*, Comp. Aided Geom. Des. 18(5) (2001) 383–396
18. G. Taubin, *Dual mesh resampling*, Proc. Pacific Graphics 2001, 2001
19. G. Umlauf, *Smooth free-form surfaces and optimized subdivision algorithms (in German)*, Shaker Verlag, 1999
20. L. Velho, *Quasi 4-8 subdivision*, Comp. Aided Geom. Des. 18(4) (2001) 345–357
21. L. Velho, D. Zorin, *4-8 subdivision*, Comp. Aided Geom. Des. 18(5) (2001) 397–422
22. D. Zorin, P. Schröder, *A unified framework for primal/dual quadrilateral subdivision surfaces*, Comp. Aided Geom. Des. 18(5) (2001) 429–454

Geometry of the Squared Distance Function to Curves and Surfaces

Helmut Pottmann and Michael Hofer

Institute of Geometry, Vienna University of Technology,
Wiedner Hauptstraße 8–10, A-1040 Wien
{*pottmann,hofer*} *@geometrie.tuwien.ac.at*

Summary. We investigate the geometry of that function in the plane or 3-space, which associates to each point the square of the shortest distance to a given curve or surface. Particular emphasis is put on second order Taylor approximants and other local quadratic approximants. Their key role in a variety of geometric optimization algorithms is illustrated at hand of registration in Computer Vision and surface approximation.

Keywords. distance function, geometric optimization, surface approximation, iterative closest point surface fitting

1 Introduction

The distance function of a curve or surface Φ, which assigns to each point $\mathbf{p}$ of the embedding space the shortest distance of $\mathbf{p}$ to Φ frequently appears in problems of geometric computing. Examples include curve and surface approximation in CAD and Geometric Modeling, registration in Computer Vision and positioning problems in Robotics [9].

A variety of contributions deals with the computation of the distance function; in many cases this computation aims towards the singular set of the function, i.e., towards points where the function is not smooth, since those points lie on the medial axis (or skeleton) of the input shape.

Early work on the geometry of the distance function comes from the classical geometric literature of the 19th century. One looks at its graph surface, which consists of developable surfaces of constant slope and applies results of classical differential geometry, line and sphere geometry (for a modern presentation, see e.g. [21]). This approach is also naturally related to the medial axis transform (MAT) which appears as cyclographic preimage in the classical language. More recent work on the MAT comes from H. I. Choi [8], especially the decomposition result for the efficient computation, which also appears in Kimmel et al. [15]. The importance of the skeleton, medial axis and medial axis transform as a shape descriptor has been realized by Blum [4], and this created a large body of literature that deals with its computation and applications in object recognition, registration and mathematical morphology [14, 24, 25]. Also in a discrete setting (pixel plane, voxel space), distance transforms and the skeleton received a lot of attention (see e.g. [18, 24, 25, 28]).

The distance function is also the (viscosity) solution of the so-called eikonal equation. Its numerical computation is not trivial because it is a hyperbolic equation and an initially smooth front may develop singularities (shocks) as it propagates. Precisely the latter belong to the medial axis and are of particular interest. The computation of viscosity solutions with the level set method of Osher and Sethian [19] proved to be a very powerful approach (see e.g. [23, 26, 27]). Distance functions also play a role in other curve and surface evolutions [23] and in the visualization of large sets of unorganized data sets [29]. The vector distance function of an object, which connects each point in space to its closest point on the object, has been used by Faugeras and Gomes [10] for the evolution of objects of arbitrary dimension and codimension.

The level sets of the distance functions are the *offsets* which are of particular importance in Computer Aided Design (see e.g. [20, 21] and the literature therein) and also appear in mathematical morphology [24]. Results on the distance function and its application to shape interrogation for CAD/CAM may also be found in the very recent book by Patrikalakis and Maekawa [20].

In the present paper, instead of the distance function we rather study the *squared distance function*. Clearly, also the latter is singular in the points of the medial axis and thus the relation to the literature just pointed out is important. However, we are mainly interested in local quadratic approximants at points different from the medial axis. A contribution dealing with local properties of the squared distance function of a manifold has been given by Ambrosio and Mantegazza [1]. Among other interesting results, the authors describe the relation between third order directional derivatives of the squared distance function (taken at points of the manifold) and the curvature invariants of the manifold. Despite the importance of the squared distance function, little effort has so far been made in better understanding, approximating and representing this function for efficient computing.

In the present paper, we will start an investigation of this subject. To get a better understanding of the nature of the problem, we first deal with planar curves. There, we can give a simple kinematic generation of the 3D graph surface of the squared distance function d^2. This is important for visualization of the behaviour of d^2 and it greatly helps us to find simple derivations for *local quadratic approximants* to d^2. The latter are of particular importance for optimization algorithms involving d^2 and therefore they are also studied for surfaces and space curves.

At hand of two examples we outline how to use the results of the present paper in certain types of geometric optimization algorithms which involve quadratic approximants to the squared distance function of a surface. The two applications we look at are registration in Computer Vision and surface approximation. An important advantage of the new approach to approximation with parametric surfaces, such as B-spline surfaces, is that we avoid the parameterization problem. The registration problem we look at is the optimal matching of a point cloud to a surface. The essential difference in our approach to standard techniques, such as the *iterative closest point (ICP) algorithm* [2] is, that we do not have to compute pairs of corresponding points.

2 Graph Surface of the Squared Distance Function to a Planar Curve

In Euclidean 3-space $\mathbb{R}^3$, we consider a planar C^2 curve $\mathbf{c}(t)$ with parameterization $(c_1(t), c_2(t), 0)$. The tangent and normal line at a curve point $\mathbf{c}(t)$ are denoted by T and N respectively. To each point $(x, y, 0)$ we compute the shortest distance $d(x, y)$ of that point to the curve $\mathbf{c}$. For visualization of the squared distance function d^2 we investigate the graph surface $\Gamma : (x, y, d^2(x, y))$ of the function d^2. By neglecting global effects of the distance function we will at first construct a surface Φ, part of which is the desired graph surface Γ. In a second step we will perform the appropriate trimming which reduces Φ to Γ.

The Frenet frame at a curve point $\mathbf{c}(t)$ consists of the unit tangent vector $\mathbf{e}_1 = \dot{\mathbf{c}}/\|\dot{\mathbf{c}}\|$ and the normal vector $\mathbf{e}_2(t)$. The two vectors form a right-handed Cartesian system in the plane. With $\mathbf{e}_3 = \mathbf{e}_1 \times \mathbf{e}_2 = (0, 0, 1)$ this system is extended to a Cartesian system Σ in $\mathbb{R}^3$. Coordinates with respect to Σ are denoted by (x_1, x_2, x_3). The system Σ depends on t and shall have the curve point $\mathbf{c}(t)$ as origin. At least locally, the shortest distance of a point $(0, u, 0)$ on the x_2-axis (curve normal) is its x_2-coordinate u. For each t, locally the graph points $(0, u, u^2)$ form a parabola p in the normal plane of $\mathbf{c}(t)$. This parabola can be considered fixed in Σ. Varying t, the positions of the parabola in the original system generate a surface Φ. A parameterization of Φ is

$$\mathbf{x}(t, u) = \mathbf{c}(t) + u\mathbf{e}_2(t) + u^2\mathbf{e}_3. \tag{1}$$

It is well-known and an immediate consequence of the Frenet equations that the instantaneous motion of the system Σ is a rotation about the axis of the osculating circle of $\mathbf{c}$. These curvature axes form a cylinder surface Λ with the evolute of $\mathbf{c}$ as orthogonal cross section and rulings parallel to $(0, 0, 1)$. Hence, the motion of Σ with respect to the fixed system can be considered as a rolling motion of the normal plane of $\mathbf{c}$ on the evolute cylinder Λ of $\mathbf{c}$. In usual terminology, the surface Φ is a *moulding surface* (cf. Fig. 1 and Color Plate 29 on page 443). Note that the contour lines of Φ at height u^2 are translates of the offsets of $\mathbf{c}$ at distance u.

The surface Φ contains a singular curve (*curve of regression*) $\mathbf{l}(t)$, which is contained in the evolute cylinder Λ. This is in accordance to the well-known result that singularities of the offsets of a curve occur at points of its evolute. The rolling motion also shows that the development of the cylinder Λ maps the curve $\mathbf{l}$ to a parabola which is congruent to the profile parabola p.

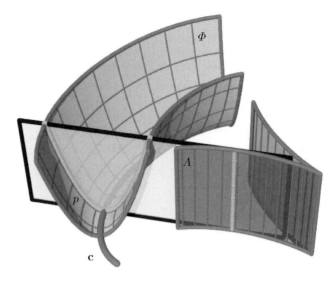

Fig. 1. When the normal plane of the planar curve **c** rolls on the evolute cylinder Λ of **c**, the parabola p generates the moulding surface Φ.

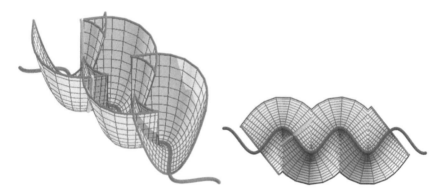

Fig. 2. A moulding surface Φ with self-intersections in an axonometric view (left) and viewed from below, where the visible points correspond to the graph Γ of the squared distance function to a sine curve (right).

The generated surface Φ is in general not the graph of a function in the horizontal xy-plane. It may happen, that a vertical line through $(x, y, 0)$ intersects Φ in several points. Among those points, the one with the smallest z-coordinate lies on the desired graph surface Γ, whereas the others do not. Thus Φ can be trimmed so as to form exactly the graph Γ of the squared distance function to **c**. The trimming can be performed with a *visibility test*. We view Φ in direction $(0, 0, 1)$, i.e., from below. Then, exactly the visible points are those closest to the xy-plane π and therefore the points of Γ (cf. Fig. 2 and Color Plate 30 on page 443). Trimming has to be performed at self-intersections of Φ. Projecting these curves orthogonally into π we obtain the *cut locus* of **c**. It is formed of those points in π, for which the shortest distance occurs at more than one normal. The limit points of this set are usually added to the cut locus: these points are curvature centers to points of **c** with locally extremal curvature. We should mention that for a closed boundary curve **c** of a planar domain D, the part of the cut locus which lies in D is also called the *medial axis* of D. It is also well-known that the trimming procedure cuts apart the singular curve **l**. Just special points of it, namely those which belong to the end points of branches of the cut locus, remain. We summarize the basically well-known results as follows.

Proposition 3. *The graph surface Γ of the squared distance function to a planar curve **c** is contained in a moulding surface Φ. The surface Φ is generated by a parabola p with parameter 1 whose plane rolls on the evolute cylinder Λ of **c**, such that p's vertex moves along **c** and p's axis remains orthogonal to the plane π of **c**. Exactly those parts of Φ lie on the graph surface Γ which are visible for an orthogonal projection onto π when viewing Φ from below (i.e., from that side of π which does not contain points of Φ).*

Remark 1. In an analogous way we may construct the graph surface of any other function $f(d)$ of the distance d. There, the profile curve has in Σ the parameterization $(0, u, f(u))$. One has to be careful with signs, however. This is easily understood at hand of the simplest example, namely the *distance function d* itself. With a signed distance function whose sign is given by the orientation of the curve normal $\mathbf{e}_2$, the profile is simply the line $(0, u, u)$ and the generated moulding surface is a developable surface of constant slope (see e.g. [21]). For a nonnegative distance, the profile is $(0, u, |u|)$ and the moulding surface is generated from the developable surface described above by reflecting the part below π at π.

In case that f is nonnegative and monotonically increasing, we perform the trimming operation as in the case of $f(d) = d^2$, i.e., with a visibility algorithm viewing the graph orthogonal to π from below. For a monotonically decreasing function, we have to view from above.

An interesting example of this type is the inverse distance function $f(d) = 1/|d|$, which is sometimes built around obstacles in robot motion planning. There, with a nonnegative distance, the profile curve of the moulding surface Φ is formed by two hyperbola segments. An image of the trimmed graph to an ellipse **c** is shown in Fig. 3. Note that for a **c** which bounds a convex domain D, there is no trimming required outside D. Regardless of f, trimming happens at the points of π which lie on the cut locus, and thus global results on cut locus, medial axis and singularities of offsets can be used to identify regions where no trimming is required (see [21]).

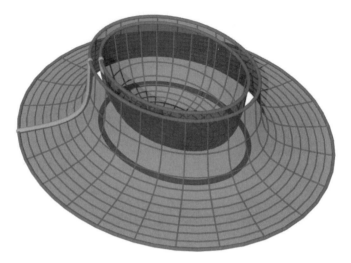

Fig. 3. Graph of the inverse distance function to an ellipse.

3 Quadratic Approximations to d^2

The previous section visualized a problem of the squared distance function, namely its non-smoothness at points of the cut locus. When we now study local quadratic (Taylor) approximants, we do not consider the global effects, and give formulae for local approximants which work on the local distance function. This means that in determining d for neighboring points of **p** we are only locally varying the footpoint of the normal to the curve **c**. In other words, at points of the medial axis we work with just one sheet of the surface Φ.

Consider a point **p** in π whose coordinates in the Frenet frame at the normal footpoint $\mathbf{c}(t_0)$ are $(0, d)$. The curvature center $\mathbf{k}(t_0)$ at $\mathbf{c}(t_0)$ has coordinates $(0, \varrho)$. Here, ϱ is the inverse curvature $1/\kappa$ and thus has the same sign as the curvature, which depends on the orientation of the curve. Since all level sets of d^2 (offsets) to points on the curve normal share **k** as curvature center, we see that *the squared distance function to the given curve and to its osculating circle at the normal footpoint agree up to second order.*

A visualization is as follows (see Fig. 4 and Color Plate 31 on page 443): rotating the profile parabola $p(t_0)$ around the curvature axis (vertical line through **k**) results in a surface of revolution Ψ which has second order contact with Φ at all points of $p(t_0)$. This is a well-known curvature property of moulding surfaces. In the Frenet frame, the function d^2 to the osculating circle is

$$f(x_1, x_2) = \left(\sqrt{x_1^2 + (x_2 - \varrho)^2} - |\varrho| \right)^2. \tag{2}$$

The graph of this function is the surface of revolution Ψ. The second order Taylor approximant F_d of f at $(0, d)$ is found to be

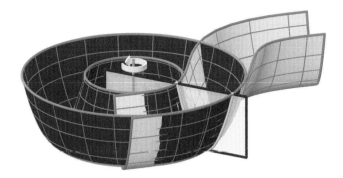

Fig. 4. Surface of revolution Ψ which has second order contact with the moulding surface Φ at all points of $p(t_0)$.

$$F_d(x_1, x_2) = \frac{d}{d - \varrho} x_1^2 + x_2^2. \tag{3}$$

Let us discuss the various cases.

- For $d = 0$ we get the Taylor approximant $F_0 = x_2^2$ at the normal footpoint. This shows the following interesting result: *At a point* **p** *of a curve* **c** *the second order approximant of the squared distance function to* **c** *and to the curve tangent* T *at* **p** *are identical.* Visually, this is not unexpected since curvature depends on the scale. Zooming closer to the curve it appears less and less curved. The graph surface Γ_0 of F_0 is a parabolic cylinder with rulings parallel to the curve tangent (see Fig. 5 and Color Plate 32(a) on page 444).
- For $d \to \infty$, the Taylor approximant tends to $F_\infty = x_1^2 + x_2^2$. This is the squared distance function to the footpoint $\mathbf{c}(t_0)$. The graph Γ_∞ of F_∞ is a paraboloid of revolution.
- For general d, it may be advantageous to view F as combination of F_0 and F_∞,

$$F_d(x_1, x_2) = \frac{d}{d - \varrho} \left(x_1^2 + x_2^2 \right) - \frac{\varrho}{d - \varrho} x_2^2 = \frac{d}{d - \varrho} F_\infty - \frac{\varrho}{d - \varrho} F_0. \tag{4}$$

This form is particularly useful for computing the second order approximant in the original x, y-system. Clearly, F_d is not defined at the curvature center $d = \varrho$, where we have a singularity. Otherwise, we see that the type of the graph surface Γ_d depends on $s = sign[d/(d - \varrho)]$. A value $s > 0$ yields an elliptic paraboloid (see Fig. 6 and Color Plate 32(b) on page 444), $s = 0$ the parabolic cylinder Γ_0 (see Fig. 5 and Color Plate 32(a) on page 444), and $s < 0$ a hyperbolic paraboloid (see Fig. 7 and Color Plate 32(c) on page 444). The latter case belongs exactly to points between the curve point and the curvature center, i.e., $0 < d < \varrho$ or $0 > d > \varrho$. Here,

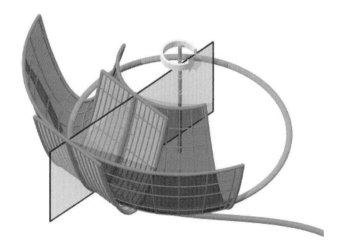

Fig. 5. The graph surface Γ_0 of F_0 is a parabolic cylinder with rulings parallel to the curve tangent.

the quadratic approximant also assumes negative values. In all other cases this does not happen.

Note also that all quadratic approximants F_d agree along the curve normal and are symmetric with respect to it. The graph paraboloids Γ_d touch the cylinder Γ_0 along the profile parabola $p(t_0)$.

– For an *inflection point* we have $\kappa = 0$ and thus $F_d = x_2^2$. This reflects the trivial fact that the squared distance function to the tangent is a second order approximant along the whole curve normal.

For the applications we have in mind, it can be important to employ *non-negative* quadratic approximants to d^2. Thus, we briefly address a convenient way to deal with those approximants.

Negative function values of the quadratic approximant F_d arise for a distance d with $0 < d < \varrho$ or $0 > d > \varrho$. Without loss of generality, we may assume an appropriate local orientation of the curve $\mathbf{c}$ such that $\varrho > 0$. Thus, only the case $0 < d < \varrho$ needs to be discussed. We fix such a distance and call it D. Our goal is to replace the local quadratic approximant F_D by a nonnegative quadratic approximant F_D^+ with the following property: For all points $\mathbf{x}$ whose distance d to $\mathbf{c}$ is less than the given value D, the local quadratic approximant F_D^+ also returns a value $< D^2$.

Let us look at the level sets of F_D, whose graph is a hyperbolic paraboloid. Those points $\mathbf{x}$ in the plane whose squared distance to $\mathbf{c}$ is less than D^2 lie between the two offset curves $\mathbf{c}_D$ and $\mathbf{c}_{-D}$ at oriented distances D and $-D$. The level set of F_D is a hyperbola h_D. Its axes agree with those of the Frenet frame at $\mathbf{c}(t_0)$ and it has second order contact with the offset $\mathbf{c}_D$ at the point $(0, D)$, where we are looking for the local quadratic approximant (see Fig. 8).

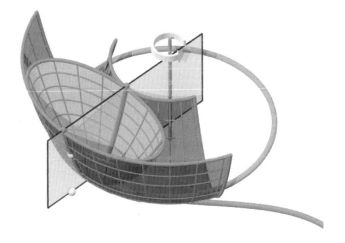

Fig. 6. The graph surface Γ_d of F_d is an elliptic paraboloid for $s > 0$.

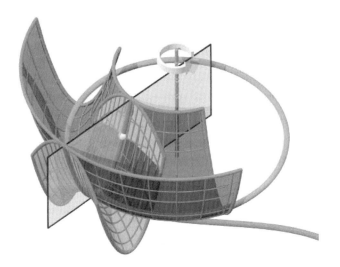

Fig. 7. The graph surface Γ_d of F_d is a hyperbolic paraboloid for $s < 0$.

The region of points $\mathbf{x}$ with $F_D(\mathbf{x}) < D^2$ is bounded by this hyperbola. Of course, there are points with $F_D(\mathbf{x}) < D^2$ and $d^2(\mathbf{x}) > D^2$.

To arrive at a practically useful result, we make a further simplification. We replace the curve by its osculating circle $\mathbf{l}$ at $\mathbf{c}(t_0)$. Points whose distance to $\mathbf{l}$ is smaller than D lie in an annulus A bounded by two concentric circles $\mathbf{l}_D$, $\mathbf{l}_{-D}$ with midpoint $(0, \varrho)$ and radii $\varrho - D$ and $\varrho + D$ (see Fig. 9). To warrant the symmetry and the precision of the approximant along the curve

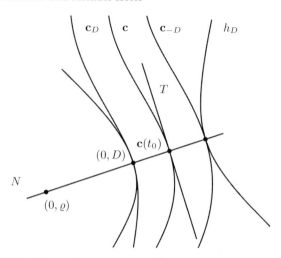

Fig. 8. Level set of F_D at height D^2 is the hyperbola h_D.

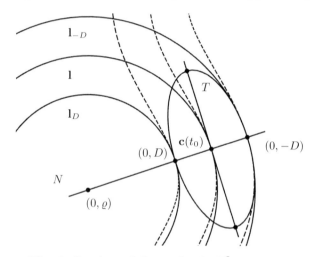

Fig. 9. Level set of F_{-D} at height D^2 is an ellipse.

normal, we have to take as F_D^+ an F_d to a point $(0, d)$ with $d < 0$. The region with $F_d(\mathbf{x}) < D^2$ is then an ellipse coaxial with the Frenet frame axes $\mathbf{e}_1, \mathbf{e}_2$, and with points $(0, D)$ and $(0, -D)$ as two vertices. The solution F_D^+ of our problem must give rise to the *largest* ellipse which still lies inside the annulus A. It is easy to prove that it has to have second order contact (in fact, fourth order contact) with the larger bounding circle of A at point $(0, -D)$. Therefore, it is the level set of the quadratic approximant F_{-D}. Thus, our solution is

$$F_D^+ = F_{-D}. \tag{5}$$

Hence, at points with an indefinite quadratic approximant we just use the nonnegative quadratic approximant at the point which is obtained by reflection at $\mathbf{c}(t_0)$. In other words, we may use the formula

$$F_d^+(x_1, x_2) = \frac{d}{d + \varrho} x_1^2 + x_2^2, \tag{6}$$

where we do not have to care about signs and always take $d > 0$ and $\varrho > 0$. This would yield 'wrong' results for distances $d > \varrho > 0$. However, a point $\mathbf{x}$ beyond the curvature center $\mathbf{k}(t)$ at $\mathbf{c}(t)$ always has another normal footpoint which is closer to it than $\mathbf{c}(t)$ and thus this case does not occur when we are looking at globally shortest distances.

4 Squared Distance Function to a Surface and its Second Order Approximants

What we said about d^2 to a planar curve can in principle be extended to the squared distance function to a surface in $\mathbb{R}^3$. However, the visualization with help of a graph in $\mathbb{R}^4$ becomes harder. Again we have a local behaviour of the distance function d^2, which we will investigate up to second order, and a global behaviour. The global effects cause non-smoothness at points of the cut locus, which in general consists of surfaces.

Consider an oriented surface $\mathbf{s}(u, v)$ with a unit normal vector field $\mathbf{n}(u, v) = \mathbf{e}_3(u, v)$. At each surface point $\mathbf{s}(u, v)$, we have a local right-handed Cartesian system whose first two vectors $\mathbf{e}_1, \mathbf{e}_2$ determine the principal curvature directions. The surface normal and the principal tangents are denoted by N, T_1 and T_2, respectively. The latter are not uniquely determined at an umbilical point. There, we can take any two orthogonal tangent vectors $\mathbf{e}_1, \mathbf{e}_2$. We will refer to the thereby defined frame as *principal frame* $\Sigma(u, v)$. Let κ_i be the (signed) principal curvature to the principal curvature direction $\mathbf{e}_i$, $i = 1, 2$, and let $\varrho_i = 1/\kappa_i$. Then, the two principal curvature centers at the considered surface point $\mathbf{s}(u, v)$ are expressed in Σ as $\mathbf{k}_i = (0, 0, \varrho_i)$. The quadratic approximant F_d to d^2 at $(0, 0, d)$ is the following.

Proposition 4. *The second order Taylor approximant of the squared distance function to a surface at a point $\mathbf{p}$ is expressed in the principal frame at the normal footpoint via*

$$F_d(x_1, x_2, x_3) = \frac{d}{d - \varrho_1} x_1^2 + \frac{d}{d - \varrho_2} x_2^2 + x_3^2. \tag{7}$$

Proof. We just give a sketch of the proof. We first have to show that it is sufficient to approximate the surface at the footpoint $\mathbf{s}(u, v)$ up to second order. This can be done with well-known results on the curvature behaviour of offset surfaces. Hence, we may replace the surface locally by an osculating torus T, which is obtained by rotating the first principal curvature circle c_1 (center $\mathbf{k}_1$, radius $|\varrho_1|$, in the plane $x_2 = 0$) around the axis of the other principal curvature circle c_2, which is defined analogous to c_1 (see Fig. 10). Clearly, by exchanging the roles of c_1 and c_2, we obtain two such tori. Any one is fine for our purposes. The level sets of the squared distance function to T are coaxial and concentric tori. Particularly, in the symmetry planes $x_1 = 0$ and $x_2 = 0$, we have the function d^2 to the principal curvature circles c_2 and c_1, respectively. Hence, almost all first and second order partial derivatives of the squared distance function with respect to x_i can be taken from the planar case. The only one, for which this is not true, is $\partial^2 f / \partial x_1 \partial x_2$. However, the second order quadratic approximant must be symmetric with respect to the planes $x_1 = 0$ and $x_2 = 0$. Hence, this mixed partial derivative vanishes. Together with (3) this proves (7). $\square$

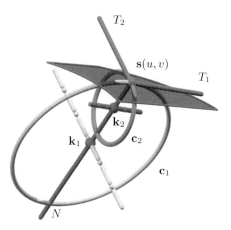

Fig. 10. Principal frame and principle circles at an elliptic surface point $\mathbf{s}(u, v)$.

The discussion of the arising cases is as in Sect. 2. At first, let us point to the case $d = 0$.

Proposition 5. *At a point $\mathbf{p}$ of a surface $\mathbf{s}$ the second order approximant of the squared distance functions to $\mathbf{s}$ and to the surface's tangent plane at $\mathbf{p}$ are identical.*

More generally, it is easy to show that if two curves or surfaces possess contact of order k at some point, their respective squared distance functions agree of order $k + 1$ there. One example of this is the fact that the second order Taylor approximation of the squared distance function at a point of a surface $\mathbf{s}$ agrees with the squared distance function of the tangent plane. A second example is the result of Ambrosio and Mantegazza [1], which says that third order derivatives of the squared distance function at points of $\mathbf{s}$ are related to the curvature invariants of $\mathbf{s}$.

For a further discussion, we distinguish between the types of surface points.

- In the case of an *elliptic point*, $\varrho_1 \varrho_2 > 0$, we assume a surface normal orientation and an indexing of the principal curvature directions that yield $\varrho_1 \geq \varrho_2 > 0$. Points of the normal, which lie on the other side of the tangent plane than the principal curvature centers, i.e., $d < 0$, cause positive factors $d/(d - \varrho_i)$, and hence a positive definite quadratic form F_d. This case always arises at points outside a closed convex surface $\mathbf{s}$. For points between the principal curvature center $\mathbf{k}_1$ and the surface point, i.e., $0 < d < \varrho_1$, we get an indefinite F_d, and for $d > \varrho_1$, we again have a positive definite second order approximant. Clearly, we exclude evaluation at the principal curvature centers, where we have a singularity.
- For a *hyperbolic point* $\mathbf{s}$, we assume w.l.o.g. $\varrho_1 < 0$ and $\varrho_2 > 0$. Here, points between the principal curvature centers belong to $\varrho_1 < d < \varrho_2$ (excluding $d = 0$) and result in an indefinite F_d, whereas for points outside the principal curvature segment $\mathbf{k}_1 \mathbf{k}_2$ the approximants F_d are positive definite. There, the level sets $F_d = c$ to any constant $c > 0$ are homothetic ellipsoids, centered at the surface point $\mathbf{s}$, with the axes of the principal frame as axes. In the indefinite cases, the level sets are hyperboloids (of one or two sheets).
- At a *parabolic point* we may assume $\kappa_1 = 0$ and $\kappa_2 > 0$. Now, the second order approximant to d^2 reads

$$F_d(x_1, x_2, x_3) = \frac{d}{d - \varrho_2} x_2^2 + x_3^2. \tag{8}$$

This shows that we never get a positive definite F_d. The level sets of F_d are, in general, cylinder surfaces with rulings parallel to the x_1-axis (principal curvature direction with vanishing curvature).
- A *flat point* is characterized by $\kappa_1 = \kappa_2 = 0$ and thus yields the obvious result $F_d = x_3^2$. Hence, the squared distance function to the tangent plane agrees along the whole surface normal up to second order with the squared distance function to the surface $\mathbf{s}$.

Analogously to the case of planar curves we derive nonnegative quadratic approximants with

$$F_d^+(x_1, x_2, x_3) = \frac{d}{d + \varrho_1} x_1^2 + \frac{d}{d + \varrho_2} x_2^2 + x_3^2, \tag{9}$$

where d, ϱ_1, ϱ_2 are taken as positive values. Again, points beyond the principal curvature centers are ruled out, but they do not arise anyway when considering global distances.

5 Squared Distance Function to a Space Curve

The study of the squared distance function to a C^2 space curve $\mathbf{c}(t)$ is also interesting in connection with surfaces, namely in regions where the closest points on the surface are on a boundary curve or at a curved edge, i.e., a surface curve with tangent plane discontinuities. Again, we are focussing on the second order approximants, and this we do from a local point of view.

Given a point $\mathbf{p}$ in $\mathbb{R}^3$, the shortest distance to the curve $\mathbf{c}$ occurs along a normal of the curve or at a boundary point of it. The latter case is trivial and thus we exclude it. At the normal footpoint $\mathbf{c}(t_0)$, with tangent line T, we form a Cartesian system with $\mathbf{e}_1$ as tangent vector and $\mathbf{e}_3$ in direction of the vector $\mathbf{p} - \mathbf{c}(t_0)$. This *canonical frame* can be viewed as limit case of the principal frame for surfaces, when interpreting the curve as pipe surface with vanishing radius. By this limit process, we can also show the following result.

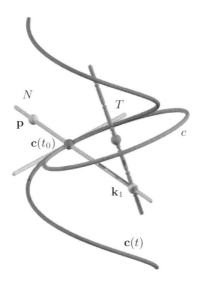

Fig. 11. A space curve $\mathbf{c}(t)$ with the osculating circle c at $\mathbf{c}(t_0)$.

Proposition 6. *The second order Taylor approximant of the squared distance function to a space curve $\mathbf{c}$ at a point $\mathbf{p}$ is expressed in the canonical frame Σ at the normal footpoint via*

$$F_d(x_1, x_2, x_3) = \frac{d}{d - \varrho_1} x_1^2 + x_2^2 + x_3^2. \tag{10}$$

Here, $(0, 0, \varrho_1)$ are the coordinates (in Σ) of the intersection point of the curvature axis of $\mathbf{c}$ at the footpoint $\mathbf{c}(t_0)$ with the perpendicular line N from $\mathbf{p}$ to $\mathbf{c}(t_0)$.

Proof. It is sufficient to consider the squared distance function to the osculating circle c of the curve at $\mathbf{c}(t_0)$. For any torus with spine circle c, the principal curvature lines are the family of parallel circles and the family of meridian circles. By Meusnier's theorem, the principal curvature centers to the parallel circles lie on the rotational axis of the torus. This is the axis of the spine circle c. If we now shrink the radius of the meridian circles to 0, we get in the limit principal curvature centers $\mathbf{k}_1 = (0, 0, \varrho_1)$ on the axis. Here,

$$\varrho_1 = \varrho / \cos \alpha, \tag{11}$$

with ϱ as curvature radius of $\mathbf{c}$ at $\mathbf{c}(t_0)$ and α as angle between the normal (x_3-axis) and the osculating plane at $\mathbf{c}(t_0)$ (see Fig. 11). With $\varrho_2 = 0$, the result follows from (7).   □

The discussion of the different cases of F_d is a limit case of the situation for surfaces and thus we omit it here.

6 Application to Geometric Optimization Problems

Optimization problems in geometric computing are frequently nonlinear and involve the squared distance function to a curve or surface. It is therefore natural to apply the results discussed above to the development of geometric optimization algorithms which are based on local quadratic approximants of the function to be minimized, such as quasi-Newton and SQP-type algorithms [5]. We will now outline this idea at hand of two important examples: registration and surface approximation.

6.1 Registration of a Point Cloud to a CAD Model

Suppose that we are given a large number of 3D data points that have been obtained by some 3D measurement device (laser scan, light sectioning, ...) from the surface of a technical object. Furthermore, let us assume that we also have got the CAD model of this workpiece. This CAD model shall describe the 'ideal' shape of the object and will be available in a coordinate system that is different from that of the 3D data point set. For the goal of shape inspection it is of interest to find the optimal Euclidean motion (translation and rotation) that aligns, or registers, the point cloud to the CAD model. This makes it possible to check the given workpiece for manufacturing errors and to classify the deviations.

A well-known standard algorithm to solve such a registration problem is the *iterative closest point (ICP) algorithm* which has been introduced by Chen and Medioni [7] and Besl and McKay [2].

In the first step of each iteration, for each point of the data point cloud the closest point in the model shape is computed. As result of this first step one obtains a point sequence $Y = (\mathbf{y}_1, \mathbf{y}_2, \dots)$ of closest model shape points to the data point sequence $X = (\mathbf{x}_1, \mathbf{x}_2, \dots)$. Each point $\mathbf{x}_i$ corresponds to the point $\mathbf{y}_i$ with the same index.

In the second step of each iteration the rigid motion M is computed such that the moved data points $M(\mathbf{x}_i)$ are closest to their corresponding points $\mathbf{y}_i$, where the objective function to be minimized is

$$\sum_i \|\mathbf{y}_i - M(\mathbf{x}_i)\|^2. \tag{12}$$

This least squares problem can be solved explicitly, cf. [2]. The translational part of M brings the center of mass of X to the center of mass of Y. The rotational part of M can be obtained as the unit eigenvector that corresponds to the maximum eigenvalue of a symmetric 4×4 matrix. The solution eigenvector is nothing but the unit quaternion description of the rotational part of M.

After this second step the positions of the data points are updated via $X_{\text{new}} = M(X_{\text{old}})$. Now step 1 and step 2 are repeated, always using the updated data points, until the change in the mean-square error falls below a preset threshold. The ICP algorithm always converges monotonically to a local minimum, since the value of the objective function is decreasing both in steps 1 and 2. A summary with new results on the acceleration of the ICP algorithm has been given by Rusinkiewicz and Levoy [22], who also suggest that *iterative corresponding point* is a better expansion for the abbreviation ICP than the original *iterative closest point*.

Actually, we want to apply that motion to the data point cloud which brings it into a position where the sum f of squared distances to the CAD-model is minimal. Minimizing f under the constraint that the applied transformation is a rigid body motion can be done with a constrained optimization algorithm. We do not directly apply a standard implementation, but find a geometrically motivated algorithm. It works with local quadratic approximants of f, and a linearization of the motion.

ICP is also based on local quadratic approximants of f, but a major disadvantage of ICP is the following: data points $\mathbf{x}_i$ are moved towards the normal footpoints $\mathbf{y}_i$. In other words, the squared distance to $\mathbf{y}_i$ is used as a quadratic approximant of the function d^2 at $\mathbf{x}_i$. As we know from the study of d^2, this approximation is only good in the 'far field', i.e., for large distances d. However, in the practical application one has to start with an initial position of the point cloud which is sufficiently close to the CAD model in order to run into the right local minimum. Thus, typically the involved distances are small. As a simple solution, we therefore may use the squared distance function to the tangent plane at the normal footpoint $\mathbf{y}_i$.

Our proposed algorithm works as follows: the first step is similar to that of the ICP algorithm. For each data point $\mathbf{x}_i \in X$ determine the nearest point $\mathbf{y}_i$ of the model surface and determine the tangent plane there. Let $\mathbf{n}_i$ denote a unit normal vector of this tangent plane in $\mathbf{y}_i$. Because $\mathbf{y}_i$ is the nearest point to $\mathbf{x}_i$ on the surface, $\mathbf{x}_i$ lies on the surface normal in $\mathbf{y}_i$, i.e., $\mathbf{x}_i = \mathbf{y}_i + d_i\mathbf{n}_i$ with d_i denoting the oriented distance of $\mathbf{x}_i$ to $\mathbf{y}_i$. Now we locally replace the function d^2 of the CAD model at $\mathbf{x}_i$ by the squared distance to the tangent plane at $\mathbf{y}_i$. Summing up these quadratic approximants results in a quadratic approximant of the function F.

The linearization of the motion is equivalent to using instantaneous kinematics, an idea that already appeared in a similar form in [6]. We estimate the displacements of points via their velocity vectors. The velocity vector field of a rigid body motion is known to have the form

$$\mathbf{v}(\mathbf{x}) = \bar{\mathbf{c}} + \mathbf{c} \times \mathbf{x}. \tag{13}$$

The distance of the point $\mathbf{x}_i + \mathbf{v}(\mathbf{x}_i)$ to the tangent plane at $\mathbf{y}_i$ with unit normal vector $\mathbf{n}_i$ is given by $d_i + \mathbf{n}_i \cdot \mathbf{v}(\mathbf{x}_i)$, where d_i again denotes the oriented distance of $\mathbf{x}_i$ to $\mathbf{y}_i$. The objective function to be minimized is

$$\sum_i \left(d_i + \mathbf{n}_i \cdot (\bar{\mathbf{c}} + \mathbf{c} \times \mathbf{x}_i) \right)^2, \tag{14}$$

which is quadratic in the unknowns $(\mathbf{c}, \bar{\mathbf{c}})$. The unique solution can be given explicitly by solving a system of linear equations.

Note that the transformation which maps $\mathbf{x}_i$ to $\mathbf{x}_i + \mathbf{v}(\mathbf{x}_i)$ is an affine map and not a rigid Euclidean motion. Nevertheless, the vector field determined by $(\mathbf{c}, \bar{\mathbf{c}})$ uniquely determines a uniform helical motion M. Axis G and pitch p of M are easily computed (see e.g. [21]). The motion we apply to $\mathbf{x}_i$ is the superposition of a rotation about this axis G through an angle of $\alpha = \arctan(\|\mathbf{c}\|)$ and a translation parallel to G by the distance of $p \cdot \alpha$.

Similar to the ICP algorithm we update the data points via $X_{\text{new}} = M(X_{\text{old}})$ and repeat the procedure until the change in mean-square error $\sum_i d_i^2$ falls below a preset threshold.

Remark 2. We iteratively minimize a quadratic approximant of f under a linearized motion. The resulting transformation would not satisfy the rigidity constraint on the moved point cloud. Thus, we project back to the constraint manifold by applying a helical motion indicated by the velocity field. Therefore, our method is a feasible point method. An SQP-algorithm [5] would work with quadratic approximants on the Lagrangian function, a linearization of the motion, but it would then directly apply the linearized transformation; it is not a feasible point method. We expect that SQP-algorithms will help in the solution of other geometric optimization algorithms where the projection onto the constraint manifold is not so simple as for the registration problem.

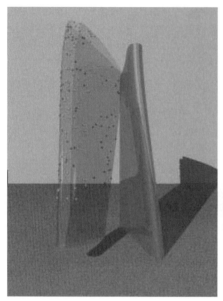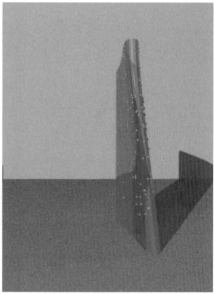

Fig. 12. Matching of a point cloud to the corresponding CAD model: (left) initial position of data points and CAD model, (right) final position after seven iterations.

An example for the registration of a point cloud to a surface is shown in Fig. 12. In order to better visualize the spatial position of the point data, a transparent surface is associated with the data point set. This transparent surface does not enter the computation in any way, it is only displayed for reasons of visualization. The pictures show the data point set in its initial position, after the first iteration step, and in its final position after 7 iterations. In this example the error tolerance reached in the final position will be obtained with the standard implementation of the ICP algorithm only after 45 iterations. This is because in the ICP algorithm the data points move towards their nearest position on the surface in each iteration step. A displacement of the data point set in tangential direction to the model surface therefore needs many iterations.

It is straightforward to extend the objective function (14) to a weighted scheme. There are 3D measurement devices that supply for each data point a tolerance for the occurring measurement errors. These can be included in the objective function to downweight outliers.

In the description above, we have approximated the squared distance function d^2 to the given model surface by the squared distance function to the tangent plane at the footpoint $\mathbf{y}_i$ of $\mathbf{x}_i$. The latter function is exactly the second order Taylor approximant of d^2 at $\mathbf{y}_i$. In a further improvement we may directly use the second order Taylor approximant of d^2 at $\mathbf{x}_i$ (if necessary, the discussed nonnegative approximant), or another quadratic approximant which is sufficiently accurate at the given position $\mathbf{x}_i$ at an appropriate level of detail. Thus, the results of the investigation on the geometry of the squared distance function can be directly applied to this registration problem and also to other types of registration and positioning problems.

6.2 Application to Surface Approximation

Approximating a given surface (in any representation) or an unstructured cloud of points by a NURBS surface is a widely investigated problem. The main approach uses a least square formulation with a regularization term that expresses the fairness of the final result (see e.g. [11]). Here, the parameterization problem is a fundamental one which largely effects the result (see e.g. [16] and the references therein). Therefore, parameter correction procedures have been suggested [12].

A different approach to the approximation of curves and surfaces are *active contour models*, which are mainly used in Computer Vision and Image Processing. The origin of this technique is the seminal paper by Kass et al. [13], where a variational formulation of parametric curves, coined *snakes*, is presented for detecting contours in images. There are various other applications and a variety of extensions of the snake model (see e.g. [3]).

Instead of a parametric representation of a curve, one may use an implicit form as zero set (level set) of a bivariate function. The formulation of active contour models via level sets goes back to Osher and Sethian [19]. The *level set method* [26] has been successfully applied to the solution of a variety of problems, e.g. for segmentation and analysis of medical images [17]. There are also extensions to surfaces. An application to the surface fitting problem has been given by Zhao et al. [30, 31].

In the registration problems outlined above, the moving object (point cloud) undergoes a rigid body motion. The motion is linearized in each iteration step and guided by the 'flow' imposed by the squared distance function. Basically the same idea applies for surface approximation. Given a very rough initial approximant, the surface is 'deformed' by moving its control points. Since the major parametric surfaces used in CAD systems are B-spline surfaces, which depend linearly on the control points, we may use the same ideas as above.

To explain the principle, we choose an example: The surface Φ in Fig. 13 (Color Plate 33 on page 444) shall be approximated by a B-spline surface patch. The initial position of the B-spline control points was chosen by linear interpolation of Φ's vertex points (Fig. 13 and Color Plate 33 on page 444). In each iteration step the contol points are recomputed, such that a sufficiently dense sample of points on the B-spline surface is moved towards the target surface Φ. The 'flow' of the sampled points is again guided by the squared distance function. In order to avoid clustering of the control points and self-intersections of the surface (folding), smoothing terms must be incorporated into the function to be minimized. In our example the resulting B-spline patch

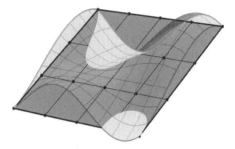

Fig. 13. Approximation of a surface patch (light colored) by a B-spline surface (dark colored): Initial position of B-spline surface.

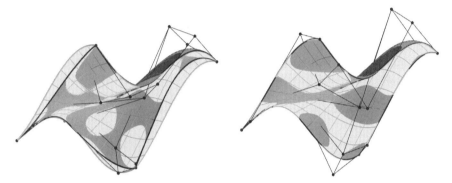

Fig. 14. Approximation of a surface patch by a B-spline surface; final position after five iterations without boundary approximation (left), and with boundary approximation (right).

after the fifth iteration step is given in Fig. 14 (Color Plate 34 on page 444), *without* boundary approximation (left), and *with* boundary approximation (right), respectively.

An important advantage of this active contour approach to surface approximation is the avoidance of the parameterization problem. Another advantage is the *applicability to subdivision surface fitting*: points at refined levels depend linearly on points of coarse levels in the subdivision procedure. This will be explored in future research.

7 Future Research

There is a large amount of work left for future research. We have to study the squared distance function d^2 to surfaces, which are just given by a dense sample of points. Moreover, computationally efficient ways of working with a piecewise quadratic approximation of d^2 need to be addressed. Furthermore, we will investigate algorithms for the solution of other geometric optimization problems. There, we believe it is important not to use an optimization algorithm as a black box, but adapt an optimization concept (Newton, quasi-Newton, SQP,...) in a geometric way to the special problem.

Acknowledgements

Helmut Pottmann is grateful for support by the Institute of Mathematics and Its Applications at the University of Minnesota; main ideas of the present work were developed during a stay at IMA in spring 2001. Special thanks go to Stefan Leopoldseder for the implementation of the applications in Sect. 6.

References

1. Ambrosio, L., Mantegazza, C. (1998): Curvature and distance function from a manifold. Journal of Geometric Analysis, **8**, 723–748
2. Besl, P. J., McKay, N. D. (1992): A method for registration of 3D shapes. IEEE Trans. Pattern Anal. and Machine Intell., **14**, 239–256
3. Blake, A., Isard, M. (1998): Active Contours. Springer New York
4. Blum, H. (1973): Biological shape and visual science, Journal of Theoretical Biology, **38**, 205–287
5. Boggs, P. T., Tolle, J. W. (1995): Sequential quadratic programming. Acta Numerica, **4**, 1–52
6. Bourdet, P., Clement, A. (1988): A study of optimal-criteria identification based on the small-displacement screw model. Annals of the CIRP, **37**, 503–506
7. Chen, Y., Medioni, G. (1992): Object modelling by registration of multiple range images. Image and Vision Computing, **10**, 145–155
8. Choi, H. I., Choi, S. W., Moon, H. P. (1997): New algorithm for medial axis transform of plane domain. Graphical Models and Image Processing, **59**, 463–483
9. Faugeras, O. (1993): Three-Dimensional Computer Vision: a Geometric Viewpoint. The MIT Press
10. Faugeras, O., Gomes, J. (2000): Dynamic shapes of arbitrary dimension: the vector distance functions. In: Martin, R. (ed.) The Mathematics of Surfaces IX. Springer, 227–262
11. Hoschek, J., Jüttler, B. (1999): Techniques for fair and shape-preserving surface fitting with tensor-product B-splines. In: Peña, J. M. (ed.) Shape Preserving Representations in Computer-Aided Geometric Design. Nova Science Publ., Commak, New York
12. Hoschek, J., Lasser, D. (1993): Fundamentals of Computer Aided Geometric Design. A. K. Peters, Wellesley, Massachusetts
13. Kass, M., Witkin, A., Terzopoulos, D. (1988): Snakes: Active contour models. International Journal of Computer Vision, **1**, 321–332
14. Kimia, B., Tannenbaum, A, Zucker, S.W. (1995): Shapes, shocks and deformations I: The components of two-dimensional shape and the reaction-diffusion space. International Journal of Computer Vision, **15**, 189–224

15. Kimmel, R., Shaked, D., Kiryati, N., Bruckstein, A. (1995): Skeletonization via distance maps and level sets. Computer Vision and Image Understanding, **62**, 382–391

16. Ma, W., Kruth, J. P. (1995): Parametrization of randomly measured points for the least squares fitting of B-spline curves and surfaces. Computer Aided Design, **27**, 663–675

17. Malladi, R., Sethian, J. A., Vemuri, B. C. (1995): Shape modeling with front propagation: A level set approach. IEEE Transactions Pattern Analysis and Machine Intelligence, **17**, 158–175

18. Nikolaidis, N., Pitas, I. (2001): 3-D Image Processing Algorithms, Wiley

19. Osher, S. J., Sethian, J. A. (1988): Fronts propagating with curvature dependent speed: Algorithms based on Hamilton-Jacobi formulation. Journal of Computational Physics, **79**, 12–49

20. Patrikalakis, N. M., Maekawa, T. (2002): Shape Interrogation for Computer Aided Design and Manufacturing. Springer

21. Pottmann, H., Wallner, J. (2001): Computational Line Geometry. Springer Berlin Heidelberg New York

22. Rusinkiewicz, S., Levoy, M. (2001): Efficient variants of the ICP algorithm, in Proc. 3rd Int. Conf. on 3D Digital Imaging and Modeling, Quebec, Springer

23. Sapiro, G. (2001): Geometric Partial Differential Equations and Image Analysis. Cambridge University Press

24. Serra, J. (1982): Image Analysis and Mathematical Morphology. Academic Press, London

25. Serra, J., Soille, J., eds. (1994): Mathematical Morphology and Its Applications to Image Processing. Kluwer Academic Publishers, Dordrecht

26. Sethian, J. A. (1999): Level Set Methods and Fast Marching Methods. Cambridge University Press

27. Siddiqi, K., Tannenbaum, A., Zucker, S.W. (1999): A Hamiltonian approach to the eikonal equation. Workshop on Energy Minimization Methods in Computer Vision and Pattern Recognition, 1–13

28. Toriwaki, J., Yokoi, S. (1981): Distance transformations and skeletons of digitized pictures with applications. In: Kanal, L. N. and Rosenfeld, A. (eds.) Progress in Pattern Recognition. North Holland, 187–264

29. Zhao, H. K. (2002): Analysis and visualization of large sets of unorganized points using the distance function. Preprint

30. Zhao, H. K., Osher, S., Fedkiw, R. (2001): Fast surface reconstruction and deformation using the level set method. Proceedings IEEE Workshop on Variational and Level Set Methods in Computer Vision, Vancouver

31. Zhao, H. K., Osher, S., Merriman, B., Kang, M. (2000): Implicit and nonparametric shape reconstruction from unorganized data using variational level set method. Computer Vision and Image Understanding, **80**, 295–314

Part IV

Image Based Visualization

A Multiscale Fairing Method for Textured Surfaces

Ulrich Clarenz, Udo Diewald, Martin Rumpf

Institut für Mathematik, Universität Duisburg, 47048 Duisburg, Germany.
{*clarenz,diewald,rumpf*}*@math.uni-duisburg.de*

Summary. Based on image processsing methodology and the theory of geometric evolution problems a novel multiscale method on textured surfaces is presented. The aim is fairing of parametric noisy surfaces coated by a noisy texture. Simultaneously features in the texture and on the surface are enhanced. Considering an appropriate coupling of the two fairing processes one can take advantage of the frequently present strong correlations between edge features in the texture and on the surface edges.

Keywords. anisotropic curvature flow, surface evolution, image processing, scale space

1 Introduction

Fairing of detailed triangulated surfaces is an important topic in computer aided geometric design and in computer graphics [1, 2, 3, 4]. Nowadays, various such surfaces are delivered from different measurement techniques [5] or derived from two- or three dimensional data sets [6]. Recent laser scanning technology for example enables very fine triangulation of real world surfaces and sculptures. Frequently, they are accompanied by grey or color valued texture maps. These surfaces and textures are usually characterized by interesting features, such as edges and corners on the geometry and in the texture intensity map. On the other hand, they are typically disturbed by noise, which is often due to local measurement errors.

The aim of this paper is to present a method which allows the fairing of discrete surfaces coupled with the smoothing of an texture coated on the surface and thus permits a drastic improvement of the signal to noise ratio. Additionally the approach is able to retain and even enhance important features such as surface and texture edges and corners. Frequently, there is a correspondence of surface and texture features. Edge features on the surface usually bound segments in the texture image, e. g. lips or hair-lines. Vice-versa jumps in the texture intensity frequently indicate geometric feature lines. Hence, we ask for a fairing method which takes advantage of this important observation and couples the fairing schemes for both quantities.

The core of the method is a geometric formulation of scale space evolution problems for surfaces. These techniques were originally developed for image processing purposes. Thus the method not only delivers a single resulting surface, but a complete scale of surfaces in time. For increasing time, we obtain successively smoother surfaces with continuously sharpened edges and a texture depending geometry.

We derive a continuous model, which leads to a nonlinear system of parabolic partial differential equations for the coordinate mapping of the surface and for the texture.

2 Image Processing Background

In physics, diffusion is known as a process that equilibrates spatial variations in concentration. If we consider some initial noisy concentration or image intensity ρ_0 on a domain $\Omega \subset \mathbb{R}^2$ and seek solutions of the linear heat equation

$$\partial_t \rho - \Delta \rho = 0 \qquad (1)$$

with initial data ρ_0 and natural boundary conditions on $\partial\Omega$, we obtain a scale of successively smoothed concentrations $\{\rho(t)\}_{t \in \mathbb{R}^+}$. For $\Omega = \mathbb{R}^2$ the solution of this parabolic problem coincides with the filtering of the initial data using a Gaussian filter $\mathcal{G}_\sigma(x) = (2\pi\sigma^2)^{-1} e^{-x^2/(2\sigma^2)}$ of width or standard deviation σ, i. e. $\rho(\sigma^2/2) = \mathcal{G}_\sigma * \rho_0$. The geometrical counterpart of the Euclidian Laplacian Δ on smooth surfaces is the Laplace Beltrami operator $\Delta_{\mathcal{M}}$ [7, 8]. Thus, one obtains the geometric diffusion $\partial_t x = \Delta_{\mathcal{M}(t)} x$ for the coordinates x on the corresponding family of surfaces $\mathcal{M}(t)$. On triangulated surfaces as they frequently appear in geometric modeling and computer graphics applications, several authors introduced appropriate discretized operators [4, 2, 3, 9]. Recently Desbrun et al. [1] considered an implicit discretization of geometric diffusion to obtain strongly stable numerical smoothing schemes.

From differential geometry [10] we know that the mean-curvature vector HN equals the Laplace Beltrami operator applied to the identity Id on a surface $\mathcal{M}$:

$$H(x)N(x) = -\Delta_{\mathcal{M}} x. \qquad (2)$$

Thus geometric diffusion is equivalent to mean curvature motion (MCM): $\partial_t x = -H(x)N(x)$, where $H(x)$ is the corresponding mean curvature (here defined as the sum of the two principal curvatures), and $N(x)$ is the normal on the surface at point x. Already in '91 Dziuk [11] presented a semi-implicit finite element scheme for MCM on triangulated surfaces. The approach by Desbrun et al. [1] is essentially identical to this earlier method.

In image processing, Perona and Malik [12] proposed a nonlinear diffusion method, which modifies the diffusion coefficient at edges. Edges are indicated by steep intensity gradients. For a given initial image ρ_0 they considered the evolution problem

$$\partial_t \rho - \mathrm{div}\left(G\left(\frac{\|\nabla\rho\|}{\lambda}\right)\nabla\rho\right) = 0 \tag{3}$$

for some parameter $\lambda \in \mathbb{R}^+$. A suitable choice for G is

$$G(s) = \left(1 + s^2\right)^{-1}. \tag{4}$$

Thus edges are classified by the involved parameter λ. Catté et al. [13] proposed a regularization of the original method where the diffusion coefficient is no longer evaluated on the exact intensity gradient. Instead they suggested to consider the gradient evaluation on a prefiltered image, i.e., they consider the equation

$$\partial_t \rho - \mathrm{div}\left(G\left(\frac{\|\nabla\rho_\sigma\|}{\lambda}\right)\nabla\rho\right) = 0 \tag{5}$$

where $\rho_\sigma = \mathcal{G}_\sigma * \rho$ with a suitable local convolution kernel $\mathcal{G}_\sigma$, e.g. an Gaussian filter kernel. The evolution including this prefiltering avoids the detection and pronouncing of initial noise as artificial edges. Weickert [14] improved this method taking into account anisotropic diffusion, where the Perona Malik type diffusion is concentrated in one direction, for instance the gradient direction of a prefiltered image. This leads to an additional tangential smoothing along edges and amplifies intensity correlations along lines. In [15] this image processing approach has been carried over for parametric surfaces. Here we in addition consider texture information on the surface. To prepare the actual discussion of the coupled model we review the basic concept as well. Furthermore we confine this to the continuous problem setting. The discretization is similar to the one presented in [15] based on the MCM discretization scheme proposed by Dziuk [11]. We also pick up the method to compute curvature quantities on discrete triangular surfaces from [15]. Finally, in [16] a finite element implementation of a level set method for anisotropic geometric diffusion is discussed which is closely related to the parametric surface evolution problem presented here.

3 Anisotropic Geometric Diffusion

Let us first review the concept of anisotropic geometric diffusion as a powerful multiscale method for the processing of surface without textures. The aim is to appropriately carry over approved methodology from scale space theory in image processing, now not only applied to image intensities *on* surfaces but to the surface itself. Here we restrict ourselves to the smoothing of a noisy surface without texture. Preliminary results on this restricted case have been presented in [15]. In the following paragraph especially a considerable implementational improvement based on the splitting in tangential and normal velocity components is presented. For the general case coupling surface and texture evolution we refer to Section 4. Let us first summarize the building blocks of the method:

• We consider a noisy initial surface $\mathcal{M}_0$ to be smoothed. Thus we replace the linear diffusion from the Euclidian case of flat images by an appropriate anisotropic diffusion of the surface geometry itself. Thereby a family of surfaces $\{\mathcal{M}(t)\}_{t \in \mathbb{R}_0^+}$ is generated, where the time t serves as the scale parameter.

• In addition to the smoothing of the surface, our aim is to maintain or even enhance sharp edges of the surface. The canonical quantity for the detection of edges is the curvature tensor, in the case of codimension 1 represented by the symmetric shape operator $S_{\mathcal{T}_x \mathcal{M}}$. Here $\mathcal{T}_x \mathcal{M}$ indicates the tangent space. An edge is supposed to be indicated by one sufficiently large eigenvalue of $S_{\mathcal{T}_x \mathcal{M}}$. Hence we consider a diffusion tensor depending on $S_{\mathcal{T}_x \mathcal{M}}$, which enables us to decrease diffusion significantly at edges indicated by $S_{\mathcal{T}_x \mathcal{M}}$. Furthermore we will introduce a threshold parameter λ as in (3) for the identification of edges.

• The evaluation of the shape operator on a noisy surface might be misleading with respect to the original but unknown surface and its edges. Thus we prefilter the current surface $\mathcal{M}(t)$ before we evaluate the shape operator. The straightforward "geometric Gaussian" filter is a short timestep of mean curvature flow. Hence, we compute a shape operator $S_{\mathcal{T}_x \mathcal{M}}^\sigma$ on the resulting prefiltered surface $\mathcal{M}_\sigma(t)$, where the parameter σ is the "geometric Gaussian" filterwidth, i.e. $\mathcal{M}_\sigma(t) = MCM(\mathcal{M}(t), \sigma^2/2)$.

• We incorporate anisotropic diffusion now based on a proper diffusion tensor $a_{\mathcal{T}_x \mathcal{M}}^\sigma$ which enables tangential smoothing along edges. Thereby, the tangential edge direction on the tangent space $\mathcal{T}_x \mathcal{M}(t)$ is indicated by the principal direction of curvature corresponding to the subdominant principal curvature. The second, perpendicular direction is considered to be the actual sharpening direction. Fig. 1 (Color Plate 3 on page 427) clearly outlines the advantage of an anisotropic diffusion tensor.

• The resulting method leads to spatial displacement and the volume enclosed by $\mathcal{M}(t)$ is changed in the evolution. Introducing an additional force f in the evolution which depends on certain integrated curvature expressions leads to volume preservation and we can further improve our multiscale method. For details we refer to [15].

We end up with the following type of parabolic surface evolution problem. Given an initial compact embedded manifold $\mathcal{M}_0$ in $\mathbb{R}^3$, we compute a one parameter family of manifolds $\{\mathcal{M}(t)\}_{t \in \mathbb{R}_0^+}$ with corresponding coordinate mappings $x(t)$ which solves the system of anisotropic geometric evolution equations given by

$$\partial_t x - \mathrm{div}_{\mathcal{M}(t)}(a_{\mathcal{T}_x \mathcal{M}}^\sigma \nabla_{\mathcal{M}(t)} x) = f \quad \text{on} \quad \mathbb{R}^+ \times \mathcal{M}(t), \qquad (6)$$

and satisfies the initial condition

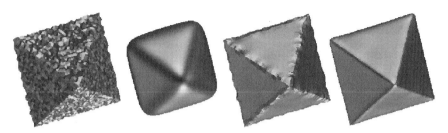

Fig. 1. *Comparison of different surface evolution models. From left to right the initial surface, the result of the mean curvature motion, the result employing an isotropic nonlinear diffusion coefficient and the resulting surface under the new edge preserving anisotropic evolution using the diffusion coefficient* (8) *are depicted. The different results are evaluated for the same time* $t = 0.0032$ *and the parameters where chosen as* $\sigma = 0.02$ *and* $\lambda = 5$. *The diameter of initial surface is chosen to be 1.*

$$\mathcal{M}(0) = \mathcal{M}_0.$$

Here, $\operatorname{div}_{\mathcal{M}(t)}$ and $\nabla_{\mathcal{M}(t)}$ are the divergence and gradient operators on the surface. Furthermore, for every point x on $\mathcal{M}(t)$ the diffusion tensor $a^{\sigma}_{\mathcal{T}_x\mathcal{M}}$ is supposed to be a symmetric, positive definite, linear mapping on the tangent space $\mathcal{T}_x\mathcal{M}$:

$$a^{\sigma}_{\mathcal{T}_x\mathcal{M}}(x) : \mathcal{T}_x\mathcal{M} \to \mathcal{T}_x\mathcal{M} .$$

Furthermore, f represents the forcing on the right-hand side that maintains the volume enclosed by $\mathcal{M}(t)$. There is an orthonormal basis $\{w^1, w^2\}$ of $\mathcal{T}_x\mathcal{M}_\sigma$ such that $S^{\sigma}_{\mathcal{T}_x\mathcal{M}_\sigma}$ is represented by

$$S^{\sigma}_{\mathcal{T}_x\mathcal{M}_\sigma} = \begin{pmatrix} \kappa^{1,\sigma} & 0 \\ 0 & \kappa^{2,\sigma} \end{pmatrix} .$$

Now we consider a diffusion tensor in equation (6) which is defined as follows with respect to the above orthonormal basis. First we introduce a diffusion on $\mathcal{T}_x\mathcal{M}_\sigma$ in the basis $\{w^1, w^2\}$ as

$$a^{\sigma}_{\mathcal{T}_x\mathcal{M}_\sigma} = a(S^{\sigma}_{\mathcal{T}_x\mathcal{M}_\sigma}) = \begin{pmatrix} G\left(\frac{\kappa^{1,\sigma}}{\lambda}\right) & 0 \\ 0 & G\left(\frac{\kappa^{2,\sigma}}{\lambda}\right) \end{pmatrix} \tag{7}$$

with the function G from above. Finally, to define the actual diffusion on $\mathcal{T}_x\mathcal{M}$ we decompose a vector $z \in \mathbb{R}^3$ in the orthogonal basis $\{w^1, w^2, N^\sigma\}$, i. e.

$$z = (z \cdot w^1)w^1 + (z \cdot w^2)w^2 + (z \cdot N^\sigma)N^\sigma$$

where $\{w^1, w^2\} \subset \mathbb{R}^3$ denotes the embedded tangent vectors corresponding to the above basis w^1, w^2 and N^σ is the surface normal of $\mathcal{M}_\sigma$. Then we define the diffusion coefficient $a^{\sigma}_{\mathcal{T}_x\mathcal{M}}$ by

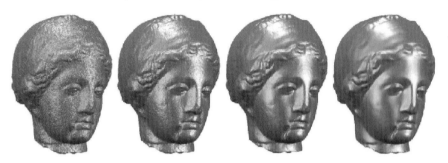

Fig. 2. *From left to right the initial surface and three timesteps of the anisotropic geometric evolution using the diffusion coefficient (8) are shown for a venus head consisting of 268714 triangles. The evolution times are 0.00005, 0.0001, and 0.0002 and the parameters are $\lambda = 10$, $\sigma = 0.02$.*

$$a^{\sigma}_{\mathcal{T}_x\mathcal{M}} z := \Pi_{TxM}\Big(G(\kappa^{1,\sigma})(z \cdot w^1)w^1$$
$$+ G(\kappa^{2,\sigma})(z \cdot w^2)w^2 + (z \cdot N^\sigma)N^\sigma \Big). \tag{8}$$

Here $\Pi_{\mathcal{T}_x\mathcal{M}}$ denotes the orthogonal projection onto the tangent space $\mathcal{T}_x\mathcal{M}$. Fig. 2 (Color Plate 5 on page 428) shows results obtained by this anisotropic geometric diffusion method.

4 Coupling Anisotropic Texture and Surface Diffusion

Up to now, we have considered anisotropic diffusion for noisy surfaces without texture (cf. Section 3). Recalling our original intention we now focus on the coupling of these two diffusion processes, making use of characteristic correlations between surface and texture features. The pure texture smoothing case has been studied by Kimmel in [17]. Usually textures are colour valued maps. For the ease of presentation we confine here to an exposition of grey valued intensities. In the vector valued case of colour textures one proceeds along the line described in [18]. Let us emphasize that the choice of a suitable colour model is of particular importance for the appropriateness of the results. For a description of the *RGB*- or *HSV*-colour model we refer to [19]. The figures depicted here already show colour textures.

As explained in the introduction we aim to intensify the modulation of the diffusivity as well in the surface as in the texture diffusion model whenever edge type features are detected not only in the one but also simultaneously in the other quantity.

Hence, for the surface evolution we decompose the texture intensity gradient with respect to the coordinate system aligned to the principal directions of curvature on the surface and vice versa for texture diffusion the curvature

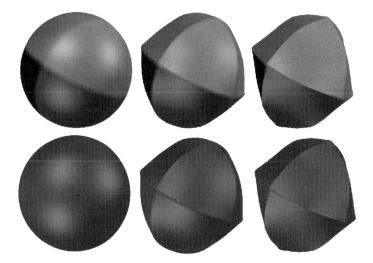

Fig. 3. *Example for the evolution of a surface with texture information under the combined diffusion (9) and (11). Because of the dependency of the diffusion coefficient a_G on the texture during the evolution geometry edges develop in areas of high texture gradients. The parameters are chosen as $\tau = 0.00001$, $\sigma = 0.0045$, $\lambda = 20$, $\mu = 2$, and the diameter of the surface is scaled to 1. On the left the initial surface is shown with and without texture information, and on the right two timesteps for $t = 0.000015$ and $t = 0.00003$ are depicted.*

directions are decomposed into the directions of the texture gradient and perpendicular to it:

If the texture gradient points in the direction of the dominant curvature we further reduce the diffusivity in this direction. Analogously we reduce the diffusion coefficient corresponding to the texture gradient direction if the geometry is significantly curved in this direction.

In what follows we will give the details separately for the surface and texture diffusion problem.

Respecting Texture Features in the Surface Evolution

Let us start considering the local situation near a geometry edge with principal curvatures $\kappa^{1,\sigma}$, $\kappa^{2,\sigma}$ and $\left|\kappa^{1,\sigma}\right| \gg \left|\kappa^{2,\sigma}\right|$. The principal direction of curvature $v^{2,\sigma}$ corresponding to the curvature $\kappa^{2,\sigma}$ points along the geometry edge, whereas $v^{1,\sigma}$ is orthogonal to the edge. The texture information on the surface will again be denoted by u. Assuming a correlation between geometry and texture we expect the angle between the texture gradient $\nabla_{\mathcal{M}} u^\epsilon$ and the direction $v^{1,\sigma}$ to be small, cf. Fig. 5. (Here, the indices ϵ and σ denote the smoothing parameters introduced in Section 2 and Section 3.)

We assume in the following that $\left|\kappa^{1,\sigma}\right| > \left|\kappa^{2,\sigma}\right|$ whenever $\kappa^{1,\sigma} \neq \kappa^{2,\sigma}$. Furthermore, we always choose the directions of $v^{i,\sigma}$ for $i = 1, 2$ such that

$$v^{i,\sigma} \cdot \nabla_{\mathcal{M}^\sigma} u^\epsilon \geq 0 \,.$$

To set up a geometric evolution which is able to take into account also texture information, we choose two orthogonal directions $\tilde{v}^{1,\sigma}, \tilde{v}^{2,\sigma} \in T_x\mathcal{M}$ – the expected directions normal and tangential to the edge – in every point $x \in \mathcal{M}$ as follows:

$$\left(\tilde{v}^{1,\sigma}, \tilde{v}^{2,\sigma}\right) := \begin{cases} \left(\dfrac{v}{\|v\|}, \dfrac{v^\perp}{\|v\|}\right), \; v \in T_x\mathcal{M} \text{ arbitrary} \\ \qquad : \quad \nabla_{\mathcal{M}_\sigma} u^\epsilon = 0 \,, \; \kappa_1 = \kappa_2 \\[2mm] \left(v, v^\perp\right), \; v = \dfrac{\alpha v^{1,\sigma} + (1-\alpha)\frac{\nabla_{\mathcal{M}_\sigma} u^\epsilon}{\|\nabla_{\mathcal{M}_\sigma} u^\epsilon\|}}{\|\alpha v^{1,\sigma} + (1-\alpha)\frac{\nabla_{\mathcal{M}_\sigma} u^\epsilon}{\|\nabla_{\mathcal{M}_\sigma} u^\epsilon\|}\|} \\ \qquad : \quad \text{otherwise} \end{cases},$$

where $v^\perp$ is a vector orthogonal to v with $\|v\| = \|v^\perp\|$ and $\alpha : \mathbb{R}_0^+ \to [0,1]$ is a monotone decreasing blending function depending on the ratio $|\kappa^{1,\sigma}/\kappa^{2,\sigma}|$. It returns 1 for $|\kappa^{1,\sigma}/\kappa^{2,\sigma}| \geq 1 + \eta$ and 0 for $|\kappa^{1,\sigma}/\kappa^{2,\sigma}| = 1$. Hence α is responsible for a blending from the frame $(v^{1,\sigma}, v^{2,\sigma})$ to the frame $(w, w^\perp)$ with $w = \nabla_{\mathcal{M}_\sigma} u^\epsilon / |\nabla_{\mathcal{M}_\sigma} u^\epsilon|$.

If the edge is already classified via the shape operator $S_{T_x\mathcal{M}}^\sigma$ we pick up directions close to those chosen in the pure geometric evolution problem. Alternatively for "softer" geometric edges - those not clearly classified via $\kappa^{1,\sigma}$ and $\kappa^{2,\sigma}$ with $\|\kappa^{1,\sigma}\| \gg \|\kappa^{2,\sigma}\|$ - we assume the texture gradient to be the canonical candidate for the direction perpendicular to a geometric edge on the surface.

The projection of the texture gradient $\nabla_{\mathcal{M}} u^\epsilon$ in the directions $\tilde{v}^{i,\sigma}$ will be denoted by

$$\nabla_{\mathcal{M}} u^{\epsilon,i} := \left\langle \nabla_{\mathcal{M}} u^\epsilon, \tilde{v}^{i,\sigma} \right\rangle \tilde{v}^{i,\sigma} \,.$$

As geometric diffusion problem for the surface coordinates $x \in \mathcal{M}(t)$ we now choose

$$\partial_t x - \Pi_{(T_x\mathcal{M})^\perp} \operatorname{div}_{\mathcal{M}} \left(a_G(S^\sigma, \nabla_{\mathcal{M}} u^\epsilon) \nabla_{\mathcal{M}} x\right) = 0$$
$$\mathcal{M}(0) = \mathcal{M}_0 \,, \tag{9}$$

where $\Pi_{(T_x\mathcal{M})^\perp}$ denotes the orthogonal projection onto the normal space $(T_x\mathcal{M})^\perp$ of the surface. This projection is essential to avoid tangential shifts of the surface coordinates which otherwise would result in a distortion of the associated texture information on the surface. The diffusion coefficient a_G in the basis $\tilde{v}^{1,\sigma}, \tilde{v}^{2,\sigma}$ is given by

$$a_G = \begin{pmatrix} G_G\left(\dfrac{\kappa^{1,\sigma}}{\lambda}, \dfrac{\|\nabla_{\mathcal{M}} u^{\epsilon,1}\|}{\mu}\right) & 0 \\ 0 & G_G\left(\dfrac{\kappa^{2,\sigma}}{\lambda}, \dfrac{\|\nabla_{\mathcal{M}} u^{\epsilon,2}\|}{\mu}\right) \end{pmatrix}, \tag{10}$$

and G_G is defined as

$$G_G(s,t) = \frac{1}{1 + s^2 + t^2} \ .$$

Hence, in the case $\nabla_{\mathcal{M}} u^\epsilon = 0$ we get back the diffusion coefficient of the pure geometric evolution problem (7), whereas in case $\kappa_1^\sigma = \kappa_2^\sigma = 0$ we obtain a diffusion coefficient which takes into account an anisotropy related to the texture gradient.

In Fig. 3 (Color Plate 4 on page 427) a test case for the evolution (9) is depicted. In time geometry edges develop on an initially smooth surface due to the dependency of the diffusion coefficient on the texture gradient.

Respecting Geometric Features in the Texture Evolution

Next, the formulation of a diffusion model for the texture u also takes into account information about the geometric features on the surface. We proceed in an analogous way like above. Again we define an orthonormal basis $\{w, w^\perp\} \subset \mathcal{T}_x\mathcal{M}$ in every point of $\mathcal{M}$ by

$$(w, w^\perp) := \begin{cases} \left(\dfrac{v}{\|v\|}, \dfrac{v^\perp}{\|v\|}\right), & v \in T_p\mathcal{M} \text{ arbitrary} \\ \qquad : \quad \kappa^{1,\sigma} = \kappa^{2,\sigma}, \ \nabla_{\mathcal{M}} u^\epsilon = 0 \\[2ex] (v, v^\perp), \quad v = \dfrac{\beta \frac{\nabla_{\mathcal{M}_\sigma} u^\epsilon}{\|\nabla_{\mathcal{M}_\sigma} u^\epsilon\|} + (1-\beta) v^{1,\sigma}}{\|\beta \frac{\nabla_{\mathcal{M}_\sigma} u^\epsilon}{\|\nabla_{\mathcal{M}_\sigma} u^\epsilon\|} + (1-\beta) v^{1,\sigma}\|} \\ \qquad : \quad \text{otherwise} \end{cases}$$

The blending function β now depends on $\|\nabla_{\mathcal{M}_\sigma} u^\epsilon\|$. It is assumed to be monotone increasing, β equals 1 for $\|\nabla_{\mathcal{M}} u^\epsilon\| \geq \eta$, and $\beta(0) = 0$.

The projections of the two principal directions of curvature $v^{i,\sigma,\pi}$ onto the directions $w, w^\perp \in \mathcal{T}_x\mathcal{M}$ are denoted by

$$v^{i,\sigma\|} := \langle v^{i,\sigma}, w \rangle w$$
$$v^{i,\sigma\perp} := \langle v^{i,\sigma}, w^\perp \rangle w^\perp ,$$

cf. Fig. 6.

Based on the pairs $(v^{1,\sigma\|}, v^{1,\sigma\perp})$ and $(v^{2,\sigma\|}, v^{2,\sigma\perp})$ we deduce directions $v^\|, v^\perp$ parallel, respectively perpendicular to w whose length is weighted by the corresponding principal curvatures:

$$v^\| := \kappa^{1,\sigma} v^{1,\sigma\|} + \kappa^{2,\sigma} v^{2,\sigma\|}$$
$$v^\perp := \kappa^{1,\sigma} v^{1,\sigma\perp} + \kappa^{2,\sigma} v^{2,\sigma\perp} ,$$

Now we prescribe the diffusion problem for the texture $u : \mathcal{M} \to \mathbb{R}$ on the surface as

$$\partial_t u - \mathrm{div}_{\mathcal{M}}\big(a_T(\nabla_{\mathcal{M}} u^\epsilon, S^\sigma)\nabla_{\mathcal{M}} u\big) = 0$$
$$u(0) = u_0 \tag{11}$$

where the diffusion coefficient in the basis $(w, w^\perp)$ is given by

$$a_T = \begin{pmatrix} G_T\left(\frac{\|v^\|\|}{\lambda}, \frac{\|\nabla_\mathcal{M} u^\epsilon\|}{\mu}\right) & 0 \\ 0 & G_T\left(\frac{\|v^\perp\|}{\lambda}, 0\right) \end{pmatrix}.$$

with

$$G_T(s, t) = \frac{1}{1 + s^2 + t^2}.$$

Let us summarize the modeling of the combined anisotropic diffusion method. The shape of the diffusion coefficients a_G and a_T results in an additional sharpening of edges in the geometry in areas of high texture gradients and vice versa of edges in the texture in areas of aligned principal directions of curvature. Here by alignment we mean nearly parallel texture gradient and principal direction of curvature corresponding to the curvature with the largest absolute value.

Furthermore, in case of a locally flat surface with $\kappa^{1,\sigma} = \kappa^{2,\sigma} = 0$ we locally obtain a pure texture diffusion. On the other hand in case of a locally constant texture indicated by $\nabla_\mathcal{M} u^\epsilon = 0$ and either $\kappa^{1,\sigma}/\kappa^{2,\sigma} > 1 + \eta$ or $\kappa^{1,\sigma} = \kappa^{2,\sigma}$ we locally get back to the pure anisotropic geometry diffusion.

Different from the pure geometric diffusion problem, the projection of the divergence term onto the normal space $(\mathcal{T}_x\mathcal{M})^\perp$ in equation (9) now suppresses a tangential shift of the surface coordinates and avoids a distortion of the associated texture information on the surface. Thus it is not only relevant for implementational purposes but already essential in the modeling.

Dependent on the application one might also be interested in other diffusion coefficients, which for example could result in an additional smoothing of the geometry in areas of low texture gradients instead of the additional sharpening of geometry edges in areas of high texture gradients. Such modifications are straightforward to incorporate based on the general approach presented here.

We have applied the combined diffusion method (9), (11) and the pure geometric diffusion method using (8) to the surface of a human head. This data set was generated by a laser scanner with an additional photography unit. Fig. 7 (Color Plate 6 on page 428) and Fig. 8 demonstrate that the combined diffusion method is characterized by a significantly better sharpening of edges, geometric ones on the surface as well as texture edges in the texture map. To underline this we study the case with (cf. Fig. 7 (Color Plate 6 on page 428)) and without (cf. Fig. 8) additional noise added to the true surface as delivered by the scanner. This is related to the fact that in this typical application we observe strong correlation of both types of features surface edges and texture edges as on most natural surfaces.

5 Comparison and Conclusions

We have presented a novel coupled multiscale method for surface fairing and texture denoising. It is able to successively smooth noisy initial surfaces with onscribed texture information while simultaneously enhancing edges and corners of the surface and edge type features of the texture maps. The evolution time plays the role of the scale parameter.

The method is based on an anisotropic curvature evolution problem. The method allows the efficient and flexible processing of arbitrary surfaces and accompanying texture maps, as they are common in geometric modeling and computer graphics applications. The user controls the surface evolution mainly by four parameters which have an intuitive meaning. Two regularization parameters σ, ϵ have to be chosen to filter out high frequency noise in the geometry and texture before the diffusion coefficients are evaluated. Furthermore, λ and μ can be regarded as user given threshold values for the edge detection on the surface and on the texture intensity map, respectively. It was not the scope of this paper to discuss the various problems involved in acquiring high quality triangulated surfaces with texture information like potential misalignement of geometry and texture or different spatial resolution of geometry and texture information. Instead we presumed to be given such a surface and restricted to a discussion of the evolution methods.

Acknowledgement

The authors thank Martin Roth from the ETH Zurich for providing the facial dataset containing texture information used in this paper.

References

1. M. Desbrun, M. Meyer, P. Schroeder, and A. Barr, "Implicit fairing of irregular meshes using diffusion and curvature flow," in *Computer Graphics (SIGGRAPH '99 Proceedings)*, 1999, pp. 317–324.
2. L. Kobbelt, "Discrete fairing," in *Proceedings of the 7th IMA Conference on the Mathematics of Surfaces*, 1997, pp. 101–131.
3. L. Kobbelt, S. Campagna, J. Vorsatz, and H.-P. Seidel, "Interactive multiresolution modeling on arbitrary meshes," in *Computer Graphics (SIGGRAPH '98 Proceedings)*, 1998, pp. 105–114.
4. G. Taubin, "A signal processing approach to fair surface design," in *Computer Graphics (SIGGRAPH '95 Proceedings)*, 1995, pp. 351–358.
5. B. Curless and M. Levoy, "A volumetric method for building complex models from range images," in *Computer Graphics (SIGGRAPH '96 Proceedings)*, 1996, pp. 303–312.
6. W.E. Lorensen and H.E. Cline, "Marching cubes: A high resolution 3d surface construction algorithm," *Computer Graphics*, vol. 21, no. 4, pp. 163–169, 1987.

7. M. P. do Carmo, *Riemannian Geometry*, Birkhäuser, Boston–Basel–Berlin, 1993.

8. I. Chavel, *Eigenvalues in Riemannian Geometry*, Academic Press, 1984.

9. I. Guskov, W. Sweldens, and P. Schroeder, "Multiresolution signal processing for meshes," in *Computer Graphics (SIGGRAPH '99 Proceedings)*, 1999.

10. U. Dierkes, S. Hildebrandt, A. Küster, and O. Wohlrab, *Minimal Surfaces*, Grundlehren der Mathematischen Wissenschaften. 295. Berlin: Springer- Verlag, 1992.

11. G. Dziuk, "An algorithm for evolutionary surfaces," *Numer. Math.*, vol. 58, pp. 603–611, 1991.

12. P. Perona and J. Malik, "Scale space and edge detection using anisotropic diffusion," in *IEEE Computer Society Workshop on Computer Vision*, 1987.

13. F. Catté, P. L. Lions, J. M. Morel, and T. Coll, "Image selective smoothing and edge detection by nonlinear diffusion," *SIAM J. Numer. Anal.*, vol. 29, pp. 182–193, 1992.

14. J. Weickert, "Foundations and applications of nonlinear anisotropic diffusion filtering," *Z. Angew. Math. Mech.*, vol. 76, pp. 283–286, 1996.

15. U. Diewald, U. Clarenz, and M. Rumpf, "Nonlinear anisotropic diffusion in surface processing," in *Proceedings of IEEE Visualization 2000*, 2000, pp. 397–405.

16. T. Preußer and M. Rumpf, "A level set method for anisotropic geometric diffusion in 3D image processing," *To appear in SIAM J. Appl.*, 2002.

17. R. Kimmel, "Intrinsic scale space for images on surfaces: The geodesic curvature flow," *Graphical Models and Image Processing*, vol. 59(5), pp. 365–372, 1997.

18. J. Weickert, "Coherence-enhancing diffusion of colour images," *Image and Vision Computing*, vol. 17, pp. 201–212, 1999.

19. J. D. Foley, A. van Dam, S. K. Feiner, and J. F. Hughes, *Computer Graphics: Principles and Practice*, Addison-Wesley, 1990.

Fig. 4. *Comparison of the pure geometric evolution (middle row) and the combined geometry and texture evolution (bottom row) applied to the noisy initial surface shown in the top row. The parameters are chosen as $\tau = 0.0005$, $\sigma = 0.0316$, $\lambda = 20$, $\mu = 1$, and the diameter of the surface is scaled to 1. The smoothed results in the middle and bottom row are evaluated for time $t = 0.0045$. On the left the surfaces are shown with color information, on the right the same surfaces are shown without color information. The noisy initial surface has been obtained from a cube with different but constant RGB color information on each of its 6 faces by adding a homogeneous white noise to the coordinate values as well as to the color information.*

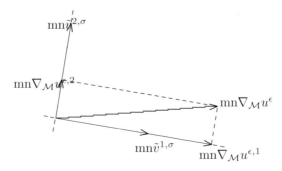

Fig. 5. *Decomposition of the texture gradient $\nabla_{\mathcal{M}} u^{\epsilon}$ into the components parallel to the two principal directions of curvature $\tilde{v}^{1,\sigma,\pi}, \tilde{v}^{2,\sigma,\pi}$.*

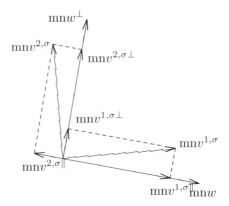

Fig. 6. *Decomposition of the two principal directions of curvature $v^{1,\sigma,\pi}, v^{2,\sigma,\pi}$ into the components parallel and orthogonal to w.*

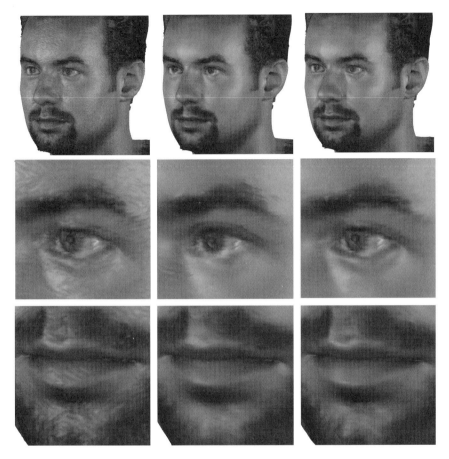

Fig. 7. *On the left a surface obtained from a laser scan with an onscribed photographic texture and moderate superimposed isotropic noise is depicted (on top the whole surface, below magnified parts of it). This surface is considered as initial surface for the pure geometric evolution and for the combined geometry and texture evolution. In the middle the result of the pure geometry evolution and on the right the result of the combined model are shown. In both cases the same evolution timestep $t = 0.00012$ is considered.*

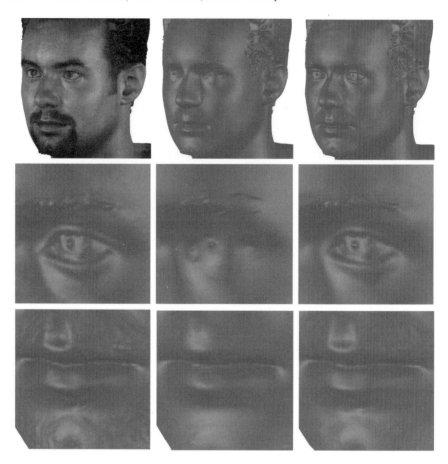

Fig. 8. *Further fairing results corresponding to Fig. 7 are shown. Now they correspond to the actual 3D scanning output without any additional noise depicted on the top left. All surfaces (expect the initial surface) are shown without color information here not to hide the significant changes in the geometry by texturing. Indeed, let us emphasis that the superimposed texture in Fig. 7 visually covers the loss of details in the pure geometry evolution (cf. especially the images in the middle of the 2nd and 3rd row of Fig. 7). The parameters are chosen as in Fig. 7.*

Generalized Block Iterative Methods

Michel Leblond[1], François Rousselle[2], and Christophe Renaud[2]

[1] Laboratoire de Mathématiques Pures et Appliquées, Laboratoire d'Informatique du Littoral Bâtiment H. Poincaré, 50 rue F. Buisson 62228 Calais Cedex. *leblond@lil.univ-littoral.fr*
[2] Laboratoire d'Informatique du Littoral Bâtiment H. Poincaré, 50 rue F. Buisson 62228 Calais Cedex. {*roussell,renaud*}*@lil.univ-littoral.fr*

Summary. In this article we present a new mathematical approach for solving the radiosity system. We introduce iterative methods using the so-called generalized block partitioning of a matrix. They are designed to improve the convergence speed of the progressive radiosity method by relaxing several components of the residual vector at the same time.

Keywords. matrix partitioning, generalized block iterative method, M-matrix, progressive radiosity, image synthesis

1 Introduction

Radiosity is an image synthesis method that simulates the diffuse global illumination of a 3D virtual scene. It takes into account the complex light interactions between all the objects that lie into the scene, restricting those interactions to purely diffuse ones (Lambertian reflection model). Many techniques have been proposed to solve the radiosity problem. Among them, the progressive radiosity method [5] gives a discrete solution by meshing the polygonal environment in n small triangular or quadrangular patches. This modeling and the purely diffuse assumption allow to express the interactions between the objects as a wavelength dependent system of linear equations called the radiosity system. Its size is $n \times n$ and its unknowns are the average values of the patches illumination. An equivalent to the Southwell iterative solver is used to compute the solution of the system. Hierarchical radiosity [10, 20, 11] is another method that allows a reduction in the number of interactions between patches to $O(n)$ with the help of a multilevel hierarchical representation of the geometry of the scene and the exchanges of energy. A modified version of the Gauss–Seidel iterative solver is used to solve the resulting hierarchical system. Monte Carlo radiosity [19, 15, 1] includes other techniques that use a stochastic iterative method to solve the radiosity system. The main advantage of these methods consists in evaluating energy transfers between patches by stochastically tracing rays through the scene. This implies no explicit computation and storage of the elements of the radiosity matrix.

Progressive, hierarchical and Monte Carlo radiosity methods all use iterative solvers. Following our mathematical study of the radiosity system and our first searches for new fast solution algorithms [12], we present in this paper the generalized block iterative method. This method computes the solution to a linear system by relaxing blocks of unknowns that are build dynamically during the iteration process. As a first attempt to speed up the solvers used in the radiosity methods, we applied the generalized block iterative method to progressive radiosity which uses a classical iterative solver. Greiner et al. proposed in [9] an iterative method also based on the idea of groups of patches: the blockwise refinement method. It solves the radiosity system by selecting a small set of important patches at each iteration. An approximate solution of the whole system is computed for this set of patches. The exact interactions between these important patches and all the other ones are taken into account. Furthermore an approximation for the interactions between the less important patches is computed. The approximated solution is refined from iteration to iteration by selecting other sets of patches that successively correct the error introduced by the approximation. This method implies the storage of an $n \times n$ matrix for each wavelength in order to keep trace of the energy exchanges and correct the error of the preceding iterations. Thus, this method is only adequate for the solution of small systems where $O(n^2)$ storage is possible. Unlike blockwise refinement the generalized block iterative methods presented in this paper make no approximation of the entries of the radiosity matrix and thus they do not introduce any error. Moreover, they require only $O(n)$ storage. Application of this technique to the hierarchical or stochastic solvers of the hierarchical radiosity and Monte Carlo radiosity methods have not been experimented yet. The potential profit in the use of generalized blocks with these state-of-the-art methods is discussed in the conclusion and perspective section.

1.1 The Radiosity System

In matrix format the standard radiosity system can be written as

$$\Phi x = (I - RF)x = b,$$

where

- I is the identity matrix of order n.
- $R = diag(\varrho_1, \ldots, \varrho_n)$. Each real number ϱ_i is the spectral reflectivity of patch i: the fraction of radiant power, incident on patch i, which is reflected back to the environment. We assume $0 < \varrho_i < 1$.
- F is the form factor matrix. The form factor F_{ij} specifies the fraction of radiant power that leaves patch i and arrives directly at patch j. Because we assume that the patches are planar or at least convex we get $F_{ii} = 0$ for $i = 1, \ldots, n$.

- b is the emission vector. $b_i \geq 0$ is the spectral irradiance emitted from patch i. $b_i \neq 0$ means that patch i is a light source.
- x is the radiosity vector. x_i is the spectral radiant exitance of patch i also called the radiosity of patch i.

1.2 Progressive Radiosity Methods

(a) after 9 iterations   (b) after 15 iterations

(c) after 100 iterations   (d) final result

Fig. 1. First iterations and final result of progressive radiosity for an office discretized in 11825 quadrangular patches (Color Plate 12 on page 433).

The geometrical complexity of scenes modeled nowadays leads to radiosity systems of very large size: hundreds of thousands or even several millions of unknowns. Usually the radiosity matrix is not sparse. Consequently all its entries must be stored. This results in a high memory cost: for example, using four bytes floats, 150 Gb are required in order to store the radiosity matrix for a scene discretized in 200, 000 patches.

If the whole radiosity matrix can be stored, standard iterative methods – such as Jacobi, Gauss–Seidel, Chebychev, conjugate gradient, ... – may be used to solve the system. The main problem of these methods is that the whole matrix has to be computed before starting to solve the system. Due to the form factors, the computation of the matrix represent the larger part of the total solution time, and it takes a long time before obtaining a first image of the scene.

Techniques named progressive radiosity methods [5, 7, 8, 18, 23] have been introduced in computer graphics. Their main advantage is to allow to solve the radiosity system when the whole matrix cannot be stored. They also provide quickly approximated solutions: they require only the computation of one column of the matrix at each iteration. These solutions may be used to display intermediate images giving some idea of the final illumination of the scene.

Among these techniques, the progressive radiosity method [5] (PR method) is known as a shooting method: at each iteration a patch shoots some energy to all other patches in the scene. In order to know how much energy remains to be shot for each patch, the PR method uses a vector called the unshot radiosity vector. It is initialized with the emission vector and it cumulates the energy received on each patch. At each shooting step the patch i having the largest unshot energy (i.e. unshot radiosity times area (A_i) of patch i) is chosen. The fraction of the radiosity leaving patch i and being reflected by all the other patches is computed and added both to the radiosities and unshot radiosities of these patches. After emission, the unshot radiosity of patch i is set to 0.

The PR method has been proved [8] to be a reformulation of the Southwell iterative solver [21]. The unshot radiosity vector corresponds to the residual vector $r = b - \Phi x$ of the Southwell solver. The algorithm of the PR method using Southwell's formulation is shown in Algorithm 1 where e_i denotes the ith vector of the canonical basis of $\mathbb{R}^n$.

Figs. 1(a), 1(b) and 1(c) show intermediate solutions and were obtained in several seconds.

Despite its ability to solve the radiosity system by using only one column of the matrix at a time and to display intermediate results, the PR method has a big drawback: it converges slowly. The time needed to obtain the final result is far longer than the one needed to compute the radiosity matrix: because some patches are selected several times, the same columns of the matrix are also recomputed several times.

As an example Fig. 1(d), showing the final result, is obtained with the PR method after about five hours but the time needed to compute the whole system matrix is about half an hour (computation with a PC at 1.7 Ghz).

Algorithm 1 The PR method

1: $x^{(0)} = 0$
2: $r^{(0)} = b - \Phi x^{(0)}$
3: **while** not converged **do**
4:     at the $(p+1)$th iteration,
5:     choose i such that $A_i r_i = \|Ar^{(p)}\|_\infty$ {$A_i r_i$ is the largest unshot energy}
6:     compute the ith column of the form factor matrix F
7:     $x^{(p+1)} = x^{(p)} + r_i^{(p)} e_i$ {update only the ith component of $x^{(p)}$}
8:     $r^{(p+1)} = r^{(p)} - r_i^{(p)} \Phi e_i$ {the ith component of $r^{(p+1)}$ is set to 0}
9:     display the scene with $x^{(p+1)} + r^{(p+1)}$ as the radiosity vector
10: **end while**

Our goal is to find a new method which converges faster than the PR method does. Our idea is to develop a similar method using more than one shooting patch at each iteration. We shall see that this new method will require the partitioning of the radiosity matrix in generalized blocks [17, pp. 100–101]. This technique allows Y. Saad to design the *general block Jacobi* and *general block Gauss–Seidel methods* [17, pp. 101–102] and methods denoted by *additive projection procedures* and *multiplicative projection procedures* [17, pp. 136–139]. For all these methods, the partitioning of the system matrix in blocks is performed before calling the iterative method. We can say that these procedures use a *static* partitioning. On the contrary, we introduce in Sect. 3 a generalized block iterative method based on a *dynamic* partitioning of the system matrix: this matrix is partitioned in generalized blocks during the solution process.

In Sect. 2 we shall recall the definition of the generalized block partitioning of a square matrix and we shall give the matrix operations which provide it.

2 Generalized Block Partitioning of a Matrix

Throughout this article we shall use the following definitions and notations:

Definition 1. For all $\alpha \in \mathbb{N}$, $\mathbb{N}_\alpha$ denotes the set of integers $\{1, 2, \ldots, \alpha\}$.
Let $n, q \in \mathbb{N}$ and let $G_1, G_2, \ldots, G_q$ form a partition of $\mathbb{N}_n$.
For any $i \in \mathbb{N}_q$, we denote by n_i the cardinal number of G_i, by Π_i the unique increasing bijection from $\mathbb{N}_{n_i}$ into G_i and by $E_{G_i} = \begin{bmatrix} e_{\Pi_i(1)} \, e_{\Pi_i(2)} \, \cdots \, e_{\Pi_i(n_i)} \end{bmatrix}$ the matrix in $\mathbb{R}^{n \times n_i}$ whose jth column is the $\Pi_i(j)$th vector of the canonical basis of $\mathbb{R}^n$.

Let $x \in \mathbb{R}^n$. The vector $x_{G_i} \in \mathbb{R}^{n_i}$ is defined by

$$x_{G_i} := E_{G_i}^T x = \begin{bmatrix} x_{\Pi_i(1)} \\ \vdots \\ x_{\Pi_i(n_i)} \end{bmatrix}, \text{i.e.}$$

the components of x_{G_i} are those of x which have their index in G_i.

Let $M \in \mathbb{R}^{n \times n}$. The matrix $M_{G_i, G_j} := E_{G_i}^T M E_{G_j} \in \mathbb{R}^{n_i \times n_j}$ is the submatrix of M whose entries have a row index in G_i and a column index in G_j. M_{G_i, G_i} is a principal submatrix of M. Its row indices and column indices run through G_i. We shall use the abbreviation M_{G_i} instead of M_{G_i, G_i}.

We shall say that the partition $(G_1, G_2, \ldots, G_q)$ carries out a partitioning of the matrix M in generalized blocks.

The following properties of the operators E_{G_i}'s are direct consequences of Definition 1.

i) $E_{G_i}^T E_{G_i} = I_{n_i}$ for $i \in \mathbb{N}_q$ and $E_{G_i}^T E_{G_j} = 0$ for $i, j \in \mathbb{N}_q, i \neq j$.

ii) Let $i \in \mathbb{N}_q$ and $x \in \mathbb{R}^{n_i}$. The vector $y \in \mathbb{R}^n$ defined by $y := E_{G_i} x$
     verifies $y_{G_j} = \begin{cases} x \text{ for } j = i \\ 0 \text{ for } j \neq i \end{cases}$ for $j \in \mathbb{N}_q$.

iii) Let $i, j \in \mathbb{N}_q$ and $X \in \mathbb{R}^{n_i \times n_j}$. The matrix $Y \in \mathbb{R}^{n \times n}$ defined by
     $$Y := E_{G_i} X E_{G_j}^T \text{ verifies } Y_{G_k, G_l} = \begin{cases} X \text{ for } k = i \text{ and } l = j \\ 0 \quad \text{otherwise} \end{cases} \text{ for } k, l \in \mathbb{N}_q.$$

iv) The partitioning in generalized blocks of the identity matrix is

$$I = \sum_{i=1}^{q} E_{G_i} E_{G_i}^T.$$

v)   For any vector $x \in \mathbb{R}^n$ we get the following partitioning in generalized blocks

$$x = \sum_{i=1}^{q} E_{G_i} E_{G_i}^T x = \sum_{i=1}^{q} E_{G_i} x_{G_i}.$$

vi) The partitioning in generalized blocks of any matrix $M \in \mathbb{R}^{n \times n}$ is

$$M = \sum_{i,j=1}^{q} E_{G_i} M_{G_i, G_j} E_{G_j}^T.$$

Remark 3. Whenever the partition is induced by $\{G, \bar{G}\}$ we shall replace G_1 and G_2 respectively by G and $\bar{G}$ in Definition 1.

Let us illustrate the partitioning of a matrix in generalized blocks.

Example 8. We consider the matrix $M = \begin{bmatrix} 1 & 2 & 3 & 4 \\ 5 & 6 & 7 & 8 \\ 9 & 10 & 11 & 12 \\ 13 & 14 & 15 & 16 \end{bmatrix}$.

Let $G_1 = \{1, 3, 4\}$, $G_2 = \{2\}$ be a partition of $\mathbb{N}_4$.

We get $E_{G_1} = \begin{bmatrix} 1 & 0 & 0 \\ 0 & 0 & 0 \\ 0 & 1 & 0 \\ 0 & 0 & 1 \end{bmatrix}$, $E_{G_2} = \begin{bmatrix} 0 \\ 1 \\ 0 \\ 0 \end{bmatrix}$.

The submatrices of the corresponding partitioning of M are

$$M_{G_1} = \begin{bmatrix} 1 & 3 & 4 \\ 9 & 11 & 12 \\ 13 & 15 & 16 \end{bmatrix}, \ M_{G_1,G_2} = \begin{bmatrix} 2 \\ 10 \\ 14 \end{bmatrix}, \ M_{G_2} = \begin{bmatrix} 6 \end{bmatrix}, \ M_{G_2,G_1} = \begin{bmatrix} 5 & 7 & 8 \end{bmatrix}.$$

If we have $x = \begin{bmatrix} 1 \\ 2 \\ 3 \\ 4 \end{bmatrix}$ then $x_{G_1} = \begin{bmatrix} 1 \\ 3 \\ 4 \end{bmatrix}$, $x_{G_2} = \begin{bmatrix} 2 \end{bmatrix}$.

If we have $x = \begin{bmatrix} 1 \\ 2 \\ 3 \end{bmatrix}$ then $E_{G_1}x = \begin{bmatrix} 1 \\ 0 \\ 2 \\ 3 \end{bmatrix}$.

Let $X = \begin{bmatrix} 1 & 2 & 3 \\ 4 & 5 & 6 \\ 7 & 8 & 9 \end{bmatrix}$. We get $E_{G_1} X E_{G_1}^T = \begin{bmatrix} 1 & 0 & 2 & 3 \\ 0 & 0 & 0 & 0 \\ 4 & 0 & 5 & 6 \\ 7 & 0 & 8 & 9 \end{bmatrix}$.

We can decompose the matrix M in generalized blocks

$$M = \sum_{i,j=1}^{2} E_{G_i} M_{G_i,G_j} E_{G_j}^T$$

$$= \begin{bmatrix} 1 & 0 & 3 & 4 \\ 0 & 0 & 0 & 0 \\ 9 & 0 & 11 & 12 \\ 13 & 0 & 15 & 16 \end{bmatrix} + \begin{bmatrix} 0 & 2 & 0 & 0 \\ 0 & 0 & 0 & 0 \\ 0 & 10 & 0 & 0 \\ 0 & 14 & 0 & 0 \end{bmatrix} + \begin{bmatrix} 0 & 0 & 0 & 0 \\ 0 & 6 & 0 & 0 \\ 0 & 0 & 0 & 0 \\ 0 & 0 & 0 & 0 \end{bmatrix} + \begin{bmatrix} 0 & 0 & 0 & 0 \\ 5 & 0 & 7 & 8 \\ 0 & 0 & 0 & 0 \\ 0 & 0 & 0 & 0 \end{bmatrix}.$$

3 Generalized Block Iterative Methods

Assume that we have to solve the linear system $Ax = b$ with $A \in \mathbb{R}^{n \times n}$, non-singular, and $b \in \mathbb{R}^n$.

Classic block iterative methods are based on a block partitioning of the matrix A carried out *before the beginning* of the solution process. We are about to describe an iterative method using a *dynamic* partitioning of the matrix A. This partitioning in generalized blocks is performed *during* the solution process. We are inspired by the PR method (cf. Algorithm 1) whose convergence speed we aim to accelerate. The PR method updates only one component of the approximation vector at each iteration. In our method we choose several of its components and we update them such that the corresponding components of the residual vector become zero. Algorithm 2 is a first formulation of our generalized block iterative methods. We use the notations given in Definition 1.

Algorithm 2 First formulation of the generalized block iterative method with dynamic partitioning of the matrix system

---

1: choose a vector $x^{(0)}$ and compute the corresponding residual $r^{(0)} = b - Ax^{(0)}$
2: **while** not converged **do**
3:     at the $(p + 1)$th iteration,
4:     choose a subset G of $\mathbb{N}_n$ and give $x_G^{(p+1)}$ a value such that:
5:         - the components of $x^{(p)}$ with indices in $\bar{G}$ are preserved,
6:             i.e. $x^{(p+1)} = E_G x_G^{(p+1)} + E_{\bar{G}} x_{\bar{G}}^{(p)}$
7:         - the components of $r^{(p+1)} = b - Ax^{(p+1)}$ with indices in G become zero,
8:             i.e. $r_G^{(p+1)} = 0$
9: **end while**

Remark 4. We emphasize that Algorithm 2 may be used with an existing partitioning of the system matrix but it does not impose this *static* partitioning. The choice of the subset G at line 4 of Algorithm 2 – number of elements and choice of these elements – may be done at each iteration: the matrix partitioning is thus *dynamic*. The criterion used to make this choice characterizes the iterative method.

Remark 5. The point Gauss–Seidel method, the PR method (using a static partitioning in blocks of size 1) are encompassed in this framework: at each step of any iteration of these methods we choose – cyclically for Gauss–Seidel – a singleton $G = \{i\}$ and we proceed exactly according to Algorithm 2.

The block Gauss–Seidel method can also be included in this framework: if the static partitioning of the system matrix is realized from the partition $\{G_1, \ldots, G_q\}$ of $\mathbb{N}_n$, then this method is obtained by choosing cyclically the G_i's at line 4 of Algorithm 2.

3.1 The Algorithm

According to Algorithm 2 we choose a subset G of $\mathbb{N}_n$ at the $(p + 1)$th iteration. Then we have to compute $x^{(p+1)} = E_G x_G^{(p+1)} + E_{\bar{G}} x_{\bar{G}}^{(p)}$ such that $r_G^{(p+1)} = 0$. Since $r_G^{(p+1)} = E_G^T r^{(p+1)}$ we can write

$$r_G^{(p+1)} = E_G^T r^{(p+1)} = 0 \iff E_G^T \left(b - Ax^{(p+1)} \right) = 0. \tag{1}$$

Replacing the vector $x^{(p+1)}$ in (1) by its imposed partitioning at line 6 of Algorithm 2 leads to

$$E_G^T \left(b - A \left(E_G x_G^{(p+1)} + E_{\bar{G}} x_{\bar{G}}^{(p)} \right) \right) = 0$$

$$\Longleftrightarrow E_G^T \left[b - A \left(\left(E_G x_G^{(p)} + E_{\bar{G}} x_{\bar{G}}^{(p)} \right) - E_G x_G^{(p)} + E_G x_G^{(p+1)} \right) \right] = 0. \qquad (2)$$

Using successively the partitioning in generalized blocks of a vector and the definition of the matrix A_G we can rewrite (2) as

$$E_G^T \left(b - A x^{(p)} + A E_G x_G^{(p)} - A E_G x_G^{(p+1)} \right) = 0$$

$$\Longleftrightarrow E_G^T r^{(p)} + \left(E_G^T A E_G \right) x_G^{(p)} - \left(E_G^T A E_G \right) x_G^{(p+1)} = 0$$

$$\Longleftrightarrow r_G^{(p)} + A_G x_G^{(p)} - A_G x_G^{(p+1)} = 0.$$

If the matrix A_G is non-singular we get

$$x_G^{(p+1)} = x_G^{(p)} + A_G^{-1} r_G^{(p)}. \qquad (3)$$

Since we required that $x_{\bar{G}}^{(p+1)} = x_{\bar{G}}^{(p)}$ by means of (3) we obtain

$$x^{(p+1)} = E_G x_G^{(p+1)} + E_{\bar{G}} x_{\bar{G}}^{(p)} = E_G \left(x_G^{(p)} + A_G^{-1} r_G^{(p)} \right) + E_{\bar{G}} x_{\bar{G}}^{(p)}$$

$$= \left(E_G x_G^{(p)} + E_{\bar{G}} x_{\bar{G}}^{(p)} \right) + E_G A_G^{-1} r_G^{(p)}. \qquad (4)$$

From (4) we derive the recurrence relation between two successive iterates

$$x^{(p+1)} = x^{(p)} + E_G A_G^{-1} E_G^T r^{(p)}. \qquad (5)$$

From (5) we easily obtain the recurrence relation which allows the computation of the iterative residual

$$r^{(p+1)} = b - A x^{(p+1)} = b - A x^{(p)} - A E_G A_G^{-1} E_G^T r^{(p)}$$

$$= \left(I - A E_G A_G^{-1} E_G^T \right) r^{(p)}. \qquad (6)$$

Multiplying the two sides of (6) by the matrix $E_{\bar{G}}^T$ gives

$$r_{\bar{G}}^{(p+1)} = r_{\bar{G}}^{(p)} - E_{\bar{G}}^T A E_G A_G^{-1} r_G^{(p)}$$

$$\Longleftrightarrow r_{\bar{G}}^{(p+1)} = r_{\bar{G}}^{(p)} - A_{\bar{G},G} A_G^{-1} r_G^{(p)}. \qquad (7)$$

From recurrence relations (3) and (7) we derive Algorithm 3 which is the general algorithm of a generalized block iterative method.

3.2 Convergence Study

We have to study three problems:

1. Which systems give non-singular matrices A_G for any G (cf. line 4 in Algorithm 3) ?
2. Under which conditions does the algorithm converge ?
3. When it converges, is its limit the solution to the system ?

Algorithm 3 Generalized block iterative method with dynamic partitioning of the system matrix

---

1: choose x and compute the residual $r = b - Ax$;
2: **while** not converged **do**
3:     choose a subset G of $\mathbb{N}_n$;
4:     solve the system $A_G s = r_G$ for s;
5:     $x_G = x_G + s$; {the components of x with indices in $\bar{G}$ remain unchanged}
6:     $r_{\bar{G}} = r_{\bar{G}} - A_{\bar{G},G} s$;
7:     $r_G = 0$; {the components of the residual with indices in G become zero}
8: **end while**

We shall study the convergence of a generalized block iterative method – defined by Algorithm 3 – when it is applied to the system $Ax = b$ where A is a non-singular M-matrix (cf. Definition 3). We are motivated by the fact that the radiosity matrix is a non-singular M-matrix (cf. [12, Proposition 3.4.7]).

For convenience we give the following definitions and notations:

Definition 2. A real matrix A is called non-negative if all its entries are non-negative. We write $A \geq 0$.
Let $A, B \in \mathbb{R}^{n \times n}$. If $A - B \geq 0$ we write $A \geq B$.
A vector $u \in \mathbb{R}^n$ is called non-negative if all its components are non-negative. We write $u \geq 0$.
Let $u, v \in \mathbb{R}^n$. If $u - v \geq 0$ we write $u \geq v$.

Definition 3. Let $A \in \mathbb{R}^n$. A is an M-matrix [2, 14] if there exists a real number s and a non-negative matrix B such that $A = sI - B$ with $s \geq \varrho(B) \geq 0$ where $\varrho(B)$ denotes the spectral radius of B.
If $s > \varrho(B)$ then A is a non-singular M-matrix.
Equivalently, a non-singular matrix A is an M-matrix if and only if its off-diagonal entries are non-positive and A^{-1} is non-negative.

The three following lemmas will be used in the proof of Theorem 8.

Lemma 5. *Let $A \in \mathbb{R}^{n \times n}$ be a non-singular M-matrix. For any subset G of $\mathbb{N}_n$ the matrix A_G is non-singular and A_G^{-1} is non-negative. Algorithm 3 can thus be applied to solve any system $Ax = b$ without any problem of inverse computation at line 4.*

Proof. Since A is a non-singular M-matrix any principal submatrix of A is a non-singular M-matrix (cf. [2, Theorem 2.3]). For any subset G of $\mathbb{N}_n$ we recall (cf. Definition 1) that A_G is a principal submatrix of A. Thus for any G the matrix A_G is a non-singular M-matrix. As a result A_G^{-1} is non-negative.

As A_G is non-singular the system $A_G s = r_G$ at line 4 of Algorithm 3 has an unique solution for any subset G. $\square$

Lemma 6. *Let $Ax = b$ be a system where $A \in \mathbb{R}^{n \times n}$ is a non-singular M-matrix and b a non-negative vector. Provided $x^{(0)}$ is equal to the null vector, Algorithm 3 gives a sequence of non-negative residuals $r^{(p)} := b - Ax^{(p)}$ and an increasing sequence of approximations $x^{(p)}$ (i.e. $p \geq q \Rightarrow x^{(p)} \geq x^{(q)}$).*

Proof. By induction we prove that the residual $r^{(p)}$ is non-negative for all $p \in \mathbb{N}$.

Since $x^{(0)}$ is the null vector we have

$$r^{(0)} = b \geq 0. \tag{8}$$

Assume $r^{(p)} \geq 0$ as an induction hypothesis. Since $G \bigcap \bar{G} = \emptyset$ the entries of the submatrix $A_{\bar{G},G}$ are off-diagonal entries of A. A is a non-singular M-matrix. Thus, its off-diagonal entries are non-positive, i.e. $A_{\bar{G},G} \leq 0$. Furthermore we proved (cf. Lemma 5) that $A_G^{-1} \geq 0$. As a result $-A_{\bar{G},G}A_G^{-1} \geq 0$. Since $r^{(p)} \geq 0$ we have $r_{\bar{G}}^{(p)} \geq 0$ and $r_G^{(p)} \geq 0$. From $-A_{\bar{G},G}A_G^{-1} \geq 0$ and (7) we deduce

$$\forall p \in \mathbb{N} \quad r^{(p)} \geq 0 \Rightarrow r^{(p+1)} \geq 0. \tag{9}$$

From (8) and (9) it follows by induction that the residual $r^{(p)}$ is non-negative for all $p \in \mathbb{N}$.

From the recurrence relation (3) we deduce that $x_G^{(p+1)} \geq x_G^{(p)}$ for all $p \in \mathbb{N}$. Since $x_{\bar{G}}^{(p+1)} = x_{\bar{G}}^{(p)}$ for all $p \in \mathbb{N}$, we conclude that $x^{(p+1)} \geq x^{(p)}$ for all $p \in \mathbb{N}$. Thus $\left(x^{(p)}\right)$ is an increasing sequence. □

Lemma 7. *Let $Ax = b$ be a system where $A \in \mathbb{R}^{n \times n}$ is a non-singular M-matrix and b a non-negative vector. Provided $x^{(0)}$ is equal to the null vector, Algorithm 3 gives a sequence $\left(x^{(p)}\right)$ bounded above by the solution x to the system. That is*

$$\forall p \in \mathbb{N} \quad e^{(p)} := x - x^{(p)} \geq 0$$

where $e^{(p)}$ is the error after the pth iteration.

Proof. We have $r^{(p)} = b - Ax^{(p)} = Ax - Ax^{(p)} = A(x - x^{(p)}) = Ae^{(p)}$. We deduce

$$e^{(p)} = A^{-1}r^{(p)}. \tag{10}$$

We proved that the residual $r^{(p)}$ is non-negative for all $p \in \mathbb{N}$ (cf. Lemma 6). Moreover, A being a non-singular M-matrix, we know that $A^{-1} \geq 0$ (cf. Definition 3). Equation (10) implies that $e^{(p)} \geq 0$ for all $p \in \mathbb{N}$. □

Now we can prove the following theorem.

Theorem 8. *Let $Ax = b$ be a system where $A \in \mathbb{R}^{n \times n}$ is a non-singular M-matrix and b a non-negative vector. If $x^{(0)} = 0$ then the sequence $\left(x^{(p)}\right)_{p \in \mathbb{N}}$ obtained from Algorithm 3 converges. Hence, there exists $c \in \mathbb{R}^n$, $c \geq 0$, such that*

$$\lim_{p \to \infty} r^{(p)} = c \tag{11}$$

with $r^{(p)} := b - Ax^{(p)}$.

Proof. Lemmas 6 and 7 state that the sequence $\left(x^{(p)}\right)_{p \in \mathbb{N}}$ is an increasing bounded above sequence in $\mathbb{R}^n$ (componentwise), hence, it is convergent. $\square$

It remains to find conditions which ensure that this limit c is equal to the solution to the system. We are about to prove a lemma which allows to conclude in some cases.

Lemma 8. *Let $Ax = b$ be a system where $A \in \mathbb{R}^{n \times n}$ is a non-singular M-matrix and b a non-negative vector. Let $\left(x^{(p)}\right)_{p \in \mathbb{N}}$ be the sequence provided by Algorithm 3 from $x^{(0)} = 0$. We denote by $G(p)$ the subset G of $\mathbb{N}_n$ chosen at the pth iteration. If $r^{(p-1)} = b - Ax^{(p-1)}$ is the $(p-1)$th residual then*

$$\lim_{p \to \infty} r_{G(p)}^{(p-1)} = 0.$$

Proof. We rewrite (5) in the following equivalent form

$$x^{(p)} - x^{(p-1)} = E_{G(p)} A_{G(p)}^{-1} E_{G(p)}^T r^{(p-1)} \text{ for } p \in \mathbb{N}. \tag{12}$$

By multiplying both sides of (12) by $E_{G(p)}^T$ we obtain

$$E_{G(p)}^T \left(x^{(p)} - x^{(p-1)}\right) = A_{G(p)}^{-1} r_{G(p)}^{(p-1)}. \tag{13}$$

Since the sequence $\left(x^{(p)}\right)_{p \in \mathbb{N}}$ converges we get

$$\lim_{p \to \infty} \left(x^{(p)} - x^{(p-1)}\right) = 0 \Rightarrow \lim_{p \to \infty} E_{G(p)}^T \left(x^{(p)} - x^{(p-1)}\right) = 0. \tag{14}$$

From (13) and (14) we deduce

$$\lim_{p \to \infty} A_{G(p)}^{-1} r_{G(p)}^{(p-1)} = 0. \tag{15}$$

The non-singular M-matrix A can be written (cf. Definition 3) $A = sI - B$ with $B \geq 0$ and $s > \varrho(B)$ where $\varrho(B) \geq 0$ denotes the spectral radius of B. Its principal submatrix $A_{G(p)}$ is also a non-singular M-matrix (cf. [2, Theorem 2.3]) and can be written

$$A_{G(p)} = E_{G(p)}^T A E_{G(p)} = E_{G(p)}^T sI E_{G(p)} - E_{G(p)}^T B E_{G(p)} = sI_{n_{G(p)}} - B_{G(p)}$$

where $n_{G(p)}$ is the cardinal number of the set $G(p)$. As a result

$$A_{G(p)} = s \left(I_{n_{G(p)}} - \frac{1}{s} B_{G(p)}\right). \tag{16}$$

$A_{G(p)}$ is a non-singular M-matrix. By means of [4, Theorem 1.4-5] we deduce from (16)

$$A_{G(p)}^{-1} = \frac{1}{s} \sum_{k=0}^{\infty} \left(\frac{1}{s} B_{G(p)}\right)^k = \frac{1}{s} I_{n_{G(p)}} + \frac{1}{s} \sum_{k=1}^{\infty} \left(\frac{1}{s} B_{G(p)}\right)^k. \qquad (17)$$

Since the matrix B is non-negative all its submatrices $B_{G(p)}$ and all their powers are non-negative. From (17) we derive

$$A_{G(p)}^{-1} \geq \frac{1}{s} I_{n_{G(p)}}. \qquad (18)$$

From Lemma 6 we know that the residual $r^{(p-1)}$ is non-negative. Therefore inequality (18) implies

$$\forall p \in \mathbb{N} \quad A_{G(p)}^{-1} r_{G(p)}^{(p-1)} \geq \frac{1}{s} r_{G(p)}^{(p-1)} \text{ with } s > 0. \qquad (19)$$

From (15) and (19) it follows immediately

$$\lim_{p \to \infty} r_{G(p)}^{(p-1)} = 0.$$

$\square$

Remark 6. Lemma 8 does not ensure that the sequence $\left(x^{(p)}\right)_{p \in \mathbb{N}}$ converges to the solution to the system for any choice of $G(p)$. For instance let G be a subset of $\mathbb{N}_n$. If we choose $G(p) = G$ for all p then we have $r_G^{(p)} = 0$ for all $p \in \mathbb{N}$. As a result the sequence $\left(x^{(p)}\right)_{p \in \mathbb{N}}$ is constant.

In this study we have not yet given a criterion for the choice of G at line 3 of Algorithm 3. In Sect. 4 we are about to define a method which uses a dynamic partitioning of the system matrix. Our goal is to define a new method for the radiosity, which

- is progressive like the PR method, i.e. after each iteration this method must allow an illumination of the scene whose quality increases with the number of iterations, and
- converges faster.

4 Application to Radiosity

4.1 The Generalized Block Progressive Radiosity Method

We recall (cf. Sect. 1.1) that the radiosity system can be written $\Phi x = b$ where Φ is a non-singular M-matrix of order n (cf. [12, Proposition 3.4.7]).

Let $k \in \mathbb{N}$ and let A be the diagonal matrix whose diagonal entries are the areas of the patches. At line 3 of Algorithm 3, at stage p, we have to choose a subset G of $\mathbb{N}_n$ of size k ($G = G(p)$ defined in Lemma 8). To stay close to the PR method for G we select the k indices corresponding to the k largest components of the vector $Ar^{(p-1)}$. We showed in Sect. 1.2 that in the radiosity context, the vector $Ar^{(p-1)}$ is interpreted as the unshot energy vector. Thus, in our algorithm, we choose the k patches which have the largest unshot energy. It may be noted that the PR method is recovered for $k = 1$.

Algorithm 4 is the algorithm of our method which we name Generalized Block Progressive Radiosity method (GBPR method).

Algorithm 4 The GBPR method

1: choose k;
2: $x = 0$ and $r = b - \Phi x = b$;
3: **while** not converged **do**
4:    for G select the k indices of the k largest components of the vector Ar;
5:    for all $i \in G$ compute the ith column of the matrix F;
6:    solve the system $\Phi_G s = r_G$ for s;
7:    $x_G = x_G + s$; {the components of x with indices in $\bar{G}$ remain unchanged}
8:    $r_{\bar{G}} = r_{\bar{G}} - \Phi_{\bar{G},G} s$;
9:    $r_G = 0$; {the components of the residual with indices in G become zero}
10:    display the scene with $x + r$ as the radiosity vector;
11: **end while**

Next we will prove

Theorem 9. *The GBPR method converges to the solution to the radiosity system $\Phi x = b$.*

Proof. As a result of our choice of $G(p)$ we have

$$\max_{j \in G(p)} \left(Ar^{(p-1)} \right)_j = \max_{j=1,\dots,n} \left(Ar^{(p-1)} \right)_j = \| Ar^{(p-1)} \|_\infty. \qquad (20)$$

As the radiosity matrix Φ is a non-singular M-matrix and as we required $x^{(0)} = 0$ and $b \geq 0$, from Lemma 8 we get

$$\lim_{p \to \infty} r^{(p-1)}_{G(p)} = 0. \qquad (21)$$

Since A is a diagonal matrix it follows from (21)

$$\lim_{p \to \infty} \left(Ar^{(p-1)} \right)_{G(p)} = 0 \Rightarrow \lim_{p \to \infty} \max_{j \in G(p)} \left(Ar^{(p-1)} \right)_j = 0. \qquad (22)$$

From (20) and (22) we deduce

$$\lim_{p \to \infty} \| Ar^{(p-1)} \|_\infty = 0 \Rightarrow \lim_{p \to \infty} Ar^{(p-1)} = 0. \qquad (23)$$

Since A is a diagonal matrix with positive entries (23) implies

$$\lim_{p \to \infty} r^{(p-1)} = 0 \Rightarrow \lim_{p \to \infty} x^{(p-1)} = x.$$

$\square$

Experimental Results. We begin with a description of our test scenes. The first one was used to illustrate the PR method in Sect. 1.2 and is showed Fig. 2(a). It is an office discretized in 11825 patches. The second one is a small empty maze (cf. Fig. 2(b)). Its purpose is to provide a scene with many occlusions implying a difficult propagation of the light between the objects. For such a scene the convergence speed of the PR method is known as very slow. This scene is discretized in 3744 patches.

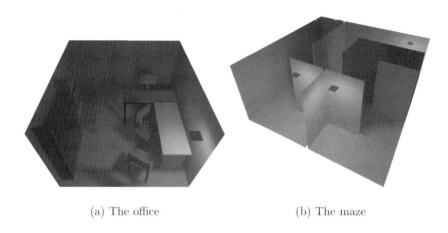

(a) The office   (b) The maze

Fig. 2. The two test scenes (Color Plate 13 on page 433)

In order to solve the system $\Phi_G s = r_G$ (cf. line 6 in Algorithm 4), we use an hybridization technique [3] (cf. Appendix 6). We showed in [13] that this technique is efficient for radiosity systems whenever their matrices can be stored.

We summarize the results of our tests in Fig. 3. The curves represent the normalized root mean square error (RMS error) as a function of the time. The RMS error after the pth iteration is given by the formula

$$RMS^{(p)} = \sqrt{\frac{\sum_{i=1}^{n} A_i \left(x_i^{(p)} - x \right)^2}{\sum_{i=1}^{n} A_i}}$$

where A_i is the area of the ith patch, $x_i^{(p)}$ is the ith component of the pth iterate and x is an accurate approximation to the solution obtained previously by any method.

The normalized RMS error is then defined by

$$RMS^{(p)}_{normalized} = \frac{RMS^{(p)}}{RMS^{(0)}}.$$

The maximal time represented in Fig. 3 is the time required for computing the whole matrix F. It would be very interesting to get a wanted minimal normalized RMS error after a time smaller than the one needed to compute F. It must be noted that a minimal normalized RMS error reduced to 5% – like the one observed on our curves – leads to images which still present some artifacts.

For all our tests we used a PC at 1.7 Ghz.

As seen on figures 3(a) and 3(b) the GBPR method does not converge really faster than the PR method does: from our experimental results we only got a 30% speedup. It appears that GBPR method cumulates some delay during its first iteration. Indeed an iteration of the GBPR method requires to compute k columns of the matrix F and to solve the corresponding subsystem. During the form factor computation none of the residual components changes and so does the convergence (see the stairs-like convergence curves for the GBPR method). Furthermore only a few patches of the scene are part of light sources. As a result only a few components of the initial residual vector are non-zero (cf. line 2 in Algorithm 4). Thus the first iteration of the GBPR method computes k form factor columns but only p of them (with $p \ll k$) correspond to non-zero components of the residual. Consequently, this reduces the efficiency of this iteration. Note that during this time, the PR method computes k form factor columns too, but for each one the corresponding component of the residual is non-zero, all the shooting being thus useful. Thus the GBPR method takes some delay from the beginning. This delay is especially important when the block size is large. This drawback appears clearly on figures 3(a) and 3(b) where small block size appears to provide the best convergence results.

Furthermore, since we have to calculate some columns of the matrix F at each iteration, it takes a while – dramatically increasing as the number of calculated columns increases –, between the display of two successive intermediate images. Thus with the GBPR method we lose a bit of progressiveness in comparison with the PR method (recall that with the PR method only one column must be calculated at each iteration).

In the next subsection, we introduce an approach whose goal is to accelerate, by means of generalized blocks, the convergence speed of the PR method without losing its progressiveness.

4.2 The Group Accelerated Shooting Method

The group accelerated shooting method (GAS method) aims to accelerate the convergence speed of the PR method. The basic idea in the GAS method is the storage of some columns of the form factor matrix F, computed during the PR process, with a later use in view.

Here is the solution process:

i)   choose k_0, $k_1 \in \mathbb{N}$,

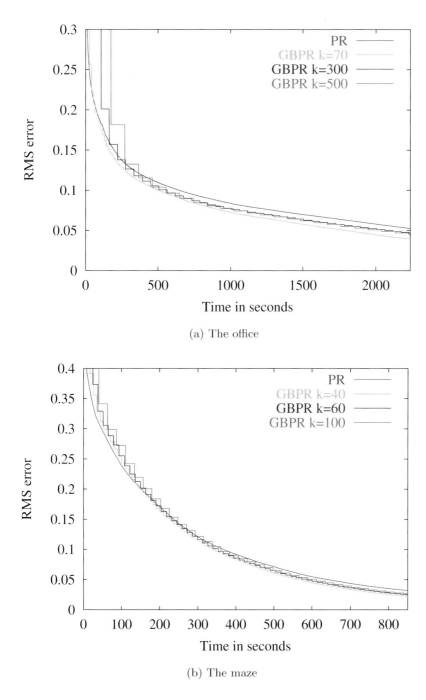

(a) The office

(b) The maze

Fig. 3. Comparison of PR and GBPR methods

ii) carry out iterations of the PR method – i.e. carry out iterations of the
     GBPR method with blocks of size 1 – and *store the computed columns of
     the matrix F*,

ii) when the number of PR iterations becomes larger than k_0, for G select
     k_1 indices among the indices of the stored columns of F and carry out
     an iteration of the GBPR method.

     Several criteria can be used for the choice of the subset G (cf. [16]).

From this description, it follows that this process is a generalized block
iterative method using cyclically two criteria at line 3 of Algorithm 3. We
describe it in Algorithm 5.

Algorithm 5 The GAS method

1: choose k_0, k_1;
2: $x = 0$ and $r = b$;
3: $k = 1$; $\{k$ is the counter of PR iterations$\}$
4: **while** not converged **do**
5:    **if** $k < k_0$ **then** $\{$PR method$\}$
6:       choose i such that $|A_i r_i| = \| Ar \|_\infty$ and set $G = \{i\}$;
7:       if not yet stored compute the ith column of F and store it;
8:       $k = k + 1$; $\{$increment the counter of PR iterations$\}$
9:    **else** $\{$GBPR method$\}$
10:      for G select k_1 indices among the indices of the stored columns of F;
11:      $k = 1$ $\{$reinitialize the counter of PR iterations$\}$
12:    **end if**
13:    solve the system $\Phi_G s = r_G$ for s;
14:    $x_G = x_G + s$; $\{$the components of x with indices in $\bar{G}$ remain unchanged$\}$
15:    $r_{\bar{G}} = r_{\bar{G}} - \Phi_{\bar{G},G} s$;
16:    $r_G = 0$; $\{$the components of the residual with indices in G are set to zero$\}$
17:    display the scene with $x + r$ as the radiosity vector
18: **end while**

We have to prove

Theorem 10. *The GAS method converges to the solution to the radiosity
system $\Phi x = b$.*

Proof. The radiosity matrix Φ is a non-singular M-matrix, hence, under the
conditions $x^{(0)} = 0$ and $b \geq 0$, the generalized block iterative method converges (cf. Theorem 8). It remains to prove that the limit of the sequence
$\left(x^{(p)}\right)$ is equal to x, the solution to the radiosity system.

The hypotheses of Lemma 8 are satisfied. This implies that

$$\lim_{p \to \infty} r_{G(p)}^{(p-1)} = 0 \tag{24}$$

where $G(p)$ denotes the subset G chosen at the pth iteration.

Let $m \in \mathbb{N}$. Algorithm 5 shows that the iterates $x^{(mk_0+1)}$ to $x^{((m+1)k_0-1)}$ are obtained from the PR method. As a result, at the $(mk_0 + 1)$th iteration the chosen subset $G = G(mk_0 + 1)$ is a singleton such that

$$\|(Ar)_{G(mk_0+1)}^{(mk_0)}\| = \max_{j=1,\ldots,n} \left|(Ar)_j^{(mk_0)}\right|. \tag{25}$$

By using (24), (25) and the fact that A is a diagonal matrix we get

$$\lim_{m\to\infty} (Ar)_{G(mk_0+1)}^{(mk_0)} = 0 \Rightarrow \lim_{m\to\infty} \left|(Ar)_{G(mk_0+1)}^{(mk_0)}\right| = 0$$

$$\Rightarrow \lim_{m\to\infty} \max_{j=1,\ldots,n} \left|(Ar)_j^{(mk_0)}\right| = 0. \tag{26}$$

Equation (26) implies

$$\forall j \in \mathbb{N}_n \quad \lim_{m\to\infty} \left|(Ar)_j^{(mk_0)}\right| = 0 \Rightarrow \lim_{m\to\infty} (Ar)^{(mk_0)} = 0. \tag{27}$$

Since A is a diagonal positive matrix it follows from (27) that

$$\lim_{m\to\infty} r^{(mk_0)} = 0. \tag{28}$$

At the beginning of this proof we showed that the sequence $\left(x^{(p)}\right)$ converges, hence there exists $c \in \mathbb{R}^n$ such that $\lim_{p\to\infty} r^{(p)} = c$. Equation (28) means that the subsequence $\left(r^{(mk_0)}\right)_{m\in\mathbb{N}}$ converges to 0. Consequently, for the whole sequence, we have $\lim_{p\to\infty} r^{(p)} = c = 0$: the sequence $\left(x^{(p)}\right)$ converges to x, the solution to the radiosity system.

$\square$

Experimental Results. In Fig. 4 we present the results obtained using the same scenes as for the GBPR method. Figures 4(a) and 4(b) include also the curves of the PR method and the curves of the GBPR method using the size of blocks that led to the better results in our previous tests.

Like we did in the tests of the GBPR method, we use the same hybridization technique to solve the system of line 13 in Algorithm 5. The criterion used to update the subset G at line 10 in Algorithm 5 is that which gives the best results as showed in [16].

The curves in Fig. 4 show that like the PR method the GAS method takes no delay at the beginning of the solution. We recall that this is a drawback of the GBPR method. In Fig. 4 we present three curves for the GAS method using different sizes of blocks (k_1 in Algorithm 5) and different frequencies for the use of GBPR iteration (k_0 in Algorithm 5). As k_1 increases, k_0 increases too in order to keep a small overhead of the GBPR iterations over the PR method. Moreover, applying too often the GBPR method with blocks of big size does not lead to good results: after an iteration of the GBPR method all the corresponding components of the residual vector are set to 0, so, if the frequency of use of the GBPR method is too high the next blocks would consist of patches with not much energy to shoot since only those in memory are used. The three curves show nearly the same results, which means that blocks of big size are not needed to obtain a good acceleration of the PR method. However the number of form factor columns stored in memory should be higher than the size of blocks to allow the building of efficient blocks.

The final RMS error of the GAS method for the office scene is 5 times lower than the one obtained with the PR method and 4 times lower than the one obtained with the GBPR method. The result obtained in the case of the maze scene is even more interesting since nearly total convergence is observed for the GAS method in the time needed to compute the whole radiosity matrix. This is explained by the fact that this scene includes many occlusions. The PR method does not obtain good results in such cases because only one patch is selected each time to shoot some energy. This implies that only few patches receive some energy at each iteration and that many iterations are needed to diffuse the light in the entire scene. On the contrary, each subsystem solution in the GAS method corresponds to the simultaneous shooting from several patches to the totality of patches seen from the block. This implies a better diffusion of light in scenes including a lot of occlusions.

With the GAS method we succeeded in cumulating the progressiveness of the PR method and the acceleration using generalized blocks. Building blocks from information stored in memory allowed us to use larger ones and more often than with the GBPR method. We have obtained a method which gives a good acceleration of the PR method especially with scenes including a lot of occlusions.

5 Conclusion and Perspectives

Due to the generally large size of the system matrix and the very high computational cost of its entries, classic methods – e.g. Gauss–Seidel, SOR, etc... – based on a static partitioning of this matrix are inadequate in the radiosity context.

That is the reason why we have introduced a new mathematical framework: generalized block iterative methods based on a dynamic partitioning of the system matrix. We proved that this approach leads to iterative methods which converge – not necessarily to the solution to the system – under some additional assumptions (A is an M-matrix, $x^{(0)} = 0$ and $b \geq 0$).

We have then presented, in the radiosity context, two methods based on this theoretical approach: GBPR and GAS methods. In GBPR method blocks are successively built from an energy criterion and solved, providing some speedup as compared to the progressive radiosity method. But this improvement is not very significant. Furthermore GBPR method reduces the

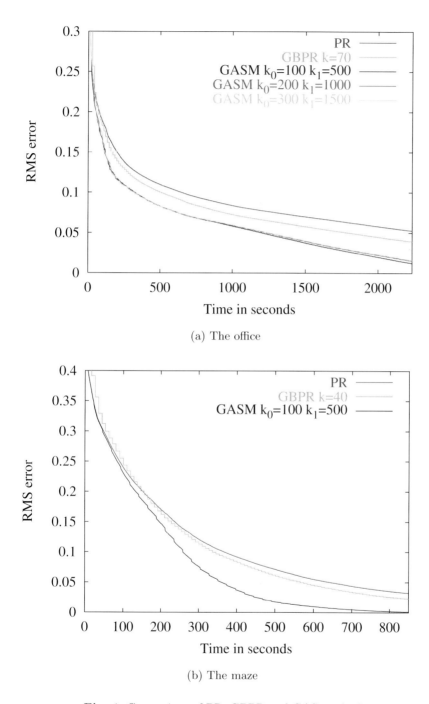

(a) The office

(b) The maze

Fig. 4. Comparison of PR, GBPR and GAS methods

frequency of the residual updates and consequently reduces the solver interactivity. The GAS method provides high speedup with the same rate of residual updates as the PR method; it alternates steps of PR method with steps of GBPR method. As a result, in some cases, it provides a nearly converged solution in a time smaller than the one that would be required for computing all the system matrix.

We are now studying the way of using dynamic sized blocks with GAS method because we observed that during the solution process the components of the residual vector decrease, reducing the efficiency of fixed size blocks. This efficiency could be improved by increasing the size of the blocks during the solution process. In this paper we restricted ourself to the classical PR method. Some enhancements of this method known as overshooting approaches [7, 23] provide generally high speedup to PR method. However it has been shown in [16] that GAS method is generally faster than overshooting methods, especially when the environment contains a large number of occlusions. We are currently investigating the ways of including overshooting in generalized block iterative methods. Our goal is to formalize this potential new approach, to prove its convergence and then to develop an efficient algorithm. The first results we obtained seems to indicate that the convergence speedup could still be increased by such a new approach.

The generalized block iterative method may be applied to hierarchical radiosity but is not interesting in all cases. If the entire hierarchical structure and all links can be stored, using generalized blocks will only speed up the solution time that represents only about 5 to 10% of the total computation time. In the case of very complex scenes where all links cannot be stored, hierarchical shooting can be used [22]. This method avoids to store all the links in memory and re-computes some of them when necessary. Since the use of generalized blocks reduces the number of iterations needed to reach the solution, in this case, it may not only reduce the solution time but also the number of links to re-compute. The use of generalized blocks may also find some interest in dynamic radiosity [6] because small modifications of the geometry of the scene generally means only small adjustments in the hierarchy. In this case the resulting solution time represents a more important part of the total computation time.

Application of generalized blocks to the stochastic iterative solvers of Monte Carlo radiosity would need further research. In these methods no explicit computation of form factors is done. Solution of the subsystems generated by the blocks would have to be adapted to the stochastic ray tracing scheme.

In conclusion, the generalized block iterative method is a new iterative solver that is not restricted to radiosity. It has been applied successfully to progressive radiosity and may be of some interest to hierarchical and Monte Carlo radiosity, but it may also be applied to any other problem using an iterative solver.

6 Acknowledgements

We would like to thank the anonymous reviewers for their constructive comments that were exceptionally detailed and insightful.

References

1. P. Bekaert. *Hierarchical and Stochastic Algorithms for Radiosity.* PhD thesis, Katholieke Universiteit Leuven, December 1999.
2. A. Berman and R. Plemmons. *Nonnegative Matrices in the Mathematical Sciences.* SIAM, Philadelphia, 1994.
3. C. Brezinski and M. Redivo Zaglia. Hybrid procedures for solving systems of linear equations. *Numerische Mathematik*, 67:1–19, 1994.
4. P. Ciarlet. *Introduction à l'analyse numérique matricielle et à l'optimisation.* Masson, 1990.
5. M. Cohen, S. Chen, J. Wallace, and D. Greenberg. A progressive refinement approach to fast radiosity image generation. *ACM Computer Graphics 22, 4(SIGGRAPH'88 Proceedings)*, pages 75–84, 1988.
6. G. Drettakis and F. Sillion. Interactive update of global illumination using a line-space hierarchy. In *Proceedings of SIGGRAPH 97*, pages 57–64, August 1997.
7. M. Feda and W. Purgathofer. Accelerating radiosity by overshooting. *Proceedings of the Third Eurographics Workshop on Rendering*, pages 21–32, May 1992.
8. S. Gortler, M. Cohen, and P. Slusallek. Radiosity and relaxation methods. *IEEE Computer Graphics and Animation*, 14(6):48–58, November 1994.
9. G. Greiner, W. Heidrich, and P. Slusallek. Blockwise refinement – a new method for solving the radiosity problem. In *Fourth Eurographics Workshop on Rendering*, pages 233–245, June 1993.
10. P. Hanrahan, D. Salzman, and L. Aupperle. A rapid hierarchical radiosity algorithm. In *Computer Graphics (SIGGRAPH'91 Proceedings)*, July 1991.
11. J. Hasenfratz, C. Damez, F. Sillion, and G. Drettakis. A practical analysis of clustering stragies for hierarchical radiosity. In R. Brunet and R. Scopigno, editors, *EUROGRAPHICS ' 99*, volume 18. Blackwell Publishers, 1999.
12. M. Leblond. *Propriétés des matrices de la radiosité. Application à la résolution du système de la radiosité.* PhD thesis, Université du Littoral Côte d'Opale, June 2001.
13. M. Leblond, C. Renaud, and F. Rousselle. Hybridization techniques for fast radiosity solvers. In IEEE Computer Society, editor, *Computer Graphics International 2000*, pages 269–278, Geneva, Switzerland, June 2000.
14. H. Minc. *Nonnegative Matrices.* Wiley Interscience, 1987.
15. L. Neumann, W. Purgathofer, R. Tobler, A. Neumann, P. Eliás, M. Feda, and X. Pueyo. The stochastic ray method for radiosity. In P. Hanrahan and W. Purgathofer, editors, *Rendering Techniques' 95*. Springer WienNewYork, 1995.
16. F. Rousselle. *Amélioration des méthodes de résolution utilisées en radiosité.* PhD thesis, Université du Littoral Côte d'Opale, December 2000.
17. Y. Saad. *Iterative Methods for Sparse Linear Systems.* PWS Publishing Company, 1996.

18. M. Shao and N. Badler. Analysis and acceleration of progressive refinement radiosity method. In *Fourth Eurographics Workshop on Rendering*, pages 247–258, Paris, France, June 1993.
19. P. Shirley. Radiosity via ray tracing. In J. Arvo, editor, *Graphics Gems II*, pages 306–310. Academic Press, San Diego, 1991.
20. F. Sillion. A unified hierarchical algorithm for global illumination with scattering volumes and object clusters. *IEEE Transactions on Visualization and Computer Graphics*, 1(3), September 1995.
21. R. Southwell. Stress-calculation in frameworks by the method of "systematic relaxation of constraints". *Proc. Roy. Soc. London*, A151 56–95;A153 41–76, 1935.
22. M. Stamminger, H. Schirmacher, P. Slusallek, and H. Seidel. Getting rid of links in hierarchical radiosity. In N. Ferreira and M. Göbel, editors, *EUROGRAPHICS ' 98*, volume 17. Blackwell Publishers, 1998.
23. W. Xu and D. Fussel. Constructing solvers for radiosity equation systems. *Proceedings of the Fifth Eurographics Workshop on Rendering*, pages 207–217, June 1994.

Hybridization

We assume that two iterative methods for solving the system $Ax = b$ respectively produce the sequences $\left(x_1^{(p)}\right)$ and $\left(x_2^{(p)}\right)$. The hybridization consists of constructing a new sequence $\left(y^{(p)}\right)$ by

$$y^{(p)} := a_p x_1^{(p)} + (1 - a_p) x_2^{(p)}.$$

If $r_1^{(p)} := b - Ax_1^{(p)}$, $r_2^{(p)} := b - Ax_2^{(p)}$ and $\varrho^{(p)} := b - Ay^{(p)}$ are the corresponding residuals we also have

$$\varrho^{(p)} = a_p r_1^{(p)} + (1 - a_p) r_2^{(p)}.$$

The parameter a_p is chosen to minimize the Euclidean norm of $\varrho^{(p)}$. Such an a_p is given by

$$a_p = -\frac{\left(r_1^{(p)} - r_2^{(p)}, r_2^{(p)}\right)}{\left(r_1^{(p)} - r_2^{(p)}, r_1^{(p)} - r_2^{(p)}\right)}$$

where $(\cdot, \cdot)$ represents the canonical innerproduct defined on $\mathbb{R}^n$.

Thanks to this minimization we get

$$\|\varrho^{(p)}\|_2 \leq \min\left(\|r_1^{(p)}\|_2, \|r_2^{(p)}\|_2\right).$$

This inequality shows that if one of the two methods converges then the hybrid method converges.

In [13] we showed that a hybridization technique derived from the Gauss–Seidel method is a good strategy in the radiosity context. At the pth iteration

$x_1^{(p)}$ is obtained by an iteration of Gauss–Seidel method using the hybrid iterate $y^{(p-1)}$ as an initial guess. For $x_2^{(p)}$ we take $y^{(p-1)}$. We may call this method hybridization of Gauss–Seidel with restart. We describe it in Fig. 5 and Algorithm 6.

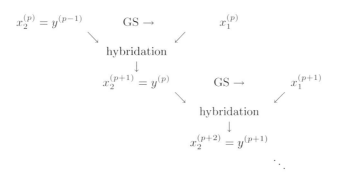

Fig. 5. Scheme of the hybridization of Gauss–Seidel with restart

---

Algorithm 6 Hybridization of Gauss–Seidel with restart

1: choose y
2: $\varrho = b - Ay$
3: **while** not converged **do**
4:    compute x_1 by means of the GS method using y as initial guess
5:    $r_1 = b - Ax_1$
6:    $z = r_1 - \varrho$
7:    $a = -\frac{(z,\varrho)}{(z,z)}$
8:    $\varrho = \varrho + az$
9:    $y = ax_1 + (1-a)y$
10: **end while**

Fast Difference Schemes for Edge Enhancing Beltrami Flow and Subjective Surfaces [*]

Ravi Malladi and Igor Ravve

Mail Stop: 50A-1148, 1 Cyclotron Road, Lawrence Berkeley National Laboratory, Computing Science Department, University of California, Berkeley, CA 94720. {*malladi,ravve*}@math.lbl.gov

Summary. Numerical integration of space-scale PDE is one of the most time consuming operation of image processing. The scale step is limited by conditional stability of explicit schemes. In this work, we introduce the unconditionally stable semi-implicit linearized difference scheme for the Beltrami flow. The Beltrami flow [14, 15] is one of the most effective denoising algorithms in image processing. For gray-level images, we show that the Beltrami flow equation can be arranged in a reaction-diffusion form. This reveals the edge-enhancing properties of the equation and suggests the application of additive operator split (AOS) methods [4, 5] for faster convergence. As we show with numerical simulations, the AOS method results in an unconditionally stable semi-implicit linearized difference scheme in $2D$ and $3D$. The values of the edge indicator function are used from the previous step in scale, while the pixel values of the next step are used to approximate the flow. The optimum ratio between the reaction and diffusion counterparts of the governing PDE is studied. We then apply this approach to fast subjective surface computation. The computational time decreases by a factor of 20 and more, as compared to the explicit scheme.

Keywords. Beltrami flow, subjective surfaces, unconditionally stable scheme, segmentation

1 Introduction

The main objective in early computer vision is to smooth images without destroying the semantic content, i.e. edges, features, corners, etc. In other words, the boundaries between objects in the image should survive as long as possible along the scale, while homogeneous regions should be simplified and flattened in a rapid way. This is particularly important since the denoising of an image is usually a precursor to segmentation and representation which often rely on edge fidelity. Improving the signal-to-noise ratio is also important in medical applications that employ the volume visualization method.

[*] This work was supported by the Director, Office of Science, Office of Advanced Scientific Research, Mathematical, Information, and Computational Sciences Division, U.S. Department of Energy under Contract No. DE-AC03-76SF00098, and LBNL Directed Research and Development Program

The use of diffusion equations for image processing in computer vision originated with the work of [1] where the authors pre-select a diffusion co-efficient function in the image that preserves the edge information. A more rigorous view was achieved with the realization that the iso-intensity contours of an image can be moved under their curvature following the work of Osher and Sethian [6]. This lead to a series of papers starting with the one by Alvarez, Lions and Morel [2], of viewing the images as a set of level contours and moving then under their curvature. Image smoothing by way of level set curvature motion [3, 8] thwarts the diffusion in the edge direction, thereby preserving the edge information. The work by Malladi and Sethian [8, 9] showed that in addition to this basic approach, a natural stopping criterion can also be chosen to prevent over smoothing a given image. For a comprehensive look at various approaches that rely on geometric diffusion, the reader is referred to [11]. Another crucial idea in [9] was also to diffuse an image by viewing it as a graph of a function and moving it under the mean curvature.

An important question still remained, what is the natural way to treat vector-valued images and images in higher dimensions? An answer to this and other questions was attempted in [14, 15] by Sochen, Kimmel, and Malladi. The result is a general mathematical framework for feature-preserving image smoothing that applies seamlessly to gray level, vector-value (color) images, volumetric images, and movies. The main idea is to view images as embedded maps between two Riemannian manifolds and to define an action potential that provides a measure on the space of these maps. The authors in [15] showed that many classical geometric flows emerge as special cases in this view as well as a new flow, the so called Beltrami flow that moves a gray level image under a scaled mean curvature, and also succeeds in finding a natural coupling between otherwise decoupled component-wise diffusion that was often used in the past in vector-valued image diffusion. In the case of gray value images, and by following a different approach, Yezzi in [22] arrived at a similar equation.

Smoothing of noisy images and edge enhancement usually presents a numerical integration of a parabolic PDE with one dimension in scale and two dimensions in space. This is often the most time consuming component of nonlinear image processing algorithms. Smoothing technique governed by the Beltrami flow equation is one of the most effective since it incorporates the edge indicator function, that minimizes diffusion at and across the edges and extensive diffusion elsewhere. On the other hand, solving the equations using explicit methods can be very time consuming due to the scaling and small time step requirement. We aim to build faster methods to solve the Beltrami flow equation in $2D$ gray level and color, and volumetric imagery. We use the method of Additive Operator Split (AOS). This technique was introduced by Weickert [4] for the nonlinear diffusion flow and later applied by Goldenberg et al. [5] to implement a fast version of the geodesic contour model.

On a different note, the Beltrami flow equation results from minimizing a (natural) generalization of $L2$ Euclidean norm to non-Euclidean manifolds, see [14] for details. This suggests that the governing equation is an "edge-preserving" in contrast to being an "edge-enhancing" flow. In other words, on grey level images, the equation simulates a mean curvature flow scaled by an edge-indicator function, thereby preserving the edge features in scale. In a recent note on the study of intermediate asymptotics of certain commonly used anisotropic diffusion equations, Barenblatt [10] reports that Beltrami flow equation forms a sharp step in the proximity of edges. In this work, an asymptotic self-similar solution was obtained for a particular case of the Beltrami equation, suggesting an "edge-sharpening" behavior. In the present context, while performing operator splitting, we confirm that Beltrami flow has both edge-preserving and edge-sharpening components.

Our goal is to build a fast and reliable method to solve the Beltrami flow equations and it is based on AOS technique. This work was first reported in Malladi and Ravve [7], where we discuss the main technique and apply the method to multidimensional gray level, and vector-valued imagery. In this paper, we first discuss the method as applied to $2D$ gray scale imagery and extend it to a related PDE, the governing equation for the subjective surface computation [23]. The paper is organized as follows. In Section 2, we rearrange the governing equation for the Beltrami flow. This approach leads to a semi-implicit linearized difference scheme. In Section 3, we present numerical simulations for $2D$ and $3D$ gray level images. We run the flow to different scales and different relative magnitudes of reaction term vs. diffusion components are considered. In Section 4 we apply the AOS technique to the subjective surface computation that is used to restore partially missing boundaries between segments. Section 5 presents the results of numerical simulation for completing missing boundaries. We summarize this work in Section 6.

2 Implicit Scheme for Beltrami Flow

Let us denote by (Σ, g) the image manifold and its metric and by (M, h) the space-feature manifold and its metric, then the map $X : \Sigma \to M$ has the following measure, [17]:

$$S[X^i, g_{\mu\nu}, h_{ij}] = \int d^m \sigma \sqrt{g} g^{\mu\nu} \partial_\mu X^i \partial_\nu X^j h_{ij}(X), \qquad (1)$$

where m is the dimension of Σ, g is the determinant of the image metric, $g^{\mu\nu}$ is the inverse of the image metric, the range of indices is $\mu, \nu = 1, \dots, \dim \Sigma$, and $i, j = 1, \dots, \dim M$, and h_{ij} is the metric of the embedding space. This is a natural generalization of the $L2$ norm to manifolds. As an example, a grey level image can be treated as a $2D$ manifold embedded in R^3, i.e. a mapping

$$X : (x, y) \rightarrow \left(X^1 = x,\ X^2 = y,\ X^3 = U(x, y)\right) \tag{2}$$

Many scale-space methods, linear and non-linear, can be shown to be a gradient descent flows of this functional with appropriately chosen metric of the image manifold. The gradient descent equation is

$$X_t^i = -\frac{1}{\sqrt{g}} \frac{\delta S}{\delta X^i}. \tag{3}$$

As shown in [14], minimizing the area action in Eq. (1), with respect to the feature coordinate U, we obtain the following Beltrami flow equation,

$$\dot{U} = \frac{U_{xx}\left(U_y^2 + 1\right) - 2U_x U_y U_{xy} + U_{yy}\left(U_x^2 + 1\right)}{\left(U_x^2 + U_y^2 + 1\right)^2}. \tag{4}$$

The nonlinear diffusion equation is the following reaction-diffusion partial differential equation [4, 5]

$$\dot{U} = \nabla \cdot \left(\frac{\nabla U}{g}\right) = \frac{\partial}{\partial x}\left(\frac{U_x}{g}\right) + \frac{\partial}{\partial y}\left(\frac{U_y}{g}\right) \tag{5}$$

where $g = (1 + U_x^2 + U_y^2)$ is the gradient magnitude.

The Beltrami equation may be reduced to a similar reaction-diffusion form, namely

$$\dot{U} = \nabla \cdot \left(\frac{\nabla U}{2g}\right) + \frac{\nabla^2 U}{2g} = h\nabla^2 U + 1/2\nabla h \cdot \nabla U \tag{6}$$

where $h = 1/g$, is the edge indicator function. In this form, the Beltrami flow equation is not a "pure" diffusion equation. It has both an (parabolic) edge-preserving and an (hyperbolic) edge-sharpening term. In addition, the reaction-diffusion form of Eq. (6) hides the mixed derivative U_{xy}, thereby making it conducive to the AOS approach. In other words, the equation can be rearranged into the form $\dot{U} = (A_x + A_y)U$, where A_x and A_y are the following differential operators:

$$A_x = \frac{\partial}{\partial x}\left(\frac{h}{2} \frac{\partial}{\partial x}\right) + \frac{h}{2} \frac{\partial^2}{\partial x^2} \qquad A_y = \frac{\partial}{\partial y}\left(\frac{h}{2} \frac{\partial}{\partial y}\right) + \frac{h}{2} \frac{\partial^2}{\partial y^2}. \tag{7}$$

Applying the backward difference formula to the above form we get,

$$\frac{U^{n+1} - U^n}{\Delta t} = (A_x + A_y)\, U^{n+1}. \tag{8}$$

The superscript n is related to the present and $n + 1$ to the next time step. The subscripts i, j index the discrete pixel location; $U_{i,j}^n$ are known values, and $U_{i,j}^{n+1}$ are to be found. Using U^{n+1} on the right side of Eq. (8) makes the integration scheme implicit and unconditionally stable, namely

$$[\mathbf{I} - \Delta t\,(\mathbf{A}_x + \mathbf{A}_y)\,]\,\mathbf{U}^{n+1} \; = \; \mathbf{U}^n \tag{9}$$

where $\mathbf{I}$ is the identity matrix. Before proceeding in time, we calculate the values of the edge indicator function h, using the known values of $\mathbf{U}^n$. Thus, the scheme is only semi-implicit. Although h depends on the gradient of U, we treat it like a given function of (x, y), making the governing PDE "quasi-linear".

Note that Eq. (9) includes a large bandwidth matrix, because all equations, related to new pixel values $\mathbf{U}^{n+1}$ are coupled. Our aim is to decouple the set (9) so that each row and each column of pixels can be handled separately. For this, we re-arrange the equations into the following form:

$$\mathbf{U}^{n+1} \; = \; [\mathbf{I} - \Delta t\,(\mathbf{A}_x + \mathbf{A}_y)\,]^{-1}\,\mathbf{U}^n. \tag{10}$$

Of course, we do not intend to invert the matrix to solve the linear set. This is only a symbolic form used for further derivation. For a small value of Δt, the matrix in the brackets on the right side of Eq. (10) is close to the identity $\mathbf{I}$. Thus, its inverse can be expanded into the Taylor series in the proximity of $\mathbf{I}$: $[\mathbf{I} - \Delta t\,(\mathbf{A}_x + \mathbf{A}_y)\,]^{-1} \approx \mathbf{I} + \Delta t\,(\mathbf{A}_x + \mathbf{A}_y)$, where the linear term is retained and the high order terms are neglected. Introducing this form into (10), we get,

$$2\mathbf{U}^{n+1} \; = \; (\mathbf{I} + 2\Delta t\mathbf{A}_x)\,\mathbf{U}^n \; + \; (\mathbf{I} + 2\Delta t\mathbf{A}_y)\,\mathbf{U}^n. \tag{11}$$

Introducing the notations $\mathbf{V} = (\mathbf{I} + 2\Delta t\mathbf{A}_x)\mathbf{U}^n$ and $\mathbf{W} = (\mathbf{I} + 2\Delta t\mathbf{A}_y)\mathbf{U}^n$ the solution is simply

$$\mathbf{U}^{n+1} \; = \; \frac{\mathbf{V} + \mathbf{W}}{2}. \tag{12}$$

In order to get an implicit scheme, we apply the differential matrix operators $\mathbf{A}_x$ and $\mathbf{A}_y$ to $\mathbf{U}^{n+1}$ (and not to $\mathbf{U}^n$), namely

$$(\mathbf{I} + 2\Delta t\mathbf{A}_x)^{-1}\,\mathbf{V} \; = \; \mathbf{U}^n \qquad\qquad (\mathbf{I} + 2\Delta t\mathbf{A}_y)^{-1}\,\mathbf{W} \; = \; \mathbf{U}^n. \tag{13}$$

Following the procedure of expanding the matrix inverses into Taylor series and applying the linearization for small Δt, we finally obtain the equation sets for $\mathbf{V}$ and $\mathbf{W}$ as follows:

$$(\mathbf{I} - 2\Delta t\mathbf{A}_x)\,\mathbf{V} \; = \; \mathbf{U}^n \qquad\qquad (\mathbf{I} - 2\Delta t\mathbf{A}_y)\,\mathbf{W} \; = \; \mathbf{U}^n \tag{14}$$

This leads to a following numerical scheme for $\Delta x = \Delta y = 1$:

$$-A_m \Delta t\,V_{i-1} + [\,1 + (A_m + A_p)\,\Delta t\,]\,V_i - A_p \Delta t\,V_{i+1} \; = \; U_i^n \tag{15}$$

where

$$A_m \; = \; \frac{h_{i-1} + 3h_i}{2} \qquad\qquad A_p \; = \; \frac{h_{i+1} + 3h_i}{2} \tag{16}$$

and a similar equation in y for W.

These equations can be solved with either Dirichlet or Neumann boundary conditions; these and other details are described in [18].

3 Simulation Results for Beltrami Flow

We ran a series of numerical simulations to demonstrate the performance of the implicit scheme for the Beltrami flow. We introduced an acceleration factor f that is defined as the ratio of the step size used in the implicit scheme to the maximum allowed step size for the explicit scheme. For a square grid, and assuming the pixels are a unit length apart, the maximum time step size, based on the CFL condition, for the explicit scheme is 0.25. We ran the scheme with values of f ranging between 1 and 200. The results are shown in Fig. 1. As we see, the implicit scheme is always stable, but for $f \gg 50$, the resulting accuracy may be insufficient for certain applications.

The next series of numerical simulations are carried out to study the edge enhancement effect on gray level images. We fix the acceleration factor to 10 and solve the following normalized reaction-diffusion equation in this series:

$$\frac{\partial U}{\partial t} = \cos\beta \ \nabla h \cdot \nabla U + \sin\beta \ h \ \nabla^2 U \tag{17}$$

The first term on the right side of Eq. (17) is a reaction term, while the second is a diffusion term and β is a parameter controlling the relative contribution of these opposing effects. The reaction term is responsible for edge enhancement, while the diffusion term smooths the noise away from the edges. Results of varying β between 0 to 90° is shown in Fig. 2. The first row in Fig. 2, presents the initial image and the results for $\beta = 0$ (pure reaction) and $\beta = 30^o$. The second row presents the results for $\beta = 45^o$ (a nonlinear diffusion flow equation), $\beta = \arctan 2 \approx 63.4^o$ (the Beltrami flow), and $\beta = 90^o$ (scaled "linear" diffusion). According to Eq. (17), the edge enhancement effect should decay with increasing β. Indeed, we see that the edge enhancement is stronger for the nonlinear diffusion flow ($\beta = 45^o$) than for the Beltrami flow ($\beta = 63.4^o$).

In Fig. 3, the implicit difference scheme in $3D$ for the Beltrami flow is applied to a volumetric image of the brain. The image consists of 124 slices of 256×256 images. In the first column of Fig. 3, we show the original slice # 20, and results of running the flow until scale $= 50$ and 100. The other two columns show exactly the same results for slice #50 and #100. An acceleration factor of 12 was applied in all the simulations. We note that the $3D$ Beltrami flow can also be arranged as follows:

$$\dot{U} = \nabla \cdot \left(\frac{h\nabla U}{2} \right) + \frac{h\nabla^2 U}{2} \tag{18}$$

where $h = 1/(1 + U_x^2 + U_y^2 + U_z^2)$ is the edge indicator function.

Note that due to the additional component U_z^2 in the denominator, the $3D$ Beltrami flow is slower than the corresponding $2D$ flow. This means that to achieve the same degree of noise reduction and edge enhancement, larger scale values should be employed in $3D$.

Fig. 1. Results of implicit difference scheme for Beltrami flow till scale $= 250$; the first row shows the initial image and results with $f = 1, 2$; second row corresponds to $f = 5, 10, 20$; third row corresponds to values of $f = 50, 100, 200$.

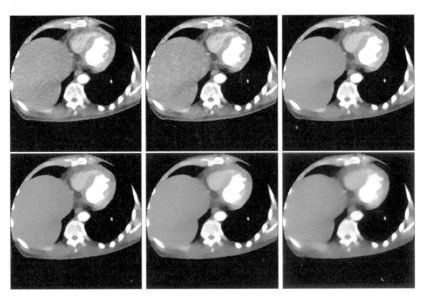

Fig. 2. Results of solving Eq. 17 until scale $= 250$ with different values of β.

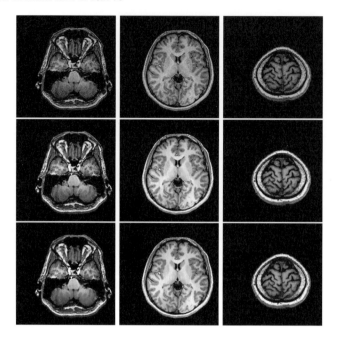

Fig. 3. Results of $3D$ Beltrami edge enhancing flow

4 Implicit Scheme for Subjective Surfaces

In this section we consider a special flow applied for segmentation of images with missing boundaries [23]

$$\dot{U} = g\,\boldsymbol{K}U + \nabla g \cdot \nabla U \tag{19}$$

The "completion" flow in Eq. (19) includes two functions to consider: an external edge function $g(x,y)$ independent of time, and a time-varying gradient function $g(x,y,t)$. External edge function is defined as

$$g(x,y) = \cfrac{1}{1 + \cfrac{G_x^2 + G_y^2}{\beta^2}} \tag{20}$$

where $G(x,y,t=\sigma)$ is the initial image with missing boundaries $G(x,y,t=0)$ subjected to a linear pre-smoothing of scale σ:

$$\dot{G} = \nabla^2 G \qquad\qquad 0 \leq t \leq \sigma \tag{21}$$

and $beta$ in Eq. (20) is a gradient scaling factor. $\boldsymbol{K}U$ is a two-dimensional differential operator, the curvature flow:

$$KU = \frac{U_{xx}\left(U_y^2 + 1\right) - 2U_xU_yU_{xy} + U_{yy}\left(U_x^2 + 1\right)}{U_x^2 + U_y^2 + 1} \tag{22}$$

With the aid of the internal gradient function

$$h = \frac{1}{\sqrt{U_x^2 + U_y^2 + 1}} \tag{23}$$

the curvature flow can be split to the additive form:

$$KU = \frac{\nabla \cdot (h\,\nabla U)}{h} = \frac{1}{h}\left[\frac{\partial\,(h\,U_x)}{\partial x} + \frac{\partial\,(h\,U_y)}{\partial y}\right] \tag{24}$$

The completion flow of Eq. (19) becomes:

$$\dot{U} = (A_x + A_y)\,U \tag{25}$$

where

$$\begin{cases} A_x\,U = \dfrac{g}{h}\dfrac{\partial\,(h\,U_x)}{\partial x} + \dfrac{\partial\,(g\,U_x)}{\partial x} - g\,\dfrac{\partial^2 U}{\partial x^2} \\[2mm] A_y\,U = \dfrac{g}{h}\dfrac{\partial\,(h\,U_y)}{\partial y} + \dfrac{\partial\,(g\,U_y)}{\partial y} - g\,\dfrac{\partial^2 U}{\partial y^2} \end{cases} \tag{26}$$

For $\Delta x = \Delta y = 1$, this leads to semi-implicit linearized numerical scheme, described by Eq. (15) with

$$A_m = g_{i-1} + g_i\,\frac{h_{i-1}}{h_i} \qquad A_p = g_{i+1} + g_i\,\frac{h_{i+1}}{h_i}. \tag{27}$$

Eq. (25) is solved with the Dirichlet boundary condition along the contour of the computational box. Initially, $U\,(x, y, t = 0)$ is an inverse distance function to a point or to a finite sector of a straight line:

$$U_o = \frac{\alpha}{\sqrt{D^2\,(x, y) + \Delta_S^2}} \tag{28}$$

where D is the distance function, α is the initial scaling factor, and Δ_S is the smoothing parameter. The value on the boundary is equal to the minimum of U_o.

To achieve the optimum accuracy for minimum computation time, we apply non-uniform scale step, assuming that more coarse steps are allowed at the initial stage of marching and finer steps are required towards the end. The steps in scale make a decreasing geometric progression. The last scale step is kept the same for all problems considered, and it corresponds to the acceleration factor 10. The average acceleration factor f changes at every step, and is constrained to be greater than 10. The average acceleration factor yields the same total number of steps N as if this factor was kept constant:

$$N = \frac{4\,S}{f} \tag{29}$$

where S the total scale of marching.

5 Simulation Results for Completing Missing Boundaries

In this Section, we consider four benchmark examples for completing partially missing edges. Dimension of images is 256×256. Recall that α is the initial scaling factor, Δ_S is the initial smoothing parameter, β is the gradient scaling factor, σ is the scale of linear pre-smoothing, and f is the acceleration factor, i.e. ratio of the *average* scale step to the "standard" step. The standard step is the stability threshold value for the explicit numerical scheme of the linear differential equation of Gaussian smoothing: $\dot{U} = \nabla^2 U$. For the unit grid in x and y, the standard step is $1/4$.

Kanizsa Triangle is presented in Fig. 4. $\alpha = 1 \cdot 10^8$, $\Delta_S = 5, \beta = 1 \cdot 10^{-3}$, $\sigma = 0$, $f = 25$. The initial surface is plotted, and the restored surface after 40, 100 and 1600 iterations.

Kanizsa Square is presented in Fig. 5. $\alpha = 1 \cdot 10^6$, $\Delta_S = 5, \beta = 1 \cdot 10^{-3}$, $\sigma = 1 \cdot 10^{-1}$, $f = 100$. The surface is plotted, after 0, 10, 20 and 300 iterations.

"Corners" are presented in Fig. 6. $\alpha = 1 \cdot 10^6$, $\Delta_S = 5, \beta = 1 \cdot 10^{-3}$, $\sigma = 1 \cdot 10^{-1}$, $f = 150$. The surface is plotted, after 0, 5, 10 and 200 iterations.

"Circle" is presented in Fig. 7. $\alpha = 1 \cdot 10^6$, $\Delta_S = 5, \beta = 1 \cdot 10^{-3}$, $\sigma = 1 \cdot 10^{-1}$, $f = 100$. The surface is plotted, after 0, 5, 10 and 400 iterations.

6 Closing Remarks

The anisotropic Beltrami operator for gray level images is reduced to reaction-diffusion form and this makes it possible to apply the Additive Operator Split (AOS) approach. Based on this approach, the unconditionally stable difference scheme is developed. The method uses the known values of the edge indicator function from the previous scale step, and incorporates the unknown pixel values from the next step, thus making the difference scheme semi-implicit and linearized. The implicit scheme leads to considerable saving computing time as compared to the explicit scheme: up to ten times and even more, depending on the value of the scale step, with no visible loss of accuracy. The approach may be applied also to the mean curvature flow in a similar way, with a different edge indicator function. A similar splitting technique is developed for the computation of subjective surfaces and applied for completing missing boundaries between segments.

Acknowledgments

We thank Dr. A. Sarti, University of Balogna, for his kind help in certain implementation details of explicit schemes for subjective surface computation.

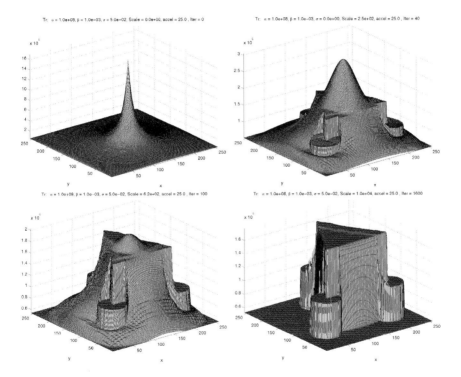

Fig. 4. Kanizsa Triangle

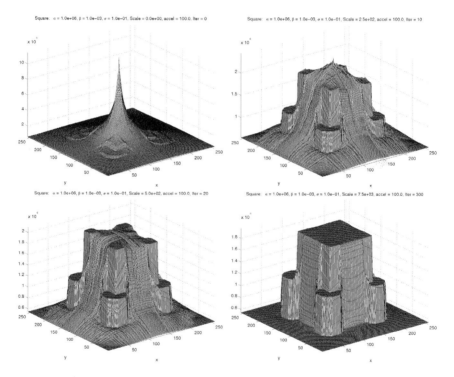

Fig. 5. Kanizsa Square

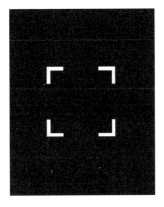

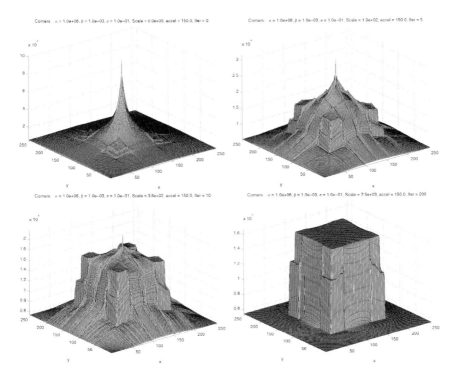

Fig. 6. Corners

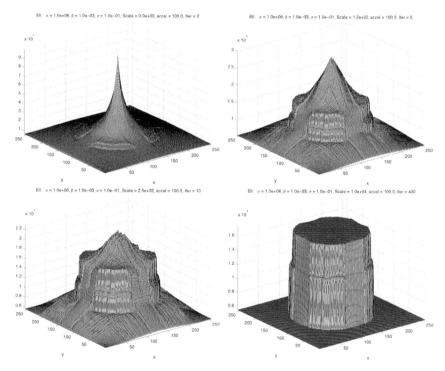

Fig. 7. Circle

References

1. P. Perona, J. Malik, "Scale-space and edge-detection using anisotropic diffusion," IEEE Trans. of PAMI, Vol. 12, pp. 629–639, 1990.
2. L. Alvarez, P.L. Lions, J.M. Morel, "Image selective smoothing and edge detection by non-linear diffusion II," SIAM Journal on Numerical Analysis, Vol. 29(3), pp. 845–866, 1992.
3. L. Rudin, S. Osher, and E. Fatemi, "Nonlinear total variation based noise removal algorithms," Physica D, Vol. 60, pp. 259–268, 1992.
4. J. Weickert, B.M. ter Haar Romeny, and M.A. Viergever. "Efficient and reliable scheme for nonlinear diffusion filtering". *IEEE Trans on Image Processing*, Vol. 7(3), pp. 398-410 (1998).
5. R. Goldenberg, R. Kimmel, E. Rivlin, and M. Rudzsky. "Fast Geodesic Active Contours". M. Nielsen, P. Johansen, O.F. Olsen, J. Weickert (Editors), Scale-space theories in computer vision, Lecture Notes in Computer Science, Vol. 1682, Springer, Berlin, 1999.
6. S. Osher and J. A. Sethian, "Fronts propagating with curvature dependent speed: Algorithms based on Hamilton-Jacobi formulation," Journal of Computational Physics, Vol. 79, pp. 12-49, 1988.
7. R. Malladi and I. Ravve, "Fast difference schemes for edge-enhancing Beltrami flow," in *Proceedings of ECCV*, LNCS 2350, pp. 343–357, Copenhagen, Denmark, May 2002.
8. R. Malladi and J. A. Sethian, "Image processing via level set curvature flow," Proc. of Natl. Acad. of Scie., Vol. 92, pp. 7046–7050, July 1995.
9. R. Malladi and J. A. Sethian, "Image Processing: Flows under Min/Max curvature and Mean curvature," Graphical Models and Image Processing, Vol. 58(2), pp. 127–141, 1996.
10. G.I. Barenblatt. "Self-Similar Intermediate Asymptotics for Nonlinear Degenerate Parabolic Free-Boundary Problems which Occur in Image Processing". to appear in *Proceedings of the National Academy of Science*, 2001.
11. B. M. ter Harr Romeny, editor, Geometry-driven diffusion in computer vision, Kluwer Academic publishers, The Netherlands, 1994.
12. R. Kimmel, R. Malladi, and N. Sochen. "Images as embedded maps and minimal surfaces: movies, color, texture, and volumetric medical images". *International Journal of Computer Vision*, 1999.
13. R. Kimmel, "A natural norm for color processing," in *Proceedings of 3-rd ACCV*, Springer-Verlag LNCS 1352, pp. 88–95, Hong-Kong, January 1998.
14. N. Sochen, R. Kimmel and R. Malladi, "From high energy physics to low level vision," LBNL Report # 39243, Lawrence Berkeley National Laboratory, University of California, Berkeley, Aug. 1996.
15. N. Sochen, R. Kimmel and R. Malladi, "A general framework for low level vision". *IEEE Transactions*, Vol. 7, No. 3, 1998.
16. G. Sapiro, "Vector-valued active contours," in *Proceedings of CVPR '96*, pp. 650–655, 1996.
17. A. M. Polyakov, "Quantum geometry of bosonic strings," in *Physics Letters B*, 103B(3), pp. 207–210, 1981.
18. R. Malladi and I. Ravve, "Fast difference scheme for anisotropic Beltrami smoothing and edge contrast enhancement of gray level and color images," LBNL report # 48796, Lawrence Berkeley National Laboratory, University of California, Berkeley, August, 2001.

19. J. Weickert. *Anisotropic Diffusion in Image Processing.* Ph. D. Thesis, Kaiser-slautern University, 1996.

20. Curtis E. Gerald and Patric O. Wheatley. *Applied Numerical Analysis*, Addison Wesley, NY, 1999.

21. William H. Press, Saul A. Tekolsky, William T. Vetterling and Brian F. Flan-nery. *Numerical Recipes in C*, Cambridge University Press, 1992

22. A. Yezzi, "Modified curvature motion for image smoothing and enhancement," IEEE Tran. of Image Processing, Vol. 7, No. 3, 1998.

23. A. Sarti, R. Malladi, and J. A. Sethian, "Subjective surfaces: A method for completing missing boundaries," Proc. Natl. Acad. Sci. USA, Vol. 97, No. 12, 2000.

Part V

Software Environments and Applications

ALICE on the Eightfold Way: Exploring Curved Spaces in an Enclosed Virtual Reality Theater

George K. Francis[1], Camille M.A. Goudeseune[2], Henry J. Kaczmarski[2], Benjamin J. Schaeffer[2], and John M. Sullivan[1]

[1] Department of Mathematics, University of Illinois, Urbana, IL, USA, *gfrancis@uiuc.edu, jms@uiuc.edu*
[2] Integrated Systems Laboratory, University of Illinois, Urbana, IL, USA, *cog@uiuc.edu, kacmarsk@uiuc.edu, schaeffr@uiuc.edu*

Summary. We describe a collaboration between mathematicians interested in visualizing curved three-dimensional spaces and researchers building next-generation virtual-reality environments such as ALICE, a six-sided, rigid-walled virtual-reality chamber. This environment integrates active-stereo imaging, wireless motion-tracking and wireless-headphone sound. To reduce cost, the display is driven by a cluster of commodity computers instead of a traditional graphics supercomputer. The mathematical application tested in this environment is an implementation of Thurston's eight-fold way; these eight three-dimensional geometries are conjectured to suffice for describing all possible three-dimensional manifolds or universes.

Keywords. cluster architecture, curved spaces, virtual reality

1 Introduction

Successful visualization projects involve two pieces, scientists with interesting content and facilities that provide the services needed to realize that content. In this paper we explore a collaboration between mathematicians interested in visualizing curved three-dimensional spaces and researchers building next-generation virtual reality environments. In many ways, the ultimate test of new visualization technology is putting it into the hands of scientists and making it useful for their projects.

The Integrated Systems Laboratory (ISL) at the University of Illinois (UIUC) provides hardware and software tools to researchers in a variety of disciplines. This particular collaboration involves a six-sided, rigid-walled virtual reality (VR) chamber, the Adaptive Laboratory for Immersive Collaborative Experiments (ALICE), which has been built by the ISL over the last year. This is a traditional VR environment, integrating active stereo, wireless tracking of a user's position and orientation, and wireless-headphone sound. To reduce cost, the display is driven by a cluster of commodity computers instead of a traditional graphics supercomputer. To meet the software challenges of the cluster environment, the lab has a toolkit for cluster-based VR, called Syzygy, under active development. A preliminary version of Syzygy was provided to the mathematicians to aid in porting their software to the cluster.

We describe several aspects of the project, starting with the challenges of constructing the physical facility. Our collaboration was a case study in porting code from a shared-memory architecture (SGI Onyx) to a cluster architecture (Syzygy). Finally, we describe the tangible benefits of a fully enclosed environment for visualizing curved three-dimensional spaces; this project could not work as well in a less immersive environment.

We gratefully acknowledge the programming assistance of Ben Bernard and Matt Woodruff, without whose work this project would not have been possible.

Fig. 1. A fisheye view of the ALICE VR theater, highlighting five of the six mirrors and projectors. The view is from near the sixth projector, which aims at the sliding door (halfway open in this photograph).

2 Other Fully Enclosed Virtual Reality Theaters

The fully enclosed systems built so far are cubes about 3 meters on a side. Binocular images are rendered via "active stereo", which rapidly alternates left-eye and right-eye projected images which are then directed to each eye by similarly flickering LCD shutter glasses.

ALICE is driven by a cluster of commodity desktop computers instead of a large Silicon Graphics Onyx computer. ALICE has rigid walls instead of fabric, which is prone to flutter and sag when made this large.

Similar theaters we know of include the following:

- The C6 at Iowa State University's Virtual Reality Applications Center [1] uses wireless motion tracking.
- The VR-CUBE at KTH Royal Institute of Technology's Center for Parallel Computers (Stockholm) [2] uses tethered motion tracking.
- The HyPI-6 system is located in the Virtual Reality Laboratory of the Fraunhofer Institute for Industrial Engineering [3]. In one configuration it uses a cluster of PC's, two per wall, to draw images with passive stereo.
- The COSMOS system at the VR Techno Center in Gifu (Japan) [4] has a motorized sliding rear wall. Motion tracking is tethered (cables go through a small hole at one of the corners of the floor). Dual CRT projectors on each face of the cube double the brightness of the images viewed.

3 Cluster Architecture

A PC cluster architecture provides a cost effective means of driving multiple graphics pipes but has disadvantages over a shared-memory SMP architecture. With SMP, shared memory is an efficient means of communication between the rendering processes, the single instance of the operating system can be used to easily manage the software, and time-tested software libraries, such as the CAVELib [5, 6], exist to aid the applications programmer. Many of these substantial advantages disappear when considering a PC cluster architecture. Communication between rendering processes can now become a bottleneck because it occurs over a network instead of a memory bus. Managing the different application components spread across the network can be a challenge, especially if they are running under different operating systems. Finally, there is no standard library for writing VR software for the cluster architecture.

Fortunately, there are several interesting software projects that may address this in the near future. Of particular note for the VR community is the NetJuggler [7] extension to VR Juggler [8]. This allows multiple copies of a VR Juggler program (satisfying certain conditions) to run synchronized across a cluster. Another set of tools to run multiple synchronized copies of an application comes from the Princeton Omnimedia group [9]. While their focus is not on VR, their software could be easily adapted. Similarly, the WireGL effort [10], which replaces the OpenGL shared library with a stub that sends data over a network for rendering, is a distributed graphics library with potential for VR applications.

The projects just described take one of two paths: distributing the application, as in the case of NetJuggler, or distributing the data, as in the case of WireGL. Distributing the application, namely running synchronized copies of an application on multiple nodes, can use far less network bandwidth than a data distribution strategy. This advantage is balanced, however, by the restrictions that typical synchronization schemes place on application design. For instance, multithreading might not be allowed, or the programmer might be constrained to change application state only in certain ways, for instance through interfaces to input devices like trackers or mouse/keyboard.

The Integrated Systems Lab is conducting ongoing research into creating a software toolkit for virtual reality on PC clusters. A very early version of the library was used to drive a head-mounted display in an human-factors study at ISL [11]. The toolkit's current incarnation is called Syzygy, is licensed under the GNU LGPL, and can be freely downloaded from the web [12]. While the software is still changing and many architectural decisions are as yet unmade, the lab has made a prototype available to help researchers port their visualizations to the PC cluster environment. This is the software backbone supporting the visualization of curved three-dimensional spaces described in this paper.

In order to produce graphics for ALICE, the six-sided VR theater at ISL, Syzygy controls six PCs running Windows NT with Wildcat 4210 graphics cards. Each PC produces an active stereo image which is viewed using standard shutter glasses. It is important that all the video images be produced in complete synchrony, with their vertical refresh locked together by means of an external *genlock* signal. The Wildcat card's ability to do this is critical for us, since active stereo across multiple displays will not work without such a feature.

Instead of trying to wrap existing libraries and hide the parallelism inherent in the PC cluster, Syzygy attempts to expose the parallelism and provide the programmer with tools to manage it. Since distributing the application tends to require fewer network resources than distributing data, the lab focused on building tools to help programmers write applications that can run synchronized across a network. As such, Syzygy does not address questions of load-balancing. The same application runs on each rendering node, with the only differences being the different point-of-view for each screen and the division of the nodes into a master and slaves. The master node is in charge of I/O and managing the synchronization of the group.

The application distribution framework for Syzygy assumes that a program can be conceptualized as a infinte loop with four phases: precomputation, data exchange between the master application instance and the slaves, post-computation, and synchronization. A simple API lets the programmer place the data-exchange and synchronization points wherever desired. The API also provides ways to determine which node is the master, to interface with a tracker, to configure the different nodes for rendering the different walls of a cube, and to compute different viewing frustums based on head position.

To use the data-exchange API, the programmer first defines the format of the packet that will be broadcast from the master to the slaves. For some applications, this can be extremely simple, possibly even a single navigation matrix, as is the case with the visualization of three-dimensional geometries described in this paper. At the data-exchange point, the master application packs this data packet and sends it to each slave. At this point, the slave applications, conversely, receive the data packet and unpack it into local storage. The data-exchange API is built on a lower layer of software that handles the details of the network connection, the data format conversion between different machine architectures, and other chores.

The underlying synchronization infrastructure needs to accomplish two goals. The computers doing the rendering not only need to produce a sequence of frames in lockstep but also need to make sure that they are drawing a consistent world at each point in time. The Syzygy synchronization primitive is used to guarantee that the data sent by the master at a given data exchange step is used by all rendering nodes to draw the next frame; this ensures consistency. The same synchronization call is used to control graphics buffer swaps. Without external genlock, the control would be imperfect. Out-of-phase vertical refreshes on the displays would mean that buffer swaps occur in a slightly staggered fashion. However, external genlock of the display video cards eliminates this problem.

Synchronization in Syzygy is accomplished by a single API call, sync(). When a slave calls sync(), it sends the master a packet via UDP and then blocks while waiting for a response. When the master calls sync(), it waits until all slaves have requested release from the barrier and then broadcasts a UDP packet to them. The master now continues, as do the slaves when they receive the broadcast packet. In GLUT-based programs, like the mathematical visualizations described here, it makes sense to put the sync() call just before the call to glutSwapBuffers(). In practice, this synchronization method yields very good coherence even on graphics hardware without genlock capabilities. Experiments with highly animated scenes on a 2×2 video wall run by non-genlocked PCs show extremely small amounts of jitter. Indeed, the jitter is only perceptible to viewers within $3\,m$ of a $2\,m \times 1.5\,m$ image. When graphics hardware with genlock capabilities is used, as in ALICE, the synchronization is perfect. No jitter whatsoever can be perceived. One should note that this level of quality is achieved using 100Mbps Ethernet and not some more exotic networking technology.

Another challenge met by the Syzygy toolkit is in management of the distributed system. Our PC cluster is heterogeneous, with the rendering nodes running Windows NT, some nodes controlling interface devices running Linux and some running Windows, and additional support nodes running Linux. Syzygy provides a remote-execution daemon that works on either Windows or Unix. It also provides an interface to send messages to running processes, which is used to provide a kill signal and also to make running applications reload their parameters. This simple interface provides a way to start, stop, and generally manage distributed applications on the PC cluster. To aid in running visualizations while enclosed in a VR theater, the lab has developed a Java interface to these functions that can run on a wireless handheld device. In this way, one can cycle through visualizations without leaving the CAVE and moving to a console; this is very convenient.

Recently, the ISL conducted an open house of its VR facility ALICE. Our visualization of three-dimensional hyperbolic space was one of the demos shown. The lab hosted almost 400 visitors over a two-day period. Eleven PC's were part of the distributed system, and that system stayed running the entire time. At any particular time, about 20 software components were running, dispersed across the system but cooperating in producing a visualization. Over the two-day period, over 10 000 seperate software components were started, connected to the system, and terminated once their work was over.

We now briefly describe the specific implementation in Syzygy of the mathematical visualizations described below. The only communication needed between the nodes is a description of the viewer's position and orientation. For the simplest geometries this is encoded as a 4×4 matrix expressing a projection from a representation of the curved space into ordinary Euclidean space (technically, into the observer's tangent space) [13]. (For the more complicated geometries slightly more information, perhaps a pair of matrices, will be needed [14].) During the pre-computation phase, the master node retrieves information from a wireless joystick and uses this to change the 4×4 matrix. The data-exchange phase distributes this matrix to all the slave nodes. In the post-computation phase, each rendering node applies the given projection and then applies the camera transformation appropriate for the point-of-view of the wall it controls. Finally, synchronization occurs and the loop repeats.

4 Mathematical Visualization of Three-dimensional Geometries

Topology is the study of deformable shapes. Topologically, a surface is anything that locally looks like a patch of the Euclidean plane. The surfaces of a round ball, a cylindrical can or a rectangular box are all topologically equivalent, since one could be deformed into another. This topological space is called the two-sphere and its most symmetric geometric realization is as a round sphere $\mathbb{S}^2$, with constant (positive) curvature.

The most interesting topological surfaces are those which, like the sphere, are compact (not infinite) but complete (with no edges). Topologically, the surfaces of a donut, an inner-tube, or a mug with a handle are also equivalent; this compact, complete space is called the two-torus. One might think that the round inner-tube gives the nicest geometry for a torus, but, in fact, by going outside the realm of geometries obtained from surfaces embedded in our ordinary three-dimensional world, we can find a much more symmetric geometry for the torus.

From video-games, we are all familiar with the notion of a wrap-around screen. When a game token flies off the top or left edge of the screen, it reappears at the bottom or right edge and proceeds at the appropriate angle. The rectangular screen, with these identifications of opposite sides, becomes topologically a torus. And it clearly has (from the original rectangle) a flat geometry, that of the Euclidean plane $\mathbb{E}^2$.

Let us ignore non-orientable surfaces, like the Möbius band, where a left-handed object can become right-handed merely by traveling around some loop on the surface. It is then a classical result [15] that any orientable, compact, complete surface other than the sphere and torus must be a multiple torus. A multiple torus is a surface with more than one handle, obtained by connecting tori together.

We have seen that the sphere and the torus each admit nice geometries (round and flat, respectively). To put an equally symmetric geometry on a multiple torus requires the use of hyperbolic geometry, $\mathbb{H}^2$, the non-Euclidean geometry discovered in the 1800s by Lobachevskii, Bolyai and Gauss. It is not hard to explicitly write down a hyperbolic geometry (with constant negative curvature) for each multiple torus. (A deep result of Poincaré and Klein, called the uniformization theorem, says that any geometric shape for a surface is in fact conformally equivalent to one of the symmetric ones we have described.)

What about three-dimensional worlds, called three-manifolds? Locally, a 3-manifold looks like a block of Euclidean three-space. Our own world is part of some 3-manifold, though we don't know which one [16]. We know from Einstein's theory of relativity that our world is curved, albeit very gently. If we could explore our universe geometrically, unconstrained by physical limitations such as the speed of light, we might find that it closes on itself like a sphere, or a torus, or some more complicated possibility. Although the classification of surfaces is relatively easy, the classification of possible 3-manifolds remains an active area of mathematical research. Many interesting open problems remain, including the Poincaré conjecture. This says there are no "fake" three-spheres, and carries a million-dollar prize for its solution [17].

Again, the three-dimensional sphere admits a nice round geometry, called $\mathbb{S}^3$, obtained when it bounds a round ball in four-dimensional space. And a three-dimensional torus admits a flat geometry modeled on $\mathbb{E}^3$, obtained by identifying opposite faces of a cubical room. Again, "most" 3-manifolds have a hyperbolic geometry, $\mathbb{H}^3$.

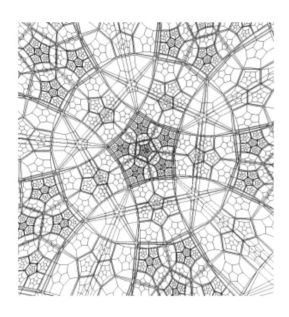

Fig. 2. The view of a sample hyperbolic manifold, with geometry $\mathbb{H}^3$, as seen on one wall of ALICE. This hyperbolic space is tiled by right-angled dodecahedra, meeting 8-to-a-vertex, just as Euclidean space can be tiled with cubes. (Color Plate 7 on page 429)

But some three-manifolds admit no nice geometry. In general, one must first perform a decomposition (along two-spheres and two-tori) into so-called irreducible and atoroidal pieces. The famous Geometrization Conjecture of Bill Thurston says that each such piece should carry a nice geometry; it has been proved for many classes of manifolds, and seems likely to be true.

Thurston's list of possible geometries [18, 19], however, is an eight-fold one. Some manifolds admit one of the isotropic geometries $\mathbb{S}^3$, $\mathbb{E}^3$ or $\mathbb{H}^3$ mentioned above. Although hyperbolic geometry is in some sense again the generic case, for 3-manifolds (unlike for surfaces) even spherical geometry can be used for infinitely many different manifolds [20], and there are several Euclidean possibilities.

For certain other 3-manifolds, however, the nicest geometry they can carry is nonisotropic, though still homogenous. There are two product geometries, $\mathbb{S}^2 \times \mathbb{E}^1$ and $\mathbb{H}^2 \times \mathbb{E}^1$, which are relatively easy to understand. The remaining three geometries on Thurston's list are the twisted geometries of the three-dimensional Lie groups Nil, Sol and $\widetilde{\mathrm{SL}_2}\mathbb{R}$. These Lie groups arise in many other areas of mathematics and physics, especially Nil, also known as the Heisenberg group.

It turns out that the isotropic geometries can all be modeled by the 4×4 projection matrices built into graphics systems like OpenGL, even though these systems were built only for $\mathbb{E}^3$. This fact was investigated by Char-

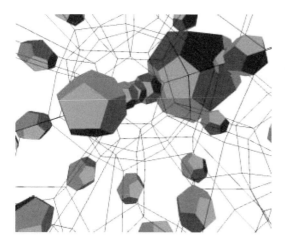

Fig. 3. The view of a sample spherical manifold, with geometry $\mathbb{S}^3$, as seen on one wall of ALICE. This space, formed by identifying opposite faces of a dodecahedron, is almost a counter-example to the Poincaré conjecture. An inhabitant of this space would see 120 repeating images of each object in the space, as in this regular tiling of the sphere by 120 dodecahedra. (Color Plate 8 on page 429)

lie Gunn and Mark Phillips at the Geometry Center [21, 13, 22], and we demonstrated it in the four-sided CAVEs at SIGGRAPH'94 [23]. Jeff Weeks has written an exposition for computer graphics programmers, describing the surprisingly simple mathematics underlying these models of the isotropic geometries [24] and his software for exploring spherical, flat, and hyperbolic spaces at home on an ordinary PC is freely available [25].

The anisotropic product geometries can be implemented without too much more difficulty, using 5×5 matrices. The three twisted geometries do have matrix representations [14], but these are not necessarily efficient. Real-time exploration of these fascinating worlds presents a challenge of considerable mathematical importance.

Our ten-year experience from frequent public demonstrations of hyperbolic and spherical geometry in a conventional (4-walled) CAVE suggests that the illusion presented in that CAVE is not adequate to truly experience non-Euclidean geometry. Too much has to be explained to visitors, and too few questions are asked. Out of the corners of their eyes, people see the empty space above, and the dimly lit room to the rear of the convential CAVE.

The exotic geometries, being anisotropic, are more subtle. Physical or geometric laws there do depend on orientation, or, to misquote George Orwell, some directions are more equal than others. We look forward to implementing these in ALICE, since the sensation of navigating these anisotropic worlds will be totally different from any previous experience.

Our experience in the conventional CAVE has, however, proved that the gravity-based sense of an absolute "down" can be fleetingly overcome by the visual suggestion to the contrary. In an illusion without "above" and "rear" to confirm the gravitational "down" we expect to overcome the Euclidean reference frame, and truly immerse visitors in the exotic three-dimensional geometries.

References

1. Virtual Reality Applications Center. C6. Iowa State University, Ames. http://www.vrac.iastate.edu
2. Center for Parallel Computers. The PDC cube. Royal Institute of Technology, Stockholm. http://www.pdc.kth.se/projects/vr-cube
3. Fraunhofer IAO. HyPI-6. Fraunhofer Institute for Industrial Engineering, Stuttgart. http://vr.iao.fhg.de/6-Side-Cave/index.en.html
4. K. Fujii, Y. Asano, N. Kubota, and H. Tanahashi. User interface device for the immersive 6-screens display COSMOS. Sixth International Conference on Virtual Systems and Multimedia, 2000. http://wv-jp.net/got/media/COSMOS/
5. C. Cruz-Neira, D.J. Sandin, T.A. DeFanti, R.V. Kenyon, and J.C. Hart. The CAVE: Audio-Visual Experience Automatic Virtual Environment. *Communications ACM*, **35**(6):65–72, 1992.
6. C. Cruz-Neira, D.J. Sandin, and T.A. DeFanti. Surround-screen projection-based virtual reality: The design and implementation of the CAVE. *Computer Graphics (Proc. SIGGRAPH '93)*, 1993.
7. J. Allard, L. Lecointre, V. Gouranton, E. Melin, and B. Raffin. Net Juggler. http://www.univ-orleans.fr/SCIENCES/LIFO/Members/raffin/SHPVR/NetJuggler.php
8. A. Bierbaum, C. Just, P. Hartling, and C. Cruz-Neira. Flexible application design using VR Juggler. *SIGGRAPH*, 2000. Conference Abstracts and Applications.
9. Y. Chen, H. Chen, D. W. Clark, Z. Liu, G. Wallace, and K. Li. Software environments for cluster-based display systems. 2001. http://www.cs.princeton.edu/omnimedia/papers.html
10. G. Humphreys and P. Hanrahan. A distributed graphics system for large tiled displays. *IEEE Visualization*, 1999.
11. B. Schaeffer. A Software System for Inexpensive VR via Graphics Clusters. 2000. http://www.isl.uiuc.edu/ClusteredVR/paper/dgdpaper.pdf
12. Integrated Systems Laboratory. Syzygy. University of Illinois, Urbana. http://www.isl.uiuc.edu/ClusteredVR/ClusteredVR.htm
13. M. Phillips and C. Gunn. Visualizing hyperbolic space: Unusual uses of 4×4 matrices. In *Symposium on Interactive 3D Graphics (SIGGRAPH)*, **25**:209–214, New York, 1992.
14. E. Molnar. The projective interpretation of the eight 3-dimensional homogogeneous geometries. *Beiträge zur Algebra und Geometrie*, **38**(2):261–288, 1997.
15. J.R. Weeks and G.K. Francis, Conway's ZIP proof. *Amer. Math. Monthly*, **106**(5):393–399, 1999.
16. J.R. Weeks. *The Shape of Space*. Dekker, 1985.
17. Clay Mathematics Institute. The Poincaré conjecture. http://www.claymath.org/prizeproblems/poincare.htm
18. P. Scott. The geometries of 3-manifolds. *Bull. London Math. Soc.*, **15**:401–487, 1983.
19. W. P. Thurston. *Three-Dimensional Geometry and Topology*. Princeton University Press, Princeton, New Jersey, 1997.
20. E. Gausmann, R. Lehouq, J.-P. Luminete, J.-P. Uzan, and J. Weeks. Topological lensing in spherical spaces. 2001. http://www.arXiv.org/abs/gr-qc/0106033

21. C. Gunn. Visualizing hyperbolic geometry. In *Computer Graphics and Mathematics*, pp 299–313. Eurographics, Springer Verlag, 1992.

22. C. Gunn. Discrete groups and visualization of three-dimensional manifolds. *Computer Graphics (Proc. SIGGRAPH '93)*, 255–262, 1993.

23. G. Francis, C. Hartman, J. Mason, U. Axen, and P. McCreary. Post-Euclidean walkabout. In *VROOM – the Virtual Reality Room*. SIGGRAPH, Orlando, 1994.

24. J. Weeks. Real-time rendering in curved spaces. *IEEE Computer Graphics and Applications* **22**(6):90–99, 2002.

25. J. Weeks. Curved spaces software.
    http://www.northnet.org/weeks/CurvedSpaces/

Computation and Visualisation in the NumLab Numerical Laboratory

Joseph M.L. Maubach and Alexandru C. Telea

Eindhoven University of Technology, Department of Mathematics and Computer Science, Postbox 513, NL-5600 MB Eindhoven, The Netherlands.
{*j.m.l.maubach, a.c.telea*} *@tue.nl*

Summary. A large range of software environments addresses numerical simulation, interactive visualisation and computational steering. Most such environments are designed to cover a limited application domain, such as Finite Elements, Finite Differences, or image processing. Their software structure rarely provides a simple and extendible mathematical model for the underlying mathematics. Assembling numerical simulations from computational and visualisation blocks, as well as building such blocks is a difficult task.

The NumLab environment, a numerical laboratory for computational and visualisation applications, offers a basic, yet generic and efficient framework for a large class of computational applications, such as partial and ordinary differential equations, non-linear systems, matrix computations and image and signal processing. Building applications which combine interactive visualisation and computations is provided in an interactive visual manner.

This paper focuses on the efficient implementation of one of the most complex NumLab components, the Finite Element assembler for systems of equations, such as Stokes or Navier-Stokes fluid-flow equations. It shows how the software framework as a whole has been targeted towards fast assemblers, and how a general purpose fast Finite Element assembler is embedded.

1 Introduction

As pointed out in [1], the NumLab (Numerical Laboratory) environment has been constructed after a thorough search through a wide range of software environments for numerical computation and data visualisation. The NumLab design goals included a seamless integration of computation and visualisation, as well as convenient computer aided application construction. Furthermore, all components should be customizable and nevertheless fast.

Paper [1] explains the design concepts in detail: The mathematics to be modeled (Sect. 2), a software framework which is one to one with the mathematical model (Sect. 3), and discusses the implementation of a Navier-Stokes Finite Element solver within the NumLab software framework (Sect. 4).

Over time, the large scope of NumLab's mathematical model – which covers numerical methods for ODEs, PDEs, non-linear systems, and all possible combinations – has induced a number of questions. Most questions have either been on mathematical framework (addressed in [1]), the technical realisation (the construction of libraries is addressed in [2]), and on overall efficient *and* customizable implementations, addressed in this paper.

This paper shows how NUMLAB's most complex component (the Finite Element assembler for non-linear systems) can be implemented such that it is both fast and customizable. On the small-detail level, introducing general purpose techniques, we demonstrate how NUMLAB's Finite Element assembler can be efficient (fast). The introduced techniques can also be used for Finite Difference and Volume assemblers. Next, we show the construction of an efficient *and* customizable Finite Element assembler for systems of equations.

The NUMLAB environment consists of two parts. The first part is its content, i.e., its c | | libraries. Standard c++ programs can be written using NUMLAB's libraries (data and operations) and can be compiled as well as interpreted (interpretation using the CINT c++ interpreter [5]). The second part of NUMLAB is its interactive program creation and simulation tool called VISSION (see [11, 12]). With this tool, standard c++ programs can be composed using NUMLAB's libraries (data and operations) in a data-flow graphical environment. This environment is well-suited for numerical and visual simulations.

There are two categories of NUMLAB libraries: *base* and *derived* libraries.
    The base NUMLAB libraries are unaltered public domain and commercial (fortran, pascal, c, c++) libraries (binaries) *adorned with*:

1. a small c++ interface for the communication of standard data types between the different libraries;
2. a smaller c++ interface for computer aided program creation and simulation (see [2]);
3. an added so-called reflection layer (see [5]) for the c++ interpreter, and;
4. a smallest data-flow simulation interface.

There are libraries for computation – LAPACK [4], NAGLIB [20], or IMSL [16], SEPRAN [17] – as well as for professional visualisation – OpenGL [8], Open Inventor [14], or VTK [9]. The (parts of) libraries which are available is research-determined, for instance OpenDX [15] is not available in NUMLAB. The NUMLAB base libraries communicate with the use of standard data types, which have been quite stable for a period of years: Open Inventor and VTK types are used for visualisation, c++-wrapped BLAS types (Basic Linear Algebra Subprograms, part of LAPACK) are used for computation with full matrices, and in-house data types are used for sparse matrix operations. Also Boost data types (http://www.boost.org/) are used. At this moment, O(1000) data types and operators are available. The data types have readers and writers for VTK, LAPACK, Matlab, and sometimes LaTeX, OpenMath, and MathML formats.

The derived NumLab libraries implement fundamental mathematical notions such as operator and derived solver, operator_pde and operator_ode, as well as the data to operate at: linear vector spaces of functions space with elements Function. Similar mathematical concepts are factored out into similar orthogonal software components, so few components can be combined to powerful solution algorithms. An example is shown in Sect. 5. The derived NumLab libraries communicate with the data types Function and Operator.

In one aspect, the NumLab graphical editor vission resembles other graphical editors such as Matlab's Simulink [18], AVS [13], IRIS Explorer [3] or Oorange [7]. In each case, data and operations from libraries are represented as rectangles to be put on a canvas, and input and output arguments can be connected.

However, in contrast with the other graphical editors, VISSION steers a c++ interpreter. Thus all kinds of c++ expressions *could be* – and in a few cases are – added in adornment-level (4), in order to implement powerful features at minimal cost. Furthermore, the use of a single language for content (libraries) and management (graphical editor) simplifies the implementation of the NumLab workbench.

For a more detailed information and a comparison of NumLab with Matlab [18], Mathematica [19], Diffpack and SciLab [6, 21], see [1].

Uniting visualisation and numerical frameworks turns out to be powerful: element and derivatives of operators are visualised without effort because (1) elements support sampling and (2) derivatives of operators are represented using matrices (see Sect. 3). All Open Inventor and VTK functions (contour surfaces, probing, Fast Fourier transforms, etc.) are available for elements (which are vector-valued functions), and even mpeg generation modules are available in NumLab. Table 1 (Color Plate 20 on page 436) shows a few examples:

1. A matrix (an image): values determine the color;
2. Interactive computation of streamlines using Open Inventor probes;
3. Axes with names and labels in 2 and 3 dimensions;
4. Computer aided design;
5. Feature detection in complex 2 and 3 dimensional vector (flow) fields;
6. Computer aided design (extrusion) post-processes numerical output;
7. A matrix (linear operator): values determine height and color;
8. A matrix: sparsity pattern;
9. Automatic generation of editors to edit network default values;
10. A VTK-generated tetrahedral grid;
11. Contour surfaces;
12. Streamlines.

Table 1. NUMLAB visualization of various computational results.

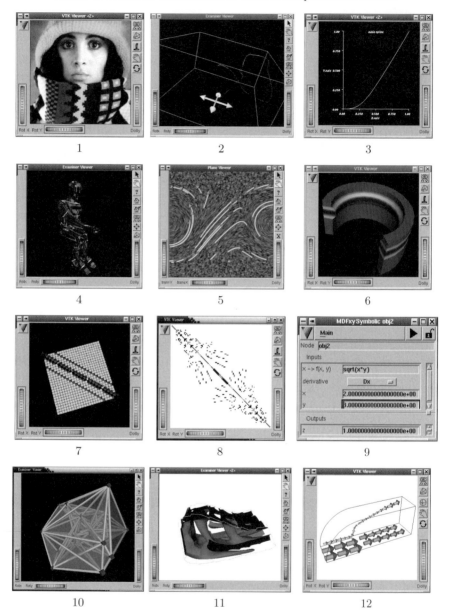

The remainder of this paper is organised as follows. Sect. 2 mentions the mathematics which can be modeled with the NUMLAB software framework. Next, Sect. 3 formulates the NUMLAB software framework. Each component and its mathematical concept is described. Further, Sect. 4 comments on the efficient implementation of a Finite Element Solver for systems of equations such as the Navier-Stokes equations. Then Sect. 5 shows how to construct simulation applications from the modules introduced in Sect. 3, using the graphical editor. Sect. 6 provides conclusions and discusses future developments.

2 The Mathematical Framework

The NUMLAB software framework models the mathematical notions *element* of a linear vector space, and *operator* on an element of such a space. In particular, elements are linear combinations (of functions), and are vector valued. The supported operations are vector space operations and function evaluations:

1. The addition of elements
2. The scaling of elements
3. The evaluation at elements $u = F(v)$ resulting in an element
4. The evaluation of the derivative $G = \partial F(v)$, resulting in an operator

The fundamental ideas behind the NUMLAB framework are:

1. all important numerical problems can be formulated as: Find the solution v of $F(v) = 0$
2. most numerical methods induce a sequence of states

$$v^{(0)} \mapsto v^{(1)} \mapsto v^{(2)} \mapsto v^{(3)} \mapsto \dots \tag{1}$$

Related to each transition $v^{(k)} \mapsto v^{(k+1)}$ is an operator which can be expressed using the four operations mentioned above. Even for complex applications (see the Finite Element solver for the Navier-Stokes equations in Sect. 5), it is possible to formulate *all* transitions from $v^{(0)}$ down to the final solution approximation $v^{(m)}$ with the use of just one transition operator F. Thus, in NUMLAB, implementations for complex problems can be composed in a simple manner.

In [1], it is shown how all of the following mathematical entities can be formulated as operators:

1. Transient boundary value problems (IBVPs);
2. Boundary value problems (BVPs)/Initial value problems (IVPs);
3. Systems of (non-)linear equations;

4. Iterative solution methods;
5. Preconditioners, etc.

The simple but powerful concept of operator makes NUMLAB a versatile workbench which is simple to use. But, it also has generated a lot of questions such as (1) how to map for instance a Finite Element assembler onto this framework and (2) how to do this in a run-time efficient *and* customizable manner. Paper [1] answers questions with respect to (1) for concrete examples of the above operators, and this paper addresses (2) for the most complex NUMLAB operator: A Finite Element BVP solver for systems of non-linear equations. An efficient implementation turns out to demand a lot of attention to small detail as well as to larger scale software design. Both aspects are addressed in Sect. 4, using the modules described in Sect. 3.

3 The Software Framework

This section describes the software framework which models two major notions: Element Function x of a set, and operator Operator F on such sets. These notions are sufficient for the solution of equations $\mathbf{F}(\mathbf{x}) = \mathbf{0}$. The names of the methods and variable passing techniques are data flow framework standards (see [1] [2]). Below, the modules introduced in [1] are presented in a form which is more suitable for the explanation of an *efficient* Finite Element implementation in Sect. 4. To this end, module Space is split into Space and Basis.

3.1 The Function Module

An instance v of Function is vector of functions from $\Omega \subset \mathbf{R}^n$ to $\mathbf{R}$:

$$v_i(\mathbf{x}) = \sum_{j=1}^{N_i} v_{ij} \phi_j^{(i)}, \tag{2}$$

where all $\phi_j^{(i)} : \Omega \mapsto \mathbf{R}$. Set $N := \sum_i N_i$. Thus, Function contains (a reference to) a vector of coefficient vectors $\mathbf{v}_i = [v_{i1}, \ldots, v_{iN_i}]$, as well as (a reference to) a set of functions, called Space. The Function module provides a few services: evaluation (*sampling*) of v, as well as of its first and second derivatives at (collections of) points of $\mathbf{R}^n$. In fact, Function contains a sequence of elements related coefficient vectors $\{\mathbf{v}^{(k)}\}_k$, either because the user or an operator (for instance a time-integrator) issues a request to this end. In retrospect, from the authors point of view, Function had better been called *LinearCombinationOfFunctions*, *Element* or just *Data*. Nevertheless, we use NumLab's name Function through the remainder of this paper.

3.2 The `Space` Module

The evaluation of a `Function` instance v at point $\mathbf{x}$ is delegated to the `Space` module because the latter "contains" all functions $\phi_j^{(i)}: \Omega \mapsto \mathbf{R}$. For an efficient implementation, $\mathbf{v}$ and $\mathbf{x}$ are passed to `Space`. The module `Space` contains a sequence of `Basis`. A `Basis` can implement pre-run-time determined functions such as $\phi_j^{(i)}(\mathbf{x}) = |\mathbf{x}|^2$, but can also contain run-time determined functions, which is the case for Finite Element implementations. Because Finite Element bases require computational grids, a `Space` module contains (a reference to) a `Grid` module. Different bases (constant, linear, etc., conforming/non-conforming), can be chosen and altered during runtime. `Space` delegates the evaluation services of `Function` to its `Basis` (plural), which is given `Grid`. In this manner, `Basis` can call on `Grid` to determine the location of the point $\mathbf{x}$ and determine the related value $v(\mathbf{x})$.

The most important service of `Space` is access to the assembler, with Finite Element computations in mind. Passed an `Equation` from an `Operator-ImplementationFiniteElement` (as in Fig. 1, box 3), `Space` delegates the assembler request – in a loop over all solution components – to each `Basis`, passing the other involved `Basis`, the `Grid`, the `Equation`, as well as the `Function` v where F or its derivative must be evaluated. For more detailed information, see 4.

The boundary conditions `BoundaryConditions` are members of module `Space`, though in retrospect a more logical place would have been module `Contour`, which describes the boundary $\partial\Omega$, used by module `Grid` for the grid generation. In retrospect, a more mathematical name for `Space` would have been *CollectionOfFunctions*.

The services of `Basis` can be provided through `Space` (see [1], where `Basis` is not mentioned). However, in order to demonstrate that an efficient but flexible implementation is possible, `Basis` is regarded as a separate module.

3.3 The `Basis` Module

Specialisations of this module implement a Finite Element basis $\{\phi_j^{(i)}\}_{j=1}^{N_i}$. The NUMLAB workbench offers the most common bases: Conforming piecewise constant, linear, quadratic, cubic, bilinear, biquadratic (etc.), and nonconforming piece-wise linear. In order to ensure that `Basis` can offer its services, `Space` invokes a method `Basis::init(Grid)`, where `Basis` generates all information it needs (amount of basis functions, support points, etc.).

There are two services: (1) Returning function values when passed a coefficient vector $\mathbf{v} = [v_1, \ldots, v_{N_i}]$ and collection of points $\mathbf{x} \in \Omega$; (2) Returning $F(v)$ or $\partial F(v)$ when passed an `Equation` (see module `Operator` below). In fact, `Basis` delegates the Finite Element assembler service in a non-trivial manner, as described in detail in Sect. 4.

3.4 The Grid Module

In order to generate (representations of) functions $\phi_j^{(i)} : \Omega \mapsto \mathbf{R}$ for Finite Element and other numerical computations, one needs a so-called computational grid. The domain of interest Ω is partitioned into a collection of non-overlapping elements e_l (spectral Finite Element methods require just one element). In the NUMLAB framework, a Grid module can contain a computational grid for $\Omega \subset \mathbf{R}^n$, where n can be adapted run-time. Available are regular grid generators – n-cube, n-simplex, or n-prism – and Delaunay grid generators – triangle, tetrahedron (see Table 1 (10) and Color Plate 20 (10) on page 436). The few services of grid are: Location of the elements in which a collection of points is situated (not public, but important intern for interpolation issues), and ensuring the element topology and data is available to module Basis. As a matter of fact, Grid takes a Contour as input, which describes the boundary $\partial\Omega$ of the region Ω.

3.5 The Operator Module

In NUMLAB, all problems such as IVPs and BVPs are expressed using Operator modules. Also (iterative) solvers and preconditioners are expressed using an Operator module. Each Operator F module provides two services: (1) Evaluation at element Function v; (2) Evaluation of the derivative of F at v. The first operation returns a Function, the latter an Operator. Operators can delegate (use) to Operators in order to solve complex problems.

As an example, consider the NUMLAB Finite Element operator F: Operator-ImplementationFiniteElementGalerkin It has (a reference to) a module Equation, which specifies the components of the involved partial differential equations, and (a reference to) a module Function, an optional predetermined solution of the BVP. Upon evaluation at element Function v, F passes its Equation member to v's Space member, which passes it to its Basis members. For further explanation, see section 4. The determination of the operator $F'(v)$ is done in a similar manner.

3.6 The Solver Module

Iterative solvers, for instance for the solution of a system of equations $Ax = b$, are described with the use of operators which are of the form

$$(v^{(0)}, b, A) \mapsto v^{(k)}, \tag{3}$$

where $v^{(k)}$ is the first iterand such that $\left|Av^{(k)} - b\right| < \epsilon$. Here both $v^{(0)}$ and b are Functions and A is an Operator. The operator in (3) need not be differentiable. Therefore, a Solver module, which represents an iterative solver, could be derived from Operator, and return a zero derivative. In fact,

Solver is derived from `OperatorIterator` which specialises `Operator` in that is has a member of `IterationControl`, which stores: maximum amount of iterations, convergence tolerance, etc. When requested for a derivative, it returns the zero `Operator`. As shown in Fig. 1, solvers can delegate (use) to solvers in order to solve complex problems.

In retrospect, the best modular approach would have been module *Operator* with evaluation service, and a derived module *OperatorDifferentiable* with added evaluation of derivative service.

4 An Efficient NumLab Finite Element Implementation

An efficient Finite Element assembler requires vector, matrix and sometimes tensor operations. To this end, in NumLab, three groups of such storage classes are required:

1. "small" storage classes (vector, matrix, tensor) for:
   a) points in a region $\Omega \subset \mathbf{R}^D$;
   b) values of T reference basis functions on Q quadrature points in $\mathbf{R}^{T \times Q}$, etc.;
2. "large" storage classes (vector, matrix) for:
   a) coefficient vectors $\mathbf{v}$ of a component of a `Function`;
   b) Jacobian matrices components of the derivative of an `Operator` evaluated at a `Function`, etc.;
3. "block" storage classes for composed items:
   a) block vectors of large vectors for the sequence of coefficient vectors of a `Function`;
   b) block Jacobian matrices of the derivative of an `Operator` evaluated at a `Function`, etc.;

In NumLab, the assembler's numerical integration routines use the small storage class, the computed entries are stored in large storage class, through a block storage class interface. The different storage classes have different services. The small storage class has just a few services which include addition, scaling, increment and specialised versions of matrix inversion for 2 x 2 and 3 x 3 matrices. The large matrix class has additional services such as for instance ILU(0) (incomplete LU factorisation).

First, we consider a scalar equation, related to a convection diffusion problem. After a step for step procedure, we arrive at a first Finite Element implementation, and discuss the reasons for its slow performance. The problem of interest is:

$$L(u) = f \quad \text{in } \Omega, \quad \text{and} \quad B(u) = g \quad \text{at } \partial\Omega, \qquad (4)$$

where

$$L(u) = -\nabla \cdot a \nabla u + \mathbf{b} \nabla u + cu, \qquad B(u) = u, \qquad (5)$$

and with the domain of interest $\Omega \subset \mathbf{R}^D$. The diffusion coefficient a, convection vector $\mathbf{b} \in \mathbf{R}^D$ and source coefficient c are all functions of $\mathbf{x} \in \Omega$. For the sake of demonstration, we assume that the domain Ω is covered with a computational grid of elements e. Assume that, on the reference element, we use Q quadrature points $\mathbf{x}_k$ and related weights w_k. Abbreviate $a_k := a(\mathbf{x}_k)$, $\mathbf{b}_k := \mathbf{b}(\mathbf{x}_k)$ and $c_k = c(\mathbf{x}_k)$.

There are two major types of Finite Element assemblers:

1. Galerkin-type Finite Element assemblers, which must integrate:

$$\int_e a\nabla\phi_s\nabla\phi_r + \phi_r\mathbf{b}\nabla\phi_s + c\phi_s\phi_r. \tag{6}$$

These integrals are approximated with the use of numerical integration:

$$\sum_k w_k \left[a_k\nabla\phi_s(\mathbf{x}_k)\nabla\phi_r(\mathbf{x}_k) + \phi_r(\mathbf{x}_k)\mathbf{b}_k\nabla\phi_s(\mathbf{x}_k) + c_k\phi_s(\mathbf{x}_k)\phi_r(\mathbf{x}_k) \right],$$
$$\tag{7}$$

for all $r, s = 1, \ldots, T$, the amount of basis functions ϕ_r on the reference element specified through `Basis`;

2. Petrov-Galerkin-type Finite Element assemblers, which must integrate:

$$\int_e a\nabla\psi_s\nabla\phi_r + \phi_r\mathbf{b}\nabla\psi_s + c\psi_s\phi_r. \tag{8}$$

These integrals are approximated with the use of numerical integration:

$$\sum_k w_k \left[a_k\nabla\psi_s(\mathbf{x}_k)\nabla\phi_r(\mathbf{x}_k) + \phi_r(\mathbf{x}_k)\mathbf{b}_k\nabla\psi_s(\mathbf{x}_k) + c_k\psi_s(\mathbf{x}_k)\phi_r(\mathbf{x}_k) \right],$$
$$\tag{9}$$

for all $r = 1, \ldots, T_1$, and $s = 1, \ldots, T_2$ the amounts of basis functions ϕ_r and ψ_s on the reference element specified through `Basis_1` and `Basis_2`.

Systems of non-linear equations such as the Navier-Stokes equations can be solved using both:

1. The Galerkin-assembler, if all different solution component bases are merged into one `Basis`, and;
2. The Petrov-assembler for each block-component of $F(v)$ and $DF'(v)$.

The `NumLab` framework uses the latter approach because it deals better with the zero entries in the derivative (as is the case for Stokes and Navier-Stokes).

Now, consider a straightforward implementation of (7). Assume we precompute and store all values $\phi_r(\mathbf{x}_k) \in \mathbf{R}^{T\times Q}$ in matrix `v_e`, $\nabla\phi_r(\mathbf{x}_k) \in \mathbf{R}^{D\times T\times Q}$ in tensor `grad_v_e`, $a(\mathbf{x}_k) \in \mathbf{R}^Q$ in vector `a`, etc. Then, the common implementation of (7) is:

```
for (int k = 0; k < Q; k++)
 for (int r = 0; r < T; r++)
  for (int s = 0; s < T; s++)
   a_rs(r, s) += w_k(k) * (
    a(k) * (grad_v_e(s, k) * grad_v_e(r, k)) +
    v_e(r, k) * (b * grad_v_e(s, k)) +
    c * v_e(s, k) * v_e(r, k)
                           );
a_rs *= abs(detAe);
```

The problem: The implementation is observed to be slow, if vector, matrix and tensor classes use run-time allocated storage (which is the case if a constructor such as `smallvector::smallvector(n)` exists). In this case, calls to `new` are slow on at least some machines, and bounds must be checked for safe selections `a(k)`. The case of multiple selections (selection of `grad_v_e(s, k)` in block `(p, q)` of a Jacobian at element face number `f`) causes the program to almost stand still. The added problem is that the *implementer* is to choose a small data storage class because c and c++ have no built-in types for such containers. Worse, because the built-in support lacks, c and c++ compilers can not optimise for it – Fortran compilers can.

The first step towards an efficient implementation: Ensure that the assembler for one component (or block component is fast). Implement the "small" storage class with type-int templates. This is *the only* manner to avoid run-time allocated storage, but bound checks would still be needed when selections are to be performed in a safe manner. However, such checks are not required if iterators are provided for the "small" storage class, and used (actual NUMLAB code):

```
const double *wk = w_k.begin();
const double *ak = a.begin();
const smallVector<T> *v = v_e.begin();
const smallMatrix<D, T> *gradv = grad_v_e.begin();
// loop over quadrature points
for ( ; wk < w_k.end(); wk++, ak++, v++, gradv++)
{
 double *ars = a_rs.value();

 const double *vs = v->begin();
 const smallVector<D> *gradvs = gradv->begin();
 // loop over trial function s
 for ( ; gradvs < gradv->end(); vs++, gradvs++)
 {
  const double *vr = v->begin();
  const smallVector<D> *gradvr = gradv->begin();
```

```
// loop over test function r
for ( ; gradvr < gradv->end(); vr++, gradvr++)
  *ars++ += (*wk) * (
                      (*ak) * ((*gradvr) * (*gradvs)) +
                      (*vr) * (b * (*gradvs)) +
                      c * (*vs) * (*vr)
                    );


  }
}
a_rs *= abs(detAe);
```

To keep the above code simple, we have assumed a constant vector field **b**, and a constant source term c. The above assembler is fast, in all aspects.

The educated reader will note that, different from usual, the iterators (supporting member `begin()` and `end()` are not new classes. For instance, invoking `begin()` on tensor `grad_v_e` in $\mathbf{R}^{D \times T \times Q}$ returns a matrix in $\mathbf{R}^{D \times T}$ and not an instance of a new iterator-class containing a reference to the matrix.

Because vectors and matrices which store the computed entries must be resisable, in NumLab, at least small and large storage classes must exist. And, because NumLab uses Petrov-assemblers, also a block storage class exists.

The Second step towards an efficient implementation: For systems, ensure that component p of F, or block component (`p`, `q`) of its derivative can be assembled using the above fast assemblers. To this end, first note that fast assemble code above is templatised, and no longer a routine: Small vectors, matrices and tensors must be declared

```
const smallMatrix<D, T> *gradv = grad_v_e.begin();
```

using parameters of the specialisation of `Basis`. In the above example, the amount of reference basis functions `T` is used – as well as the dimension of the region of interest `D`.

The assembler code is put into two functions, a general one related to (9, and a specialised one related to 7. For the sake of demonstration we consider the assembler for the derivative, and for now focus on the case of a system of *linear differential equations*:

```
template<class Psi, class Phi>
int assembleDF(const Psi &basis_psi, const Phi &basis_phi,
               const Grid &grid,
               largeMatrix &DF, ... );
template<class Phi>
int assembleDF(const Phi &basis_phi,
               const Grid &grid,
               largeMatrix &DF, ... );
```

In each specialisation `assembleDF<Basis_1, Basis_2, Grid>(...)`, all calls to methods of `Basis` (such as `Basis::get_element_dofs()`) are **non-virtual** (see below), *and* all calls to methods of `Grid` (e.g., `get_element_vertices()`) are *non-virtual*. The latter methods are non-virtual because different `Grid` specialisations share data members of the base. `assembleDF()` just accesses these base data members. This technique is called *virtual through data member information*. For the `Grid`, `assembleDF()` must make use of such – or standard virtual methods – because `Grid` can be re-attached to `Space` during run-time.

It remains to be explained how all calls to `Basis` members can be *non-virtual*, and how `Space` delegates the assembling to `Basis`. First, observe that the user can run-time alter the collection of `Basis` (plural) (see figure 1, where `SpaceReferenceTriangleLinear` and `SpaceReferenceTriangleQuadratic` are input to `Space` for the pressure and velocities). Next, observe that `assembleDF()` is parametrised with *two* `Basis`. Based on these two observations, NumLab uses a technique called *double dispatch*. Each module derived from `Basis`, such as `SpaceReferenceTriangleLinear`, implements a list of virtual methods (we omit the `SpaceReference-` part):

```
class TriangleLinear: public Basis
{
  ...

  int assembleDF(const Basis &basis, const Grid &g, ...)
   { return basis->assembleDF_execute(*this, g, ...) }

  int assembleDF_execute(const TriangleLinear &basis,
                  const Grid &g, ...)
   { return assembleDF<TriangleLinear>(*this, g, ...); }


  int assembleDF_execute(const TriangleQuadratic &basis,
                  const Grid &g, ...)
   { return assembleDF<TriangleQuadratic,
               TriangleLinear>(basis, *this, g, ...); }

  ...
};
```

This shows how the finite element assemblers in NumLab can be fast and customizable: The `Basis` (plural) can be exchanged to different bases at run-time because the actual assembler is not present in `Space` nor in `Basis`:

```
class Space
{
  ...
```

```
int assembleDF(blockMatrix &DF, ... )
{
 for (int p = 0; p < nb_basis; p++)
 {
  for (int q = 0; q < nb_basis; q++)
  {
   DF(p, q) = basis[p]->assembleDF(basis[q], *grid,
                              DF, ...);
  }
 }
}

...
};
```

The NumLab implementation is also fast because the assembler defacto runs with the routines assemble<., .>(...) which are specialised for the combination of Basis chosen at run-time. It should be noted that one assemble<b1, b2>(...) instance is needed for each basis b1 $\neq$ b2, a small price to be paid for the ensured fast execution.

Next, the above approach is refined for system of *non-linear differential equations*, where as an added problem, in each block component (p, q), *all or a few solution components of v are required* – as a rule. Because (Navier-Stokes) v consists of three components, it would seem that a two-Basis parametrisation of assemble() is not sufficient. However, a two-Basis parametrisation of assemble() still works fine, when the value of the solution components v is passed as a vector of values at the quadrature points, just as for instance the diffusion a, or convection b could be passed to assemble(). This turns out to the solution: At all quadrature points x_k, Space ensures the pre-computation (sampling) of all values which are not related to a basis function $\phi_j^{(i)}$. For solution components – which do depend on basis functions – this task is delegated to Basis. This information is passed to assembleDF().

For the sake of a lucid exhibition of the software framework implementation on module level, a range of details has been not discussed. The reader will note that for the non-linear case, Space must negotiate a cross-basis set of quadrature points, and, that in fact assemble() must be parametrised with this amount of points. An alternative is that different bases use the same set of quadrature points, suited for the highest degree of polynomial basis. The best software framework solution to such problems is continuously under investigation. Other not-commented on issues are the use of Equation inside the assemble(), specialised Navier-Stokes assemblers, the sampling of elements v required for multiple grid computations (in NumLab, also delegated by Space to Basis), etc., etc.

Readers of [1], [2], [11], [12] and other NumLab related papers have pointed out that the implementation of this powerful NumLab software framework would be far from trivial. This paper shows that this has been true indeed. Each small detail of the NumLab framework has been designed based on years of experience in: (1) iterative solution methods for complex non-linear systems of transient BVPs – with Finite Element, Difference, Volume discretisations, grid generation, etc; (2) visualisation of the resulting complex data-sets; (3) modern computer-language software design techniques.

The above modular software framework makes NumLab Finite Element computations highly customizable: The NumLab workbench contains a module called SpaceReferenceTriangle, which is a specialisation of Basis. This specialisation has a switch, which allows run-time switching between the different available basis specialisations. The implementation is standard: Module SpaceReferenceTriangle has a pointer to a specialisation of a member of Basis, which can be reassigned to at run-time.

Thus, summarising the facts which ensure that NumLab modules are both fast *and* customizable:

1. customization: a software framework with sophisticated delegation-model;
2. discretisation mixes: dispatch techniques in combinations with templates;
3. fast execution: different storage classes for different circumstances.

Because there are just two Finite Element assembler templates (Galerkin and Petrov type) in NumLAB, NumLab modules are not just both fast *and* customizable, but *also* simple to maintain.

5 Application Design and Use

This section demonstrates how the NumLAB modules introduced in Sect. 3 are combined to form a numerical simulation.

Fig. 1 shows a NumLAB c++ Navier-Stokes simulation program composed with, and represented using, the NumLAB graphical editor VISSION. The involved modules occur in quite standard groups:

1. A computational Grid specialisation;
2. A Space specialisation, involving specification of Finite Element bases Basis, and BoundaryConditions;
3. The BVP operator:
   a) A Finite Element Galerkin Operator specialisation, using;
   b) A Navier-Stokes Equation, and an optional;
   c) Predetermined solution, a specialisation of type Function.

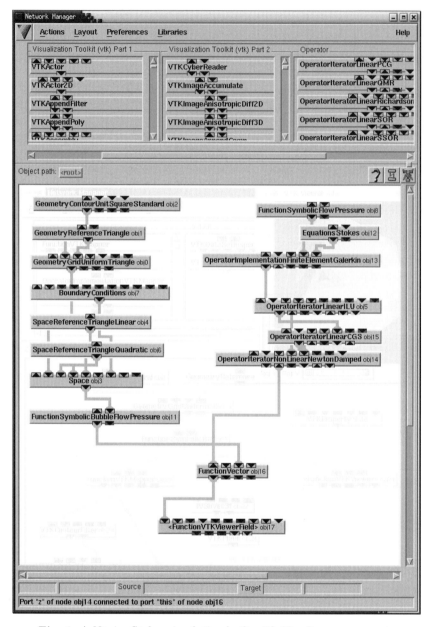

Fig. 1. A Navier-Stokes simulation built with NUMLAB components

4. The composed solver:
   a) A Non-linear `Solver` specialisation, using;
   b) A Linear `Solver` specialisation, using;
   c) A Preconditioner `Solver` specialisation.
5. The initial guess $v^{(0)}$ (velocities and pressure);
6. The function $v^{(k)}$ containing all subsequent iterands – velocities and pressure – (see the explanation below (1));
7. A visualisation group `FunctionVTKViewerField` for $v^{(k)}$.

Module group (2) defines the linear vector space, which contains the targeted solution (6). The operator composed in group (4) acts on this solution, i.e., it performs all transitions $v^{(k)} \mapsto v^{(k+1)}$, until all criteria are met.

The application can be extended or altered, taking modules from the libraries right at the top of Fig. 1, or taking modules from libraries which can be loaded during run-time. Eigenvalues of Jacobians can be visualised, etc. In this example, several NUMLAB libraries have been loaded, but visible are just three:

- `Visualization ToolKit 1`;
- `Visualization ToolKit 2`;
- `NumLab Operator`.

In fact, the visualisation module in Fig. 1 is not a single module, it is a group which contains other connected modules. Thus, when simulations get more complex, it is possible to group (hide) less important parts into one module. Non-connected input and output ports of the hidden modules become input and output ports of the new group-module. The contents of module `FunctionVTKViewerField` in Fig. 1 is shown if Fig. 2 (Color Plate 21 on page 437), together with the resulting output. The content of the `FunctionVTKViewerField` group is accessible after a simple double-click on the `FunctionVTKViewerField` icon. This group can be replaced with a range of available visualisation groups (stream lines, contour surfaces, contour lines, etc.).

Because NUMLAB adorns the original unaltered VTK and Open Inventor libraries with an interface (see Sect. 1), its VTK and Open Inventor modules inter-mix. NUMLAB combines the strength of Open Inventor (superb rendering engine) with the strength of VTK (lots of high level modules, such as Image Transformations).

6 Conclusions and Future Work

As was concluded in [1], NUMLAB addresses two categories of limitations of current computational environments. First, NUMLAB builds on a few fun-

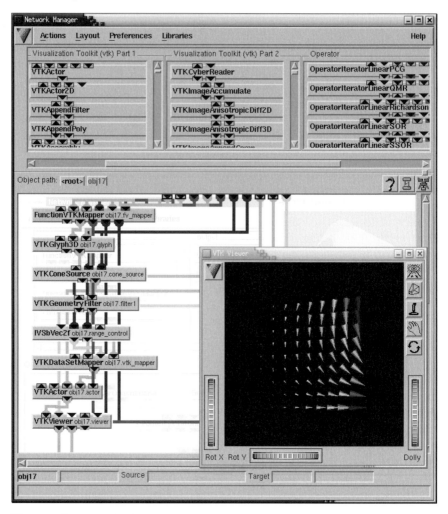

Fig. 2. The modules hidden inside group module `FunctionVTKViewerField` of NUMLAB. The viewer at the end of the pipe-line visualizes the result.

damental numerical notions. All entities such as iterative solver, preconditioners, time integration, Finite Element assemblers, etc., are formulated as operators which cause state transitions $v^{(k)} \mapsto v^{(k+1)}$. Each new operation is simple to integrate because it is of the same nature, i.e., a transition operation. Visualisation and numerical operations follow the same concepts, whence in NUMLAB, computations and visualisation mixes without problems.

Next, in detail demonstrated in this paper, NUMLAB combines a high level of *customizable features with nevertheless fast implementations*. We have shown that this is possible with a minimal amount of code, making targeted use of templates in c++. Though the amount of code is minimal – and thus the maintenance costs are low – the larger scale required module-coworking turns out to be non-trivial pattern.

The next steps on the road towards an even more complete workbench are: merging available parallel Finite Element assemblers, integrating problem-specialised iterative solvers and preconditioners, and, started a few month ago, separating the address space of the graphical editor from the address space where all modules execute. Then failing contributed research-modules will no longer crash entire NumLab applications. The visualisation viewer module and computational modules should be run on different threads so data can be visualised while computations continue.

References

1. Maubach, J.M.L., Telea A. (accepted): The NumLab numerical laboratory for computational and visualisation. Computing and Visualisation in Science. Springer Verlag

2. Maubach J.M.L., Drenth, W.(2002): Data-Flow Oriented Visual Programming Libraries for Scientific Computing. In: Sloot, P.M.A., Tan, C.J.K., Dongarra, J.J., Hoekstra, A.G. (eds) Computational Science – ICCS 2002 (LNCS 2329). Springer Verlag

3. Abram, G., Treinish, L. (1995): An Extended Data-Flow Architecture for Data Analysis and Visualisation. In: 6th Proc. IEEE Visualisation 1995, ACM Press, 263–270.

4. Anderson, E., Bai, Z., Bischof C., et al. (1995): LAPACK user's guide. SIAM, Philadelphia

5. Brun, R., Goto, M., Rademakers, F.: The CINT c/c++ interpreter http://root.cern.ch/root/Cint.html

6. Bruaset, A.M., Langtangen, H.P. (1996): A Comprehensive Set of Tools for Solving Partial Differential Equations: Diffpack. In: Daehlen, M., Tveito, A. (eds) Numerical Methods and Software Tools in Industrial Mathematics. Springer Verlag

7. Gunn, C., Ortmann, A., Pinkall, U., Polthier, K., Schwarz, U. (1996): Oorange: A Virtual Laboratory for Experimental Mathematics, Sonderforschungsbereich 288, Technical University Berlin. http://www-sfb288.math.tu-berlin.de/oorange/OorangeDoc.html

8. Jackie, N., Davis, T., Woo, M. (1993): OpenGL Programming Guide. Addison-Wesley

9. Schroeder, W., Martin, K., Lorensen, B. (1995): The Visualisation Toolkit: An Object-Oriented Approach to 3D Graphics. Prentice Hall

10. Stroustrup, B. (1997): The c++ Programming Manual (3rd edition). Addison-Wesley

11. Telea, A., van Wijk, J.J. (1999): VISSION: An Object Oriented Dataflow System for Simulation and Visualisation. In: Gröller, E., Löffelmann, H., Ribarsky, W. (eds) Proceedings of IEEE VisSym 1999. Springer Verlag

12. Telea, A. (1999): Combining Object Orientation and Dataflow Modeling in the VISSION Simulation System. In: Proceedings of TOOLS'99 Europe. IEEE Computer Society Press
13. Upson, C., Faulhaber, T., Kamins, D., Laidlaw, D., Schlegel, D., Vroom, J., Gurwitz, R., van Dam, A. (1989): The Application Visualisation System: A Computational Environment for Scientific Visualisation. IEEE Computer Graphics and Applications, 30–42
14. Wernecke, J. (1993): The Inventor Mentor: Programming Object-Oriented 3D Graphics with Open Inventor. Addison-Wesley
15. The Open Source Software Project Based on IBM's Visualisation Data Explorer. http://www.opendx.org/
16. IMSL (1987): FORTRAN Subroutines for Mathematical Applications, User's Manual. IMSL
17. SEPRA Analysis http://ta.twi.tudelft.nl/sepran/sepran.html
18. Matlab (1992): Matlab Reference Guide. The Math Works Inc.
19. Wolfram, S. (1999): The Mathematica Book 4-th edition. Cambridge University Press
20. NAG (1990): FORTRAN Library, Introductory Guide, Mark 14. Numerical Analysis Group Limited and Inc.
21. INRIA-Rocquencourt (2000): Scilab Documentation for release 2.4.1. http://www-rocq.inria.fr/scilab/doc.html

A Generic Programming Approach to Multiresolution Spatial Decompositions

Vinícius Mello[1], Luiz Velho[1], Paulo Roma Cavalcanti[2], and Cláudio T. Silva[3]

[1] IMPA – Instituto de Matemática Pura e Aplicada, Estrada Dona Castorina 110, Rio de Janeiro, RJ 22460-320, Brazil. *vinicius,lvelho@visgraf.impa.br*
[2] Federal University of Rio de Janeiro - UFRJ. *roma@lcg.ufrj.br*
[3] AT&T Labs, 180 Park Ave., Florham Park, NJ 07932, USA. *csilva@research.att.com*

Summary. We present a generic programming approach to the implementation of multiresolution spatial decompositions. From a set of simple and necessary requirements, we arrive at the Binary Multitriangulation (BMT) concept. We also describe a data structure that models the BMT concept in its full generality. Finally, we discuss applications of the BMT to visualization of volumetric datasets.

1 Introduction

Generic programming was born from the observation that most algorithms rely on a few basic semantic assumptions about the data structures, and not on any particular implementation of these structures. Given a problem, the generic programming basic task is to isolate these essential concepts, framing them in a well defined interface where semantic requirements and computational complexity guarantees are clearly posed. The algorithms that comply with that interface are free from idiosyncrasies of data structures, which can be changed or even replaced by procedural schemes. The great success of C++ Standard Template Library is the main proof on behalf of that methodology [29].

The main contribution of this paper is to show how generic programming techniques can be used to build a computational framework to deal with multiresolution spatial decompositions. We have studied from combinatorial topology classics like [1] to modern works on multiresolution modeling [12] in order to identify the meaningful concepts. As a result, we arrive at a new concept called Binary Multitriangulation (BMT) that is a particular case of the Multiresolution Simplicial Model (MSM) described in [12], but more manageable and closer in spirit to well stablished procedures of combinatorial topology. The BMT concept can also be regarded as a 3-dimensional extension of variable resolution structures like [31].

Currently, generic programming methodology is being used as design philosophy of many libraries in several areas: computational geometry (CGAL [10]), combinatorics (BGL [26]) and scientific computing (MTL [27]), for instance. Concerning to the specific problem of spatial decompositions, our work resembles the GrAL library [2], although we were unaware of that until just before the publication of this paper. The similarities can be ascribed to the application of the same paradigm (generic programming) to a same problem (spatial decompositions) with the same search for conceptual rigor. More important, nonetheless, are the dissimilarities: while the GrAL library concentrates in fixed resolution cell decompositions, we focus on multiresolution simplicial decompositions.

A common point of all those libraries is the omnipresence of C++ language, the most fitted language to generic programming (specially after the recent ISO/ANSI standardization). This fact justifies the extensive use of C++ code in this paper, but we must emphasize that there is no intrinsic dependence to specific programming languages in the concepts we describe. In a sense, genericity can be achived in almost any language, with more or less effort. We even can regard the pioneering work of Guibas and Stolfi [14], Laszlo and Dobkin [7] and Mäntylä [18] as first attempts to attain genericity in surface and solid modeling. The C++ language just provides a set of built in facilities to do it.

The paper is organized as follows. In Sect. 2, we set the context where multiresolution spatial decompositions are needed and discuss informally some of their advantages. Sect. 3 is dedicated to the detailed examination of the concepts we isolate, in increasing order of complexity. In Sect. 4, we describe data structures that models the previously defined concepts. Some applications are presented in Sect. 5. Finally, Sect. 6 contains concluding remarks and indications of future works.

2 Background

If multiresolution methods are significant in the processing of triangle meshes [9, 8], they are indispensable in the case of tetrahedra meshes, or spatial decompositions, as we prefer to call it, since the complexity of the mesh increases with the power of its dimension. Therefore, most applications dealing with three-dimensional data, like scientific visualization, medical imaging, geoprocessing etc., will benefit of techniques that allow the user to extract from the original data an equivalent representation, in a sense will be made precise subsequently, but in a resolution more adequate to the task at hand. Because the user often doesn't know *a priori* the desired resolution, the solution is to store a good set of possible resolutions and to give him the ability of browsing between them. That is the essence of the multiresolution methods.

A cost-benefit analysis of multiresolution methods with respect to the current technology is presented in [4]. The main conclusion of this analysis is that, in the case of direct volume rendering applications (DVR), the graphics constraints are stronger than memory constraints. In fact, one can store a mesh three orders of magnitude larger than that one can visualize with DVR. It is reasonable to assume that similar conclusions can be extended to other applications dealing with volume data. Thus, the extra cost to store a multiresolution mesh structure is compensated by the flexibility to choose the most adequate resolution.

It remains to define more precisely what is a tetrahedral mesh. Although the basic intuitive concept is clear, it is not so clear that some properties of a mesh are independent of the geometry of its constituent tetrahedra or, more specifically, of their spatial embedding. The area of mathematics that studies such properties is called *combinatorial topology*. A *manifold* is a well known mathematical object and it is very useful in combinatorial topology. A combinatorial manifold is characterized by a certain uniformity in the way its parts are glued. Despite of the fact that in some applications it is necessary to consider "non-manifold" structures, the concept of a manifold is sufficient for most applications.

In the next session, we will describe a series of concepts progressively more complex, until we arrive at a definition of a combinatorial manifold. Each concept will be followed by an API which defines the operations required to work with the concept. We will postpone the introduction of geometric concepts as much as possible, in order to clearly isolate the topological properties.

However, geometric concepts, such as volume, area, aspect ratio, etc., are of fundamental importance in applications and are deeply related with the mechanisms that are employed to select a particular mesh resolution from a multiresolution structure. In general, we deal with functions whose domain is the combinatorial manifold, and we would like to have a more refined mesh in regions where these functions exhibit high variations. This property is called *adaptivity*. We intend to discuss techniques for generation and processing of multiresolution adaptive tetrahedral meshes in a future paper.

3 Concepts

In this section, we adopt the following strategy in describing the concepts: we will define the mathematical objects involved, the name and type signature of the requirements, and let the semantics be derived from them and from hints in the text body. Auxiliary tables containing associated types and notations fulfill the description. Concerning to type signatures, each concept has a *trait class* where all type information is encoded. Trait classes are essentially a mechanism to ensure algorithm independence of data structure implementation [30]. In order to clarify the programs presentation to the noninitiated in the C++ technicalities, we will apply the notation defined in the auxiliary tables to the sample code and we will omit the **template** clause (see program 1 and program 2).

Some definitions bellow differs from the usual (see [6], for instance), but this happens because we have choosen equivalent definitions easily translatable to algorithm requirements.

3.1 Abstract Simplicial 3-Complex

Definition 1 *Given a finite set V, called* vertex set, *an abstract simplicial complex on V is a set K of subsets of V verifying the following properties:*

1. *For each $\nu \in V$, $\{\nu\} \in K$;*
2. *If $\sigma \in K$ and $\phi \subset \sigma$, them $\phi \in K$. An element σ of K is called* simplex *and the subsets of σ are called* faces;
3. *There is a total ordering on the vertices of each simplex of K such that the ordering on the vertices on any face of a simplex σ is the ordering induced from the ordering on the vertices of σ.*

If $n+1$ is the cardinality of a simplex $\sigma \in K$, we say that σ is a n-simplex and K is an *abstract simplicial n-complex* if the largest simplex of K is a n-simplex. We will use the same name simplex to mean the subcomplex of K formed by the faces of a simplex σ, and we define $\partial\sigma = \{\tau \in K : \tau \subsetneq \sigma\}$, that is, the *boundary* of σ.

We are interested here in abstract simplicial 3-complexes, AS3C for short. In this case, we adopt the terminology *vertex, edge, face* and *simplex* for 0, 1, 2 and 3-simplex, respectively (Tables 1 and 2).

description	type
Vertex descriptor	as3c_traits<T>::vertex_descriptor
Edge descriptor	as3c_traits<T>::edge_descriptor
Face descriptor	as3c_traits<T>::face_descriptor
Simplex descriptor	as3c_traits<T>::simplex_descriptor
Complex vertices iterator	as3c_traits<T>::vertex_iterator
Complex edges iterator	as3c_traits<T>::edge_iterator
Complex faces iterator	as3c_traits<T>::face_iterator
Complex simplices iterator	as3c_traits<T>::simplex_iterator

Table 1. Associated types of an AS3C. T is a type that models an AS3C. The *_descriptor types are intended to mean "small types", that is, types which objects can be passed by value without overhead. The *_iterator types must be at least forward iterators.

Item 3 from definition 1 has a twofold propose: it provides a "canonical form" for each simplex and enables us to define the *face operator* d_i, that assigns for each simplex σ the face of σ obtained by removing the i-th vertex. The face operator satisfies

symbol	definition
vertex	typedef as3c_traits<T>::vertex_descriptor vertex;
v	a object of type vertex
edge	typedef as3c_traits<T>::edge_descriptor edge;
e	a object of type edge
face	typedef as3c_traits<T>::face_descriptor face;
f	a object of type face
simplex	typedef as3c_traits<T>::simplex_descriptor simplex;
s	a object of type simplex
vi	typedef as3c_traits<T>::vertex_iterator vi;
ei	typedef as3c_traits<T>::edge_iterator ei;
fi	typedef as3c_traits<T>::face_iterator fi;
si	typedef as3c_traits<T>::simplex_iterator si;

Table 2. AS3C related notation.

$$d_i d_j = d_{j-1} d_i, \text{ if } i < j. \tag{1}$$

The existence of operators satisfying (1) is sufficient to recover all relations between the faces of a simplex[1]. Figure 1 shows all those relations. We exploit this fact to define a minimum set of requirements on an AS3C (see table 3).

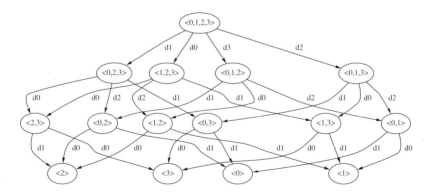

Fig. 1. Face operator graph. The nodes are the simplex's faces and the edges are the face operators d_i.

As we mentioned previously, we are going to apply the definitions on the requirement tables above to simplify the program's code. Programs 1 and 2 show two versions of the same algorithm, with and without this simplification. This simplified notation will be used in all subsequent programs in the paper.

---

[1] Face operators appear in algebraic topology in the definition of simplicial sets [19].

expression	return type
empty_vertex(t)	vertex
empty_edge(t)	edge
empty_face(t)	face
empty_simplex(t)	simplex
face_op(t, s, i)	face
face_op(t, f, i)	edge
face_op(t, e, i)	vertex
vertices(t)	pair<vi, vi>
edges(t)	pair<ei, ei>
faces(t)	pair<fi, fi>
simplices(t)	pair<si, si>

Table 3. Requirements of an AS3C. Some remarks about the notation: t is a object which type models an AS3C; empty_vertex(t) returns a null vertex descriptor; vi is the type of a iterator which traverses the vertex container; vertices(t) return a pair of iterators: the first points to the first vertex and the second is a "past-the-end" iterator. The other operators work analogously.

Program 1 i-th vertex of a simplex. Note how the trait class as3c_traits isolates the algorithm from the data structure implementation.

```
template <typename T>
as3c_traits<T>::vertex_descriptor ith_vertex(T t,
                                    as3c_traits<T>::simplex_descriptor s, int i) {
    int table[][3]={ 1,1,1, 1,1,0, 1,0,0, 0,0,0 };
    return face_op(t, face_op(t, face_op(t, s, table[i][2]), table[i][1]), table[i][0]);
}
```

Program 2 Simplified version of program 1

```
vertex ith_vertex(T t, simplex s, int i) {
    int table[][3]={ 1,1,1, 1,1,0, 1,0,0, 0,0,0 };
    return face_op(t, face_op(t, face_op(t, s, table[i][2]), table[i][1]), table[i][0]);
}
```

Program 2 shows how to get the i-th vertex of a simplex and program 3, the i-th edge (with respect to the lexicographic ordering over its vertices). The correctness can be verified by looking to figure 1. Of course, we can specialize both programs if more information about the underlying data structure is known.

Program 3 i-th edge of a simplex.

```
edge ith_edge(T t, simplex s, int i) {
    int table[][2]={2,2, 2,1, 1,1, 2,0, 1,0, 0,0};
    return face_op(t, face_op(t, s, table[i][1]), table[i][0]);
}
```

3.2 Abstract 3-Manifold

In principle, the requirements on an AS3C are enough to answer incidence queries like "get all faces meeting an edge" or "get all edges meeting a vertex". But, in many cases, we have more information about the local structure of the complex in each vertex. That information can be used to speed-up those queries. To describe precisely that local structure, we need some definitions.

Two simplices σ_1, σ_2 are *independent* if $\sigma_1 \cap \sigma_2 = \emptyset$. The *join* $\sigma_1 \star \sigma_2$ of independent simplices σ_1, σ_2 is the set $\sigma_1 \cup \sigma_2$. The join of complexes K and L, written $K \star L$, is $\{\sigma \star \tau : \sigma \in K, \tau \in L\}$. The *link* of simplex $\sigma \in K$, denoted $\mathrm{link}(\sigma, K)$, is defined by

$$\mathrm{link}(\sigma, K) = \{\tau \in K : \sigma \star \tau \in K\}.$$

And finally, the *star* of σ in K, $\mathrm{star}(\sigma, K)$, is the join $\sigma \star \mathrm{link}(\sigma, K)$.

The link and star operators provides a combinatorial description of a neighborhood of a simplex. We can use them also to define certain changes in a complex, but care must be taken to not modify essentially ("topologically") that neighborhood.

The *stellar moves* are a such change. Indeed, many concepts of combinatorial topology are founded on stellar moves [17]. Let K be a complex on the vertex set V, K' a complex on V', σ a simplex in K and ν a vertex in V'. The operation that changes K into K' by removing $\mathrm{star}(\sigma, K)$ and replacing it with $\nu \star \partial\sigma \star \mathrm{link}(\sigma, K)$ is called a *stellar subdivision* and is written $K' = (\sigma, \nu)K$. The inverse operation $(\sigma, \nu)^{-1}$ that changes K' into K is called a *stellar weld*. These operations are depicted in figure 2.

Two complexes are *stellar equivalent* if they are related by a sequence of stellar moves. A (abstract) *n-ball* is a complex stellar equivalent to a n-simplex and a (abstract) *n-sphere* is a complex stellar equivalent to the boundary of a $(n+1)$-simplex.

We can now define a special kind of abstract simplicial complex that has nice local properties.

Definition 2 *An* abstract *n*-manifold M *is an abstract simplicial n-complex such that for each vertex* $\nu \in M$, $\mathrm{link}(\nu, M)$ *is a* $(n-1)$-*ball or a* $(n-1)$-*sphere.*

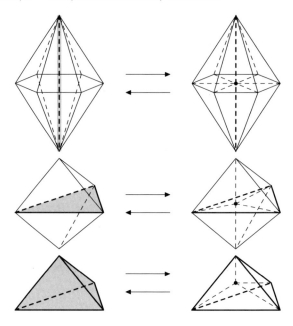

Fig. 2. Stellar moves applied to edge, face and simplex, from top to bottom. ($\rightarrow$) indicates subdivision and ($\leftarrow$) indicates welding.

The *boundary of* M, denoted by ∂M, is the subcomplex $\partial M = \{\sigma \in M : \mathrm{link}(\sigma, M)$ is a ball$\}$. One can proof that ∂M is a $(n-1)$-manifold.

Many properties follows from definition 2. In our particular case, we want properties that help us to speed-up local queries over abstract 3-manifolds (A3M). Each boundary face of a 3-manifold, for example, is incident to a unique simplex and internal faces (faces not on boundary) are shared by exactly two simplices. Therefore, we can require operators that, in constant time, retrieve all simplices meeting at a face.

Tables 4, 5 and 6 list a set of additional requirements to the AS3M ones that are sufficient to formalize the A3M concept. With that additional requirements, we can implement program 4 which, given an internal face f and an incident simplex s, returns the simplex that shares f with s.

description	type
Incident faces iterator	a3m_traits<T>::radial_face_iterator
Incident simplices iterator	a3m_traits<T>::radial_simplex_iterator

Table 4. Associated types of an A3M.

symbol	definition
rfi	typedef a3m_traits<T>::radial_face_iterator rfi;
rsi	typedef a3m_traits<T>::radial_simplex_iterator rsi;

Table 5. A3M related notation.

Refinement of abstract simplicial 3-complex	
expression	return type
on_boundary(t, v)	bool
on_boundary(t, e)	bool
on_boundary(t, f)	bool
incident_simplex(t, f)	simplex
incident_simplices(t, f)	pair<simplex, simplex>
a_incident_face(t, e)	face
boundary_faces(t, e)	pair<face, face>
a_incident_edge(t, v)	edge
radial_simplices(t, e)	pair<rsi, rsi>
radial_faces(t, e)	pair<rfi, rfi>

Table 6. Requirements of an A3M. Type t models an A3M. Types rsi and rfi model a radial simplex iterator and radial face iterator, respectively. Some pre-conditions must hold: boundary_faces(t, e) can be used only if on_boundary(t, e)==true, for instance.

Program 4 Opposite simplex. The tie function is just a compact way of assign a pair of values to two variables.

```
simplex opposite_simplex(T t, simplex s, face f) {
  simplex s1, s2;
  tie(s1, s2)=incident_simplices(t, f);
  if(s1==s) return s2 else return s1;
}
```

That modest operation is the key component of a *radial iterator*, that is, an iterator that traverses all faces or simplices meeting an edge. Radial iterators are used in algorithms that compute the star of vertices and edges, for instance, and are reminiscent of the Weiler's radial edge structure (RED) [32]. Again, nothing prevents users from implementing *ad hoc* iterators, perhaps based on some reliable implementation of the facet-edge structure of Dobkin and Laszlo [7].

3.3 Oriented Abstract 3-Manifold

Orientation is another notion we want capture. Since orientation can be defined in a purely combinatorial way, without reference to geometrical concepts, we choose to place the *oriented abstract 3-manifold* concept as a refinement of abstract 3-manifold.

An *orientation* on a n-manifold M is a function s that assigns for each n-simplex $\sigma \in M$, an integer in the set $\{+1, -1\}$. The choice of orientation in σ induces an orientation in its faces in the following way:

$$s(d_i(\sigma)) = (-1)^i s(\sigma), i = 0, \ldots, n. \tag{2}$$

An orientation is *coherent* if *contiguous* n-simplices, i.e., simplices sharing an $(n-1)$-simplex, induces opposites orientations in its common face, that is,

$$d_i(\sigma_1) = d_j(\sigma_2) \Rightarrow s(d_i(\sigma_1)) = -s(d_j(\sigma_2)),$$

where σ_1 and σ_2 are n-simplices in M. Now, we can define another basic object.

Definition 3 *An* oriented abstract n-manifold *is an abstract n-manifold plus a coherent orientation.*

In the 3-dimensional case, the additional requirement on an A3M is just the operator simplex_orientation that takes a manifold and a simplex and returns an int in the set $\{-1, 1\}$ (Table 7). Program 5 is the obvious implementation of the equation 2.

Refinement of abstract 3-manifold	
expression	return type
simplex_orientation(t, s)	int

Table 7. Requirements of an OA3M.

Program 5 Induced orientation of the faces.

```
int face_orientation(T t, simplex s, face f) {
    int o=simplex_orientation(t, s);
    for(int i=0; i<4; ++i, o*=-1)
        if(face_op(t, s, i)==f) return o;
}
```

3.4 3-Polyhedron and Combinatorial 3-Manifold

Until now, we discussed only combinatorial concepts. Let's introduce the geometrical counterpart of the previously defined concepts.

We call an *euclidean embedding* of an abstract simplicial complex K a function g from the vertex set V to an euclidean space E^m that maps a vertex $\nu \in V$ to a euclidean point $g(\nu) \in E^m$, such that $g(\sigma)$ is a set in general position in E^m, for all $\sigma \in K$. A subset $\mathcal{P}$ of E^m is a *geometric realization* of K if there is an embedding g satisfying

$$x \in \mathcal{P} \Leftrightarrow x \in \mathrm{ConvHull}(g(\sigma)), \text{ for some } \sigma \in K.$$

Below, we define the geometrical objects corresponding to abstract simplicial n-complex and abstract n-manifold.

Definition 4 *A n-polyhedron is a set $\mathcal{P} \subset E^m$ for which exists an abstract simplicial n-complex K and an euclidean embedding g such that $\mathcal{P} = |K|_g$.*

Definition 5 *A combinatorial n-manifold is a set $\mathcal{M} \subset E^m$ for which exists an abstract n-manifold M and an euclidean embedding g such that $\mathcal{M} = |M|_g$.*

From the computational side, a polyhedrom concept (Poly3) is just a refinement of AS3C with an additional requirement **euclidean_point** that takes a manifold and a vertex and return an euclidean point, see tables (8, 9 and 10). A combinatorial 3-manifold (C3M) is a Poly3 plus the requirements of an A3M and an oriented combinatorial 3-manifold (OC3M) is a Poly3 plus the requirements of an OA3M.

description	type
Point type	poly3_traits<T>::point_type

Table 8. Associated types of a Poly3.

symbol	definition
point	typedef poly3_traits<T>::point_type point;

Table 9. Poly3 related notation.

Refinement of abstract simplicial 3-complex	
expression	return type
euclidean_point(t, v)	point

Table 10. Requirements of a Poly3.

3.5 Binary Multitriangulation

Now, we'll investigate the interplay between combinatorial and geometrical concepts related to subdivision process and how this leads naturaly to the concept of binary multitriangulations.

A polyhedron $\mathcal{P}' = |K'|_g$ is a *subdivision* of the polyhedron $\mathcal{P} = |K|_h$, denoted by $\mathcal{P}' < \mathcal{P}$, if $\mathcal{P}' = \mathcal{P}$ and for each $\sigma' \in K'$ exists a $\sigma \in K$ such that

$$\text{ConvHull}(g(\sigma')) \subset \text{ConvHull}(h(\sigma)).$$

The above definition uses geometrical concepts like euclidean embeddings. Therefore, we can not assert *a priori* anything about how the complexes K and K' are related. However, a theorem of Newman, presented in modern form in [17], shows that $P' < P$ if, and only if, K' is stellar equivalent to K. Moreover, the stellar equivalence can be chosen in such a way that only stellar moves on 1-simplices ("edges") are used.

There is a good reason to restrict the stellar moves to moves on edges. Whenever a stellar subdivision happens in an edge ε, all simplices containing ε are splitted in two. Accordingly, a sequence of stellar subdivision induces a binary tree structure in the simplices. And binary trees often leads to simpler algorithms.

In order to define the binary multitriangulation concept (BMT), we need some auxiliary definitions. We follow closely the definitions in [12]. A *partially ordered set* (poset) $(C, <)$ is a set C with an antisymmetric and transitive relation $<$ defined on its elements. Given $c, c' \in C$, notation $c \prec c'$ means $c < c'$ and there in no $c'' \in C$ such that $c < c'' < c'$. An element $c \in C$, such that for all $c' \in C$, $c \leq c'$, is called a *minimal* element in C. If there is a unique minimal element $c \in C$, then c is called the *minimum* of C. Analogously are defined *maximal* and *maximum* elements.

Definition 6 *A* binary multitriangulation *is a poset* $(\mathcal{T}, <)$, *where* $\mathcal{T}$ *is a finite set of abstract 3-manifolds (named* triangulations*) and the order* $<$ *satisfies:*

1. *$M' \prec M$ if, and only if, $M' = (\varepsilon, \nu)M$, for some edge $\varepsilon \in M$.*
2. *There is maximum and minimum abstract 3-manifolds in $\mathcal{T}$, called* base triangulation *and* full triangulation*, respectively* [2];

Property 2 says, in fact, that a BMT is a *lattice*. Other fact which follows from the definition is that every two triangulations in $\mathcal{T}$ are stellar equivalent. As usual, a BMT can be thought as a directed acyclic graph (DAG), with one drain and one source, whose arrows are labeled with stellar subdivisions

---

[2] One can replace $M' \prec M$ by $M \prec M'$ in property 1. In this case, we must interchange base triangulation and full triangulation. This is a transformation from an *increasing* to a *decreasing* BMT. In [22], Puppo demonstrates that increasing and decreasing multitriangulations are equivalent.

on edges. From an algorithimic perspective, the key idea is to use the above mentioned binary tree structure in the simplices to encode the DAG.

To describe the requirements on a BMT, we need to do a little digression about *state changes* in a data structure. The formerly defined requirements, like incident_simplices, are deterministic functions without side effects, at least from the user viewpoint. In other words, incident_simplices must return the same value in sucessive invocations. The situation changes in the BMT case, because we want to be able to move from a triangulation in $\mathcal{T}$ to another. We can regard this move as a state change in the underlying data structure modeling a BMT. The point is that, between state changes, the functions like incident_simplices behave deterministically.

The BMT requirements in table 11 are divided in two groups: operators that changes the state (subdivide, weld and base_triangulation) and the others. The operator base_triangulation set the current triangulation in $\mathcal{T}$ to the base triangulation, while subdivide(t, e) applies a stellar subdivision to the edge e and weld(t, v) applies a stellar weld "removing" the vertex v. The predicate is_current is usefull to check if a simplex belongs to the current triangulation.

Refinement of abstract 3-manifold	
expression	return type
was_subdivided(t, e)	bool
subdivide(t, e)	void
in_base_triangulation(t, v)	bool
weld(t, v)	void
base_triangulation(t)	void
has_children(t, s)	bool
children(t, s)	pair<simplex, simplex>
has_parent(t, s)	bool
parent(t, s)	simplex
subdivided_edge(t, s)	edge
welded_vertex(t, s)	vertex
is_current(t, {v, e, f, s})	bool

Table 11. Requirements of a BMT. Type t models a BMT. The binary tree structure in the simplices can be traversed with children and parent.

We must remark that operators subdivide and weld implements just "local" transitions in the DAG, that is, if $\mathcal{T}$ and $\mathcal{T}'$ are the triangulations before and after subdivide be called, respectively, then $\mathcal{T}' \prec \mathcal{T}$. Program 6 illustrates how subdivide can be used to achieve non-local transitions. Note that the order of subdivision of the incident simplices to the subdivision edge is relevant: they are subdivided from the lowest to the highest level, where the level function is defined in program 7.

Program 6 Non-local subdivide. This program also ilustrates how the encoding of traversal capabilities in radial iterators makes the algorithm more generic and readable at no extra cost.

```
void non_local_subdivide(T t, edge e) {
    typedef set<pair<int, edge> > edge_set;
    edge_set sub_edges;
    if((e==empty_edge(t))||(!was_subdivided(t, e))) return;
    do {
        sub_edges.clear(); rsi i, end;
        for(tie(i, end)=radial_simplices(t, e); i!=end; ++i) {
            edge se=subdivided_edge(t, *i);
            if(se!=e) sub_edges.insert(make_pair(level(t, *i), se));
        }
        for(edge_set::iterator j=sub_edges.begin(); j!=sub_edges.end(); ++j)
            non_local_subdivide(t, j->second);
    } while(!sub_edges.empty());
    subdivide(t, e);
}
```

Program 7 The simplex level.

```
int level(T t, simplex s) {
    if(!has_parent(t, s)) return 0; else return level(t, parent(t, s))+1;
}
```

4 Models

In this section, we present data structures which are models of the concepts defined above, in the sense that they fill all necessary requirements. We have absolutely no pretension of describing "the best" data structure, because we think that data structures are somehow application dependent. But, if we did a good analysis in Sect. 3, most algorithms can be used in different applications without change.

Figure 3 resumes the prototypical data structure which models a BMT. Most operators are easily inferred by inspection. Stripping out some data, we obtain models to simpler concepts like AS3C and A3M. It remains to clarify certain points:

- Each element has sufficient information to recover its incident elements (incident_* fields) and its star (star fields);
- The subdivided edge of a simplex s is given by ith_edge(t, s, s.subdivided_edge);
- The array edge.star stores the boundary faces incident to a boundary edge, or stores a face incident to an internal edge;

- A face f is on boundary if, and only if, f.star[1]==0, and a edge e is on boundary if, and only if, e.star[0]!=e.star[1].
- A face f is current if, and only if, f.star[0].current==true. And a edge e is current if, and only if, is_current(t, e.star[0])==true.
- A ordering in adopted in the vertices of the simplex in such a way that the welded vertex is always the last vertex.

Actually, the data structure exhibited in figure 3 it's not quite the same we use in our implementation. The main difference is that the **struct bmt** is parametrized by an *attribute class*, which enables that new attributes be added to any element. One can, for instance, add the plane equation to each face or a scalar value to each vertex. This is done in compile time and causes no storage overhead. A similar technique is described in [3].

The creation process of binary multitriangulations in our implementation is straightforward. There is a **add_simplex** function that takes a triangulation t and four vertices v0, v1, v2 and v3 as arguments and adds to t all faces and edges not yet in t, updating the incidence and star data. Thus, all simplices of the base triangulation are entered. Afterward, according to applications needs, a sequence of calls to **edge_split** is dispatched. The **edge_split** function takes a triangulation t, an edge e and a *visitor object* **vis** as arguments. It applies a stellar move on t that splits the edge e. Each time a new element is created, a internal face of a incident simplex for example, a corresponding function in **vis** is called. This allows that user attributes, as the above mentioned face plane equation, be updated. The Visitor concept is explained in [26].

We note that this data structure is too general. Once more, the application guides the real implementation. In dealing with regular spatial decompositions, for example, most of work can be done procedurally (see, for example, [16]). Even out-of-core techniques (e.g., [8, 11]) can be implemented without changing the interface.

5 Applications

The development of multi-resolution techniques for volume visualization is one of our driving applications. Although multi-resolution techniques for 3D surfaces is a well-developed research area [13], the same is not true for multi-resolution techniques for 3D volumes. This is particularly true for multi-resolution techniques for unstructured volumetric grids (see [5] for a recent survey). In fact, compared to the surface case, it is possible to argue that multi-resolution work in volumetric grids is still at its infancy.

The data structures presented in this paper are aimed at providing a more formal and disciplined way of handling unstructured volumetric grids, which we hope will aid the further development of this area. To prove the usefulness of our concepts, we have implemented two simple applications:

```
struct vertex;
struct edge;
struct face;
struct simplex;
typedef vertex * vertex_descriptor;
typedef edge * edge_descriptor;
typedef face * face_descriptor;
typedef vertex * simplex_descriptor;

struct vertex {
  edge_descriptor star;
  bool current;
  bool boundary;
  bool in_base_triangulation;
};

struct edge {
  vertex_descriptor incident_vertices[2];
  face_descriptor star[2];
  bool was_subdivided;
};

struct face {
  edge_descriptor incident_edges[3];
  simplex_descriptor star[2];
};

struct simplex {
  face_descriptor incident_faces[4];
  simplex_descriptor child[2];
  simplex_descriptor parent;
  int subdivided_edge;
  bool current;
};

struct bmt {
  list<simplex_descriptor> simplex_list;
  list<face_descriptor> face_list;
  list<edge_descriptor> edge_list;
  list<vertex_descriptor> vertex_list;
};
```

Fig. 3. Modeling a BMT.

a simple unstructured grids volume renderer based on the ideas presented in [33, 25]; and a progressive volume approximation system similar in some respects to the one described by Roxborough and Nielson [24].

Direct Volume Rendering. An efficient technique for exploring graphics hardware for volume rendering is the Projected Tetrahedra (PT) algorithm of Shirley and Tuchman [25], which uses the traditional 3D polygon-rendering pipeline. This technique renders a volumetric grid by breaking the volumetric grid into a collection of tetrahedra. Then, each tetrahedra is rendered by *splatting* its faces on the screen. A key idea of the PT algorithm is exploit the fact that the aspect graph [21] of a tetrahedron consists of a few simple cases which can be encoded in a small table which depends completely on the dot products of the normal of the faces of the tetrahedron with the viewing direction. On average, one needs to render 3.4 triangles per tetrahedron [34].

In order to apply PT, one needs to compute a visibility-ordering of the cells. Williams' Meshed Polyhedra Visibility Ordering (MPVO) algorithm [33] developed in the early 1990s provides a very fast visibility-ordering algorithm. The MPVO algorithm explores the topological adjacencies of a convex model to produce a visibility ordering of the cells with respect to a given viewpoint. Given a tetrahedral mesh S, it produces an adjacency graph G of the mesh S, with vertices v of the graph corresponding to each tetrahedral cell c of the mesh, and edges e representing a common face between cells. A directed version of this graph is obtained when a viewpoint or viewing direction is specified, which is then used to define the direction of each edge by a simple dot product with the face normals (see Figure 4). For every new viewpoint, a directed graph as described is formed, and a topological sorting procedure outputs cells in visibility ordering. MPVO, which runs in linear time, works well for well-behaved meshes (acyclic and convex). General acyclic meshes can be handled with more complex algorithms as shown in [28].

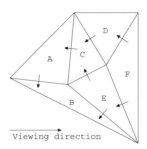

(a) MPVO phase 1

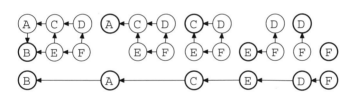

(b) MPVO phase 2

Fig. 4. Depiction of the MPVO algorithm [33]. MPVO works by first determining pairwise ordering relations between cells that share a face (a). Then, in a second phase, the ordering is determined by a topological sort of the induced visibility graph (b).

Our implementation of the PT part of the algorithm is quite simple, and it is completely driven by a set of tables for classifying the different cases (including degenerate cases). Figure 5 shows the six cases of projected tetrahedra.

Implementing the sorting is more interesting, since we are able to completely avoid building a separate graph or actually writing any sorting code. Our implementation for computing the visibility orders works by developing a *visibility graph* concept that is an adaptor over the C3M concept, i.e., it

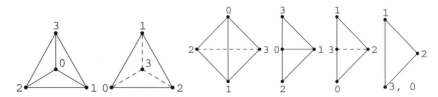

Fig. 5. Cases of the Projected Tetrahedra (PT) algorithm [25]. Depending on the particular case, between one to four triangles are rendered.

provides the appropriate graph interface to a combinatorial manifold. So, we can use all graph machinery built in the Boost Graph Library [26], for example, to solve the visibility problem. More specifically, in the simpler regular convex case, the visibility is computed by a single call to the BGL function topological_sort.

Figure 6 (Color Plate 22 on page 438) gives an example of direct volume rendering using the algorithm described above. Figure 6 (a) (Color Plate 22 (a) on page 438) shows the underlying space decomposition and the tetrahedral mesh. Figure 6 (b) (Color Plate 22 (b) on page 438) shows the volume visualization of a lung tumor dataset.

Progressive Tetrahedral Approximation of 3D Functions. A more interesting example of the use of our framework is our system for the hierarchical approximation and rendering of 3D scalar functions (see [20] for a survey). We use a simple approximation algorithm which recursively refines the triangulation until a user-defined error bound has been achieved. The exact procedure we implemented works by generating a collection of random points inside a given tetrahedron [23]. Then finding the best linear least-squares fit for the scalar function inside the simplex. If the error bound of the approximation is above the user-defined threshold, we continue to subdivide.

The approximation algorithm we implemented is somewhat similar to the described in [24] for computing a progressive tetrahedral approximation of ultrasound data, and it could potentially be used for the same purpose. In their system, a global least-squares fit is used, while in our system we handle each tetrahedron separately. Note that for a given vertex of the triangulation, each incident simplex determines one (possibly different) scalar value. We set the value at the vertex as the average value. Alternatively, we could perform an analysis of the discontinuities in the scalar function and represent this information as radial wedges [15].

For rendering purposes, we use a very simple sorting procedure shown in Program 8. The basic idea is the same used for traversing a BSP in back-to-front order. We always traverse the space region that does not contain the viewpoint (i.e., the "back"), before traversing the region that contains the viewpoint. This leads to a very efficient rendering algorithm.

Program 8 Hierarchical splatting.

```
void hier_splat(T t, simplex s) {
  if(is_current(t, s))
    splat(t, s);
  else {
    face f=internal_face(t, s);
    simplex s0, s1;
    tie(s0, s1)=children(t, s);
    if(visibility_test(t, f)) {
      hier_splat(t, s1);
      hier_splat(t, s0);
    } else {
      hier_splat(t, s0);
      hier_splat(t, s1);
    }
  }
}
```

Figure 7 demonstrates the BMT adaptation capabilities. These adapted decompositions were computed using least-squares fitting described above. Figure 7 (a) shows the mesh corresponding to a function that varies linearly in one direction and has a sharp discontinuity. Figure 7 (b) shows the mesh corresponding to the characteristic function of a sphere. Observe that we compute the full hierarchical decomposition, where the top level is a subdivision of a cube into six tetrahedra. In these images, we have removed one of the top level tetrahedra to reveal the internal structure of the mesh.

6 Conclusion

We have presented a generic programming framework for multiresolution spatial decompositions which was formulated through a rigorous mathematical analysis of the concepts involved. Indeed, our current implementation contains numerous generic algorithms for the extraction of topological information, as well non-generic functions to build a multiresolution mesh and execute other operations such as input/output. It is certainly possible to employ generic programming techniques in the creation of multiresolution meshes. Nonetheless, the problem is more involved and we plan to consider it in a future paper.

Concerning to future works, there are essentially two lines to follow. The first line is theoretical and consists in to investigate the BMT properties, its n-dimensional extension, its applicability in non-manifold settings and the incorporation of all stellar moves, not only moves on edges. The second line is to prove, by means of applications, mainly in the context of visualization of large datasets and finite element analysis, the advantages of our approach. We think that our work can be classified in the confluence of two new trends in graphics. On one hand, our concern to clearly separate geometric and topological concepts, emphasizing the later ones, brings us closer to *Computational Topology* as posed in [6]. This new branch is as promising today as Computational Geometry was thirty years ago. On the other hand, generic programming is a powerful methodology for computer programming, which holds the promise to complete, specially in relation to algorithm abstraction, the revolution started twenty years ago by object-oriented programming. Our hope is that this work will become the basis of a library called CTAL, that is, Computational Topology Algorithm Library.

References

1. J. Alexander. The combinatorial theory of complexes. *Ann. Math.*, 31:294–322, 1930.
2. G. Berti. *Generic software components for Scientific Computing*. PhD thesis, BTU Cottbus, 2000.
3. Mario Botsch, Stephan Steinberg, Stephan Bischoff, and Leif Kobbelt. Openmesh - a generic and efficient polygon mesh data structure. In *OpenSG PLUS Symposium*, 2002.
4. P. Cignoni, L. De Floriani, P. Magillo, E. Puppo, and R. Scopigno. TAn2 - visualization of large irregular volume datasets. Technical Report DISI-TR-00-07, University of Genova (Italy), 2000.
5. E. Danovaro, L. De Floriani, M. Lee, and H. Samet. Multiresolution tetrahedral meshes: an analysis and a comparison. In *Proceedings International Conference on Shape Modeling*, 2002.
6. T. Dey, H. Edelsbrunner, and S. Guha. Computational topology. In B. Chazelle, J. E. Goodman, and R. Pollack, editors, *Advances in Discrete and Computational Geometry (Contemporary mathematics 223)*, pages 109–143. American Mathematical Society, 1999.
7. D. P. Dobkin and M. J. Laszlo. Primitives for the manipulation of threedimensional subdivisions. *Algorithmica*, 4:3–32, 1989.
8. Jihad El-Sana and Yi-Jen Chiang. External memory view-dependent simplification. *Computer Graphics Forum*, 19(3):139–150, August 2000.
9. Jihad El-Sana and Amitabh Varshney. Generalized view-dependent simplification. *Computer Graphics Forum*, 18(3):83–94, September 1999.
10. Andreas Fabri, Geert-Jan Giezeman, Lutz Kettner, Stefan Schirra, and Sven Schonherr. On the design of CGAL a computational geometry algorithms library. *SP&E*, 30(11):1167–1202, 2000.
11. Ricardo Farias and Cláudio T. Silva. Out-of-core rendering of large, unstructured grids. *IEEE Computer Graphics & Applications*, 21(4):42–51, July / August 2001.
12. L. De Floriani, E. Puppo, and P. Magillo. A formal approach to multiresolution modeling. In W. Straßer, R. Klein, and R. Rau, editors, *Theory and Practice of Geometric Modeling*. SpringerVerlag, 1996.

13. M. Garland. Multiresolution modeling: Survey & future opportunities. In *Eurographics '99, State of the Art Report (STAR)*, 1999.
14. L. J. Guibas and J. Stolfi. Primitives for the manipulation of general subdivisions and the computation of voronoi diagrams. *ACM Trans. Graph.*, 4:74–123, 1985.
15. Hugues Hoppe. Efficient implementation of progressive meshes. *Computers & Graphics*, 22(1):27–36, February 1998.
16. M. Lee, L. De Floriani, and H. Samet. Constant time neighbor finding in hierarchical tetrahedral meshes. In *Proceedings International Conference on Shape Modeling*, pages 286–295, 2001.
17. W. B. R. Lickorish. Simplicial moves on complexes and manifolds. In *Proceedings of the Kirbyfest*, volume 2, pages 299–320, 1999.
18. M. Mäntylä. *An Introduction to Solid Modeling.* Computer Science Press, Rockville, Maryland, 1988.
19. J. Peter May. *Simplicial Objects in Algebraic Topology*, volume 11. D. Van Nostrand Company, Inc., Princeton, 1967.
20. Greg Nielson. Tools for triangulations and tetrahedrizations and constructing functions defined over them. In *Scientific Visualization: Overviews, Methodologies, and Techniques*, pages 419–515. IEEE CS Press, 1997.
21. H. Plantinga and C. R. Dyer. Visibility, occlusion, and the aspect graph. *Internat. J. Comput. Vision*, 5(2):137–160, 1990.
22. E. Puppo. Variable resolution triangulations. *Computational Geometry Theory and Applications*, 11(34):219–238, 1998.
23. C. Rocchini and P. Cignoni. Generating random points in a tetrahedron. *Journal of Graphics Tools*, 5(4):9–12, 2000.
24. T. Roxborough and Gregory M. Nielson. Tetrahedron based, least squares, progressive volume models with application to freehand ultrasound data. In *IEEE Visualization 2000*, pages 93–100, October 2000.
25. P. Shirley and A. A. Tuchman. Polygonal approximation to direct scalar volume rendering. *Computer Graphics*, 24(5):63–70, 1990.
26. Jeremy G. Siek, Lie-Quan Lee, and Andrew Lumsdaine. *Boost Graph Library, The: User Guide and Reference Manual.* Addison-Wesley, 2002.
27. Jeremy G. Siek and Andrew Lumsdaine. The matrix template library: A generic programming approach to high performance numerical linear algebra. In *IS-COPE*, pages 59–70, 1998.
28. C. T. Silva, J. S. B. Mitchell, and P. Williams. An exact interactive time visibility ordering algorithm for polyhedral cell complexes. In *1998 Volume Visualization Symposium*, pages 87–94, October 1998.
29. A. A. Stepanov and M. Lee. The Standard Template Library. Technical Report X3J16/94-0095, WG21/N0482, ISO Programming Language C++ Project, 1994.
30. B. Stroustrup. *The C++ Programming Language: Third Edition.* Addison-Wesley, 1997.
31. Luiz Velho and Jonas Gomes. Variable resolution 4-k meshes: Concepts and applications. *Computer Graphics forum*, 19:195–212, 2000.
32. Kevin Weiler. Edge-Based Data Structures for Solid Modeling in Curved-Surface Environments. *IEEE Computer Graphics and Applications*, 5(1):21–40, 1985.
33. P. Williams. Visibility ordering meshed polyhedra. *ACM Transactions on Graphics*, 11(2):103–126, 1992.

34. C. Wittenbrink. Cellfast: Interactive unstructured volume rendering. In *Proceedings IEEE Visualization'99, Late Breaking Hot Topics*, pages 21–24, 1999. Also available as Technical Report, HPL-1999-81R1, Hewlett-Packard Laboratories.

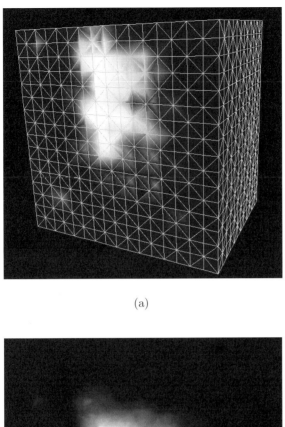

(a)

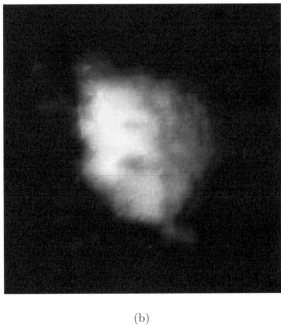

(b)

Fig. 6. Volume visualization of tumor dataset.

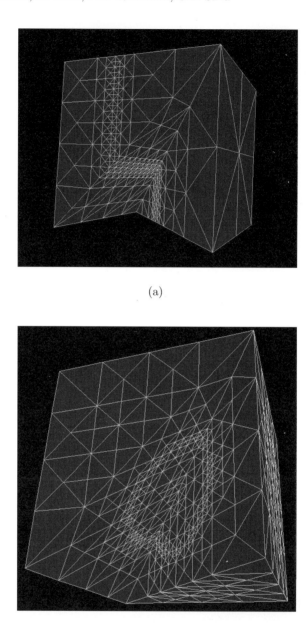

(a)

(b)

Fig. 7. Adapted decompositions of two different scalar functions.

Mathematical Modelling and Visualisation of Complex Three-dimensional Flows

Alfio Quarteroni[1,2], Marzio Sala[1], M.L. Sawley[1,3], N. Parolini[1], and G. Cowles[1]

[1] MOCS, SB, École Polytechnique Fédérale de Lausanne, Switzerland,
   {*alfio.quarteroni,marzio.sala*}*@epfl.ch*
[2] MOX, Dipartimento di Matematica, Politecnico di Milano, Italy
[3] Granulair Technologies, Lausanne, Switzerland

Summary. Three-dimensional fluid flows are characterised by the presence of complex physical phenomena. Numerical algorithms that provide accurate approximations to the governing flow equations and visualisation to enable the detection and analysis of particular flow features, both play important roles in the mathematical modelling of such flows. Two specific three-dimensional flow applications are presented to illustrate the use of appropriate visualisation techniques.

1 Introduction

Using mathematical modelling, a problem arising in a particular discipline (e.g. physics, chemistry, biology, engineering, medicine or economics) is analysed via a suitable set of mathematical equations. The original "real world" problem is thus pre-processed into a mathematical frame for numerical resolution of the governing equations, and then transformed back to the real world frame for post-processing of the results (Figure 1). The determination of the governing equations, the analysis of their qualitative properties, the development of numerical methods for their approximation, their efficient resolution by computer, and the visualisation, analysis and validation of the results obtained, form the essential steps of the mathematical modelling process.

The numerical simulation of complex fluid flows is currently performed for a wide variety of scientific and engineering problems. Not only can such simulations provide detailed insights into fundamental flow behaviour, but they can also be used as a *virtual prototyping* tool for industrial product design and manufacturing. Certain industrial sectors – such as aeronautical, mechanical, civil, chemical and nuclear engineering – have for a number of years benefited from the use of numerical flow simulation for the design, control, optimisation and management of critical processes. More recently, numerical flow simulation has also become increasingly common in fields such as micro-electronics, environment, finance and medicine (Figure 2).

Mathematical modelling is inspired by fundamental physical principles. For fluid flows these consist of the conservation of mass, momentum and energy of the continuous medium. The description of the fluid properties, the

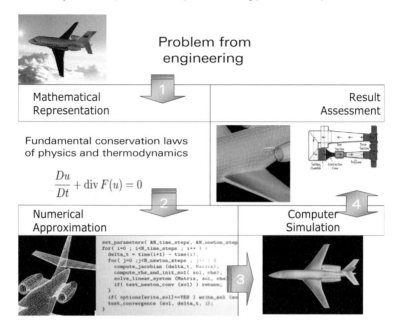

Fig. 1. The process of mathematical modelling.

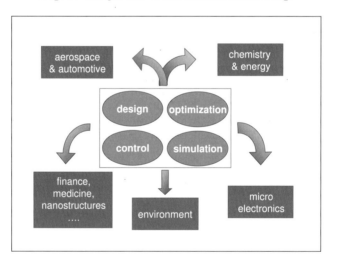

Fig. 2. Scope of mathematical models and fields of application.

determination of boundary conditions, and the dimensionalisation and measurement of the problem parameters are provided by an initial engineering analysis.

Engineers use in a complementary manner experimental data and numerical simulation. The example outlined in Figure 3 refers to the design of a vehicle (e.g. a car or an aircraft) and shows the interplay between experimental tests in wind tunnel and numerical simulation before obtaining the final design. The mathematical modelling process, therefore, is sometimes referred to as a *virtual wind tunnel*.

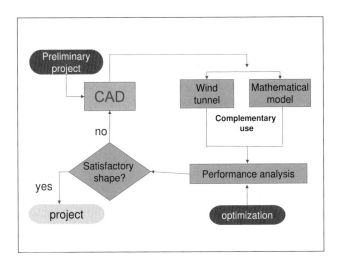

Fig. 3. Typical project cycle.

Mathematical modelling of complex three-dimensional flows involves the generation of a large solution database consisting of the values of the flow variables at selected locations throughout the computational domain. Visualisation plays an important role in the extraction of useful information from this database. Although on-line (or runtime) visualisation can be effectively used in a number of situations (e.g. for the debugging of the solution procedure), post-processing is generally undertaken through a detailed *a-posteriori* analysis.

In this paper, several important aspects of the mathematical modelling of complex three-dimensional flows are presented. Techniques are described that have been developed to enable the resolution of the governing mathematical equations and the visual representation of the obtained solutions. These techniques are demonstrated by two different applications, the aerodynamic flow around a complete aircraft, and the hydrodynamic and aerodynamic flow around a racing yacht. The first application was calculated using the THOR code (developed at the Von Karman Institute, Belgium) and visualised with

Visual3 [3], while the second application was calculated and visualised using the commercial software package FLUENT.

2 Mathematical Formulation

The primary goal of an applied mathematician is the effective solution of the problem which is expressed by a mathematical model. Most often, these problems are governed by complex partial differential equations, whose solution in explicit form is seldom available from pure analysis. In some other cases, the solution can be represented only implicitly through complex expression involving series or integrals, which is of no constructive use or its representation is not constructive. Numerical algorithms thus have to be developed to allow the construction of an approximate solution of the mathematical model at hand. Mathematical modelling aims at guaranteeing that not only the approximate solution can be efficiently computed, but also that the error between the physical and numerical solution is small and controllable. For a better insight of the concepts outlined in this Section, the reader is referred to [9, 7, 5] and the references therein.

In an abstract framework, the mathematical formulation of a physical problem can be represented as

$$u = P(d), \tag{1}$$

where d represents the data (e.g. the internal forces and the boundary conditions), u the solution and P an operator which represents the mathematical model, typically based on partial or ordinary differential equations.

Very often the quantities of physical interest are continuously varying functions, and therefore (1) has infinite dimension. In this case, the problem is discretised by dividing the computational domain into small elements of simple shape (e.g. triangles or quadrilaterals in 2D, tetrahedrons or prisms in 3D). An example of the resulting computational mesh is given in Figure 4.

A numerical approximation u_h of the solution u of (1) can be obtained by solving the family of numerical problems

$$u_h = N(d_h). \tag{2}$$

Here d_h represents the "numerical" data, N the discretised operator and h a parameter which is related to the dimension of the numerical problem (typically, h is the mesh size). On each element the unknown solution is assumed to have, for example, constant, linear or quadratic variations with respect to the spatial (and temporal) coordinates. Smaller values of h indicate better approximations.

The strategy for choosing N distinguishes the various families of discretisation methods. For instance, for finite element methods (2) is a suitable orthogonal projection of (1) upon piecewise polynomial subspaces, while for

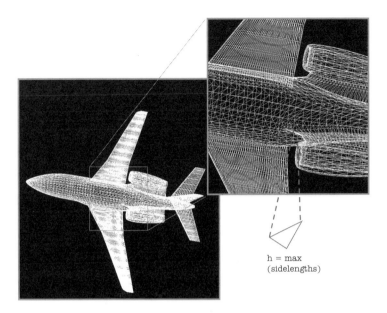

h = max
(sidelengths)

Fig. 4. Example of a computational mesh (only the surface mesh is shown).

finite volume methods (2) is obtained by applying the Gauss divergence theorem to (1) on each element.

In order to be well-posed, the discretised operator N must be *consistent*, that is

$$P^{-1}(u) - N^{-1}(u) \to 0 \text{ as } h \to 0. \tag{3}$$

Moreover, it must be *stable*: given two discrete sets of data d_h and δ_h "close" enough, the corresponding numerical solutions $u_h = N(d_h)$ and $\omega_h = N(\delta_h)$ must also be "close", i.e.

$$\exists C > 0 : \quad |u_h - \omega_h| \approx C|d_h - \delta_h| \quad \forall h > 0. \tag{4}$$

The fundamental property is that consistency and stability imply *convergence*, that is

$$|u - u_h| \to 0 \text{ as } h \to 0. \tag{5}$$

Typically it can be shown that the norm of the error can be bounded *a-priori* by a formula of the following form

$$|u - u_h| \leq K(u)h^p \tag{6}$$

for a suitable $p \in \mathbb{N}$ which is the *order* of the method. It is therefore clear that the parameter h plays a key role in the numerical model.

One can also develop an *a-posteriori* analysis, yielding

$$|u - u_h| \approx \sum_T h_T \sigma_T(u_h), \tag{7}$$

where T represents a generic element of the mesh associated to problem (2), h_T its diameter, and σ_T a given function of the numerical solution u_h.

3 Complexity

The algorithm *efficiency* is measured, among other factors, by its complexity, which is related to the amount of resources (typically, computation time and memory storage) needed for its implementation. The belief that available computer systems now allow the solution of problems of arbitrary complexity is illusory. For a number of application areas, resource limitations will continue to exist inhibiting the resolution of the complete set of mathematical equations. This is particularly true for computational fluid dynamics considered here. The numerical resolution of the complete set of partial differential equations governing viscous, compressible, rotational flows – the Navier-Stokes equations – provide substantial challenges for the mathematical modelling of complex flows in 3D geometries.

When a problem is too complex, a suitable model reduction must be considered. A list of potential reduction schemes includes:

– *physical reduction*: based on physical considerations, a simpler mathematical model can often be employed. As shown in Figure 5, the complete Navier-Stokes equations can be simplified under certain flow conditions, leading to a hierarchy of mathematical equations of different complexity. The appropriate choice of governing equations is determined by the physical flow properties.

  In principle, it is possible to combine different equations sets to resolve in an effective manner problems containing regions of different flow behaviour [10]. An example is provided in Figure 6, where the transonic flow over an aircraft wing is divided into two flow domains; the full potential equation is applied where there are no shock waves and the fluid is irrotational, whereas the Euler (or the Navier-Stokes) equations are used near the wing, where the interaction of shock waves with the body requires a more complete model.

  Another important application of physical reduction is its application to turbulent flows. The direct simulation of turbulence is not computationally feasible for flows with high Reynolds number (defined as the ratio of inertial to viscous forces acting on the fluid). It is then necessary to incorporate turbulence models into the governing equations to account for the effect of small-scale motions on the larger scales that are actually simulated (see Figures 7 and 8).

– *geometrical reduction*: in some circumstances 2D, 1D and even 0D (that is, point-wise) models can furnish enough information. Again, a judicious

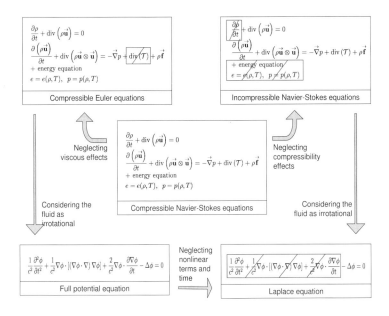

Fig. 5. Mathematical models for fluid flows. Starting from the (compressible) Navier-Stokes equations, one can derive simpler models, depending on the physics of the problem. Suitable initial and boundary conditions must be added to the equations in order to make the mathematical problem well-posed.

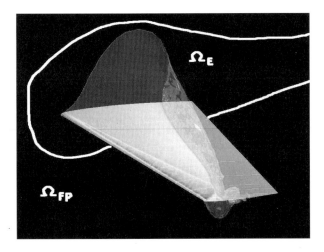

Fig. 6. Example of physical model reduction. The numerical solution shows the Mach number contours on the wing, and the iso-surface at unitary Mach number. Ω_{FP} is the far-field region where the full-potential equation is used, whereas in Ω_E, near the wing, the Euler equations are solved.

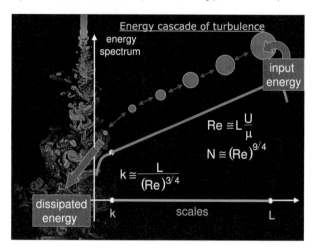

Fig. 7. A cascade of energy can be used to explain the turbulence scales. However, a rough estimation of the computational resources needed for a direct simulation suggests that they are excessive for real-world computations.

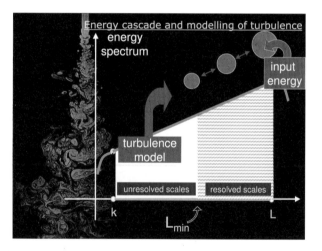

Fig. 8. Example of model reduction arising from turbulence. Only the large scales are simulated, while the small scales are treated using a simpler model.

combination of these reduced models can often provide the most appropriate choice. For example, for the human cardiovascular system, point-wise models describing the pressure and flux distribution have been available since the 1950s. These models can be enriched by treating veins and arteries as 1D domains, and 3D geometries considered if further flow details require investigation (e.g. oscillation of shear stresses that may be responsible for artherosclerotic plaque formation). See, for example, [6, 8].

– *algorithmic reduction*: once a suitable model has been obtained after physical and geometrical reduction, it may be prohibitive to solve due to the computational cost. Efficient numerical algorithms must be employed to minimise these costs. Of particular interest in recent years has been the use of high-performance parallel computer systems. Most currently available parallel computers have distributed memory, and thus require specific algorithm development to extract the required performance. For computational fluid dynamics, algorithms based on domain decomposition are generally employed [10, 12]; the computational domain is partitioned into a number of subdomains, the flow in which is computed by different processors (Figure 9).

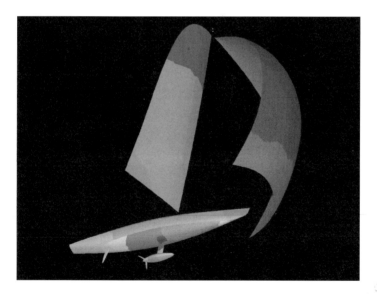

Fig. 9. Partitioning of the domain into 32 subdomains for parallel computation of the flow around a racing yacht. Different colours represent subdomains that are treated by different processors. For the sake of clarity, only the surface is shown.

– *mesh adaption* [2, 1, 13]: Equation (6) states that, for a given numerical model (and henceforth for a given p), one should use a value of h as small as possible in order to reduce the error. However, since $K(u)$ is a function of the unknown solution u, no quantitative information can be extracted from (6). To reduce the total error one possibility is to modify the local value of h, accordingly to the local behaviour of the numerical solution. For instance, given a solution u_h obtained using a certain mesh, Formula (7) suggests that smaller values of h_T should be used where the local error is large, while larger h_T can be adopted for small values of the local error. This can be implemented by a mesh adaption (refinement-derefinement) procedure to be applied locally. An example of meshes and solutions for non-adapted and adapted meshes is given in Figure 10.

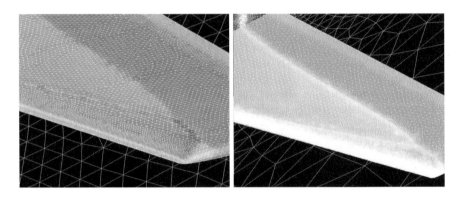

Fig. 10. Mach number contours for a non-adapted mesh (left) and for an adapted mesh (right) on an ONERA M6 wing.

4 Visualisation Techniques

Visualisation has become an integral part of mathematical modelling, and especially for complex 3D numerical flow simulation [4]. A number of different goals can motivate the use of visualisation tools, which include:

– analysis of the solution to determine its correctness,
– qualitative analysis of the solution to extract physical insights into the flow behaviour,
– quantitative analysis of specific details of the solution, for example for comparison with experimental data (Figure 11 and Color Plate 35 on page 445),
– global presentation of the simulation results to non-specialists, for example in the framework of a global optimisation procedure.

A wide variety of visualisation techniques have been developed over the years to capture the wealth of physical phenomena that are inherent in complex 3D flows. Of particular importance to our investigations have been:

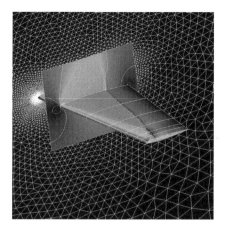
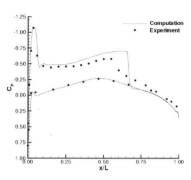

Fig. 11. On the left, the Mach number contours for an ONERA M6 wing. On the right, comparison of the between numerical (on continuous line) and experimental (dotted) pressure distributions on a section of the wing.

- contours of physical quantities (e.g. pressure, velocity),
- iso-surfaces, indicating surfaces of constant values of physical quantities,
- cutting planes, to highlight details within the flow domain,
- streamlines, which provide a more "dynamic" view of certain physical phenomena, such as recirculating flow.

Combining the above standard visualisation techniques has enabled a wide variety of physical phenomena to be illustrated and analysed. Here we will concentrate on two representative applications.

The aeronautical industry has long been a strong advocate of the use of mathematical modelling. In this application area, it is critical for numerical flow simulation to produce reliable estimates, for example, for the aerodynamic forces exerted on an aircraft. A full range of mathematical models are employed, ranging from the potential flow equations to the full Navier-Stokes equations, with each model having its appropriate domain of applicability (see Figure 5). Figure 12 (Color Plate 36 on page 445) provides an illustration of the results of an Euler calculation around a complete Falcon aircraft. Contours of Mach number in critical areas of the flow field provide useful information on the flow characteristics. Iso-surfaces of Mach number over a wing are shown in Figure 13. The red zone on the upper part of the wing represents a supersonic area. The sudden transition from red to green indicates the presence of a strong shock wave; moreover, the concentration of iso-surfaces behind it reveals that a further (weaker) shock wave develops.

Mathematical modelling is commonly employed to gain a competitive edge in the design of racing yachts for the America's Cup. While traditionally, reduced models such as the potential flow equation have been employed, there is

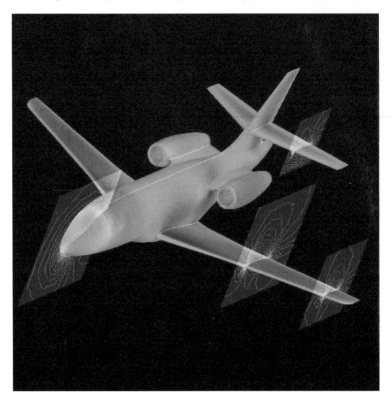

Fig. 12. Mach number contours on the surface of a Falcon aircraft and on selected cutting planes. The asymptotic Mach number is 0.45, with one degree of angle of attack.

now considerable effort in the resolution of the (incompressible) Navier-Stokes equations. Numerical simulations have been performed for the hydrodynamic flow around the hull and its appendages (keel, bulb, winglets and rudder), the aerodynamic flow around the sails, mast and exposed section of the hull, as well as the waves generated at the water-air free surface. Figure 14 (Color Plate 37 on page 446) shows an example of the aerodynamic flow around the sails of a downward sailing boat. Visualisation of streamlines demonstrates that the sails act as a combination of a parachute (indicated by the presence of separated flow on their leeward surface) and a vertical wing (indicated by the generation of trailing vortices) [11]. The total pressure iso-surfaces indicate the extent of the perturbed flow behind the sails. By considering two identical boats at different spatial locations, the surface pressure contours shown in Figure 15 can be used to evaluate the effect of the resulting aerodynamic shadowing. Further information on the aerodynamic interaction between the two boats can be deduced by displaying streamlines around the

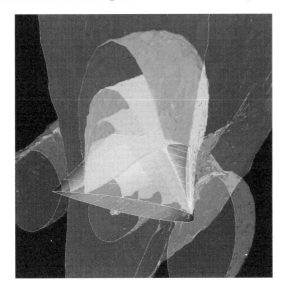

Fig. 13. Mach number iso-surfaces for the ONERA M6 wing. The asymptotic Mach number is 0.84 and the angle of attach 3.06. A second order scheme with 585725 nodes and 3477090 cells has been used in this computation.

sails, as in Figure 16. Finally, visualisation of the height of the free surface surrounding a boat (shown in Figure 17 and Color Plate 38 on page 446) reveals the typical pattern of surface waves generated by the hull, which are responsible for a significant fraction of the total drag on the boat.

5 Conclusions

Practical experiments will continue to play a key role in the study of fluid dynamics. However, due to their flexibility, which provides rapidity in the evaluation of different designs, numerical simulations using a virtual wind tunnel (Figure 18) are expected to become the main tool for problems involving complex fluid flows.

In this note we have briefly outlined the process of deriving a mathematical model of fluid flows. Firstly, the problem has to be discretised in space and time; then, a (convergent) numerical scheme has to be introduced to obtain an approximate solution. Unfortunately, a complete model (like the compressible Navier-Stokes equations) is often too complex to be solved in its entirety. Most often, model reduction must be applied, based on physical, geometrical or algorithmic considerations. In particular, we have focused our attention on simulations of aerodynamic and hydrodynamic flow examples.

We have shown that appropriate visualisation tools for complex simulations can greatly help in both monitoring the solution and analysing it.

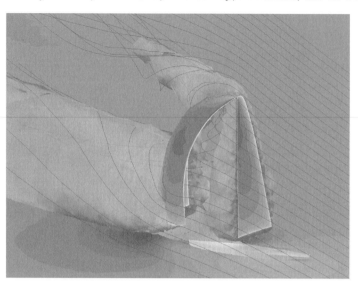

Fig. 14. Streamlines and total pressure iso-surfaces around the downwind sails of a racing yacht, illustrating the flow separation behind the sails and the generation of trailing vortices. The boat speed is 10 knots (1 knot = 0.514 m/s), and the true wind speed is 15 knots at an angle of 160 degrees.

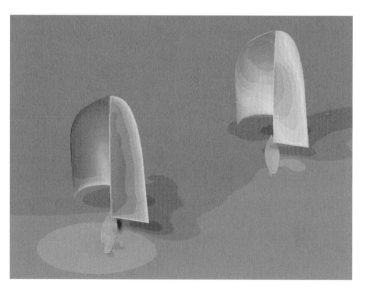

Fig. 15. Surface pressure on the downwind sails of two identical racing yachts, indicating the influence of the aerodynamic shadow.

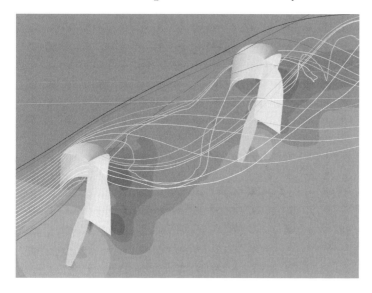

Fig. 16. Streamlines around two identical racing yachts sailing downwind, indicating the aerodynamic interaction between the two boats.

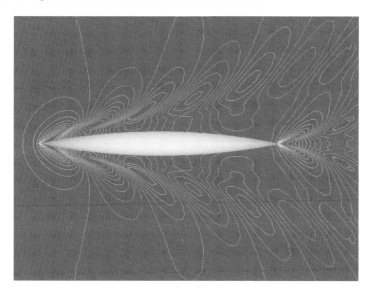

Fig. 17. Contours of surface height indicating the waves generated around a Wigley hull at the water-air interface.

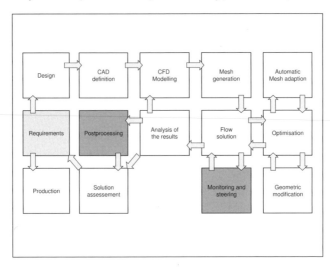

Fig. 18. A virtual wind tunnel. Starting from given requirements (the yellow box), the modelling phase leads to the definition of a mathematical model. A preliminary flow field solution is then computed using an initial mesh. Model reduction, mesh adaption and optimisation cycles usually follow. Visualisation has an established role, in both monitoring the solution and in interpreting the results.

Streamlines, iso-surfaces and particle tracking can provide a global understanding of the flow field that may be of paramount importance during the whole simulation process.

Acknowledgement

The results related to the aeronautical simulations were obtained during the European project BRITE/EURAM III IDeMAS under contract number BRPR-CT97-0591. L. Formaggia is acknowledged for many helpful discussions, and the OFES (Office Fédérale de l'Éducation et de la Science) is acknowledged for the financial support. The simulations around a racing yacht was undertaken within the framework of the engagement of the Ecole Polytechnique Fédérale de Lausanne as Official Scientific Advisor to the Alinghi Challenge for the America's Cup 2003. The interest and interaction of the Alinghi Design Team are gratefully acknowledged.

References

1. M. Ainsworth and J.T. Oden. *A Posteriori Error Estimation in Finite Element Analysis*, John Wiley & Sons, 2000.

2. I. Babuška and T. Stroboulis. *The Finite Element Method and its Reliability*, Oxford University Press, 2001.
3. R. Haimes. Visual3: Interactive Unsteady Unstructured 3D Visualization. *AIAA Paper 91-0794*, 1991.
4. H.-C. Hege and K. Polthier (Eds.). *Mathematical Visualization*, Springer-Verlag, 1998.
5. C. Hirsch. *Numerical Computation of Internal and External Flows*, John Wiley & Sons, 1990.
6. A. Quarteroni and L. Formaggia. Mathematical Modelling and Numerical Simulation of the Cardiovascular System, in *Modelling of Living Systems, Handbook of Numerical Analysis Series*, (P.G. Ciarlet and J.L. Lions Eds.) (to appear), 2002.
7. A. Quarteroni, R. Sacco, and F. Saleri. *Numerical Mathematics*, Springer-Verlag, 2000.
8. A. Quarteroni. Modeling the Cardiovascular System: a Mathematical Challenge, in *Mathematics Unlimited - 2001 and Beyond* (B. Engquist and W. Schmid Eds.), Springer-Verlag, 2001.
9. A. Quarteroni and A. Valli. *Numerical Approximation of Partial Differential Equations*, Springer-Verlag, 1994.
10. A. Quarteroni and A. Valli. *Domain Decomposition Methods for Partial Differential Equations*, Oxford University Press, 1999.
11. P. J. Richards, A. Johnson, and A. Stanton. America's Cup Downwind Sails – Vertical Wings or Horizontal Parachutes? *Journal of Wind Engineering*, Vol. 89, pp. 1569–1577, 2001.
12. B. Smith, P. Bøjrstad and W. Gropp. *Domain Decomposition: Parallel Multilevel Methods for Elliptic Partial Differential Equations*, Cambridge University Press, 1996.
13. R. Verfürth. *A Review of a Posteriori Error Estimation and Adaptive Mesh Refinement Techniques*, Wiley-Teubner, 1996.

web *Mathematica*

Tom Wickham-Jones

Wolfram Research Inc., 100 Trade Center Drive, Champaign Illinois, USA.
twj@wolfram.com

Summary. This article describes web*Mathematica*, a technology that allows the building of web sites that carry out computations and visualizations. It will discuss the basics of web*Mathematica* and how to develop material for web based visualization and computation. It will also discuss the use of MathML for visualization of mathematical notation on the web.

1 Introduction

web*Mathematica* provides server based web computation and visualization tools using *Mathematica* as its engine. Some of the features that make *Mathematica* particularly suitable include: numerical and symbolic computation, visualization, programming, connectivity to other languages such as Java, the *Mathematica* notebook user interface and mathematical typesetting. The features of *Mathematica* are explored further on the Wolfram Research web site, `http://www.wolfram.com`.

There are a number of standard benefits that derive from the use of web based technology. One important issue is the ease of use. In the modern world people typically learn to use web technology at a very early stage in their adoption of computer technology and hence they need less training with web based applications. In addition, web applications often provide a very structured interface. Configuration of the application can benefit from the server arrangement; everything is held on the server and so updates and fixes need only take place in one central location. In addition, installation of software is considerably simpler, delivery of material such as HTML pages, JavaScript functions or Java applets are all taken care with standard well understood solutions. Another important feature is that access to sites can be controlled and measured with standard technologies.

In this paper it is not possible to give a broad and deep description of web*Mathematica*, instead it will give an overview. It will start with a very brief review of the underlying technology provided by web*Mathematica* itself. It will then consider some of the extension technologies that are available for building web applications and summarize how they can be used with web*Mathematica*. Easy interaction and cooperation between different technologies is particularly important since the web, by definition, is a highly distributed and often extremely heterogeneous environment. This is followed by a section that reviews actual applications that incorporate web*Mathematica* technology.

2 web*Mathematica* Technology

The web interaction of web*Mathematica* is provided by a Java web technology called Java servlets. Servlets are special Java programs that run on a web server machine. Support is provided by a separate program called a servlet container (or sometimes a "servlet engine") that connects to the web server. One popular servlet container is Apache Tomcat, `http://jakarta.apache.org/tomcat/`, which is the reference implementation. Essentially all modern web servers support servlets natively or through a plug-in servlet container. This includes Apache, Microsoft's IIS and PWS, Netscape Enterprise Server, iPlanet, and application servers (such as IBM WebSphere). Closely related to Java servlets are Java Server Pages (JSPs); both servlets and JSPs integrate very closely with web*Mathematica*.

2.1 *Mathematica Server Pages*

web*Mathematica* is driven by a technology called *Mathematica Server Pages*. This involves inserting *Mathematica* commands into HTML pages, a form of HTML templating. These pages, called MSP scripts, are easy to write and fit well with HTML development tools such as editors. Part of a sample MSP script is shown below.

```
<form action="Expand" method="post">
  Enter a polynomial (eg x+y):
  <input type="text" name="expr" size="10">
  Enter a positive integer (eg 4):
  <input type="text" name="num" size="3">
  <br/>

<%Mathlet
        MSPBlock[{$$expr,$$num},
                Expand[$$expr^$$num]] %>

  <br/>
  <input type="submit" name="button" value="Evaluate">
</form>
```

This is all HTML except for the Mathlet tag. Its contents are computed by *Mathematica* on the server and the result inserted into the page. The generated page is then returned to the browser.

There are many *Mathematica* commands for working with web requests and responses. Support is built in for plotting, formatting and typesetting, embedding applets, and returning general content. In the example above, the command `MSPBlock`, takes care of converting the input variables `$$expr` and `$$num`, to *Mathematica* input and then evaluating the `Expand` command. If there is any security problem with the input, this will be detected.

Some examples are available at: `http://www.wolfram.com/products/ webmathematica/examples`.

2.2 Technology Details

To set up the web*Mathematica* site, a number of components are added to the servlet container. These follow a standard layout to form what is called a web application. The features that are supported by these components are summarized here.

One important feature is the session manager which launches and initializes *Mathematica* and shuts down sessions that exceed some preset limit for computations. Input to the server uses a standard HTTP mechanism, with name/value pairs being sent, either on the URL or in an HTTP header. The result from the server can be any result that *Mathematica* can generate. This could be, for example, HTML, MathML, XML, images, or *Mathematica* notebooks.

Since web*Mathematica* is built on top of web standards it can make use of standard web technology for security. This allows a web*Mathematica* site, if desired, to limit access to users from certain domains or to make use of a variety of web authentication systems. In addition a security system is built in to process all inputs to a web*Mathematica* site, an input that does not pass the security system will be rejected. The security system is configurable and by default will reject all except fundamental mathematical objects such as those that are necessary for basic arithmetic, as well as elementary, transcendental, and special functions in addition to numbers and names that are known not to have any meaning in *Mathematica*. This means that potentially dangerous *Mathematica* commands, such as those for launching process, or inspecting or modifying file systems cannot be used. In addition, programming constructs and other manipulation functions, that might be used to programmatically bypass the security system, are also rejected.

An overview of the steps in processing a page is shown below. For the sake of illustration, this has a *Mathematica* session pool with three sessions.

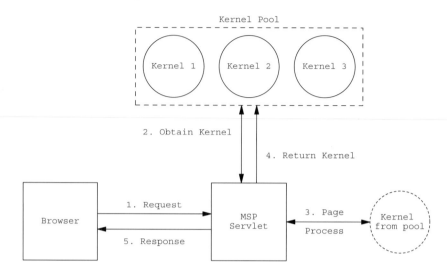

3 Web Extension Technologies

There are a number of ways to extend a web*Mathematica* site. Since it is fundamentally a server technology it is worth using client technologies to build a more sophisticated web application. Of course, web*Mathematica* can be used with a basic HTML client. However, it can also be used with more sophisticated client technology. In addition, there are a number of important server features that can be used with web*Mathematica*.

3.1 Client Technologies

Some of the client technologies for extending a web application include the following.

JavaScript, which is a scripting language useful for manipulating documents and other features of browsers, can be used. Scripting the browser is very useful providing a richer user experience. For example, checking input values before they are sent to the server. The web*Mathematica* distribution contains a number of examples that show how to use JavaScript in a web*Mathematica* site.

Plug-ins, Active-X controls, and Java applets are often used to provide extra functionality to the browser. They fit well into the world of HTML, Java, and JavaScript. One useful functionality is the rendering of MathML, this is described in a following section. Another involves the use of JavaView, also described in a following section.

A final example of client technology is Microsoft Excel^{TM} which has a built-in mechanism to get data from a web site. This can be readily deployed to call on a web*Mathematica* server.

3.2 Server Technologies

Since web*Mathematica* is based on a Java web technology, these are the server technologies with which web*Mathematica* is most easily integrated. These Java technologies, Java Servlets and Java Server Pages (JSPs) are frequently used for many different types of web site, including very large scale web sites for applications such as e-commerce, online banking and frequent-flyer programs. This Java web technology integrates very well with web*Mathematica* because web*Mathematica* is built on the same technology and because *J/Link*, the *Mathematica* Java toolkit, has been carefully designed to facilitate exactly this type of interoperation.

As a consequence, many important features are readily available to web*Mathematica*. Examples include, HTTP session support, file upload, database connectivity, XML integration, load balancing and automatic failover.

4 web*Mathematica* Services

There are many different services that can be provided by a web*Mathematica* server. This section will review a selection and some of the web sites that they allow to be constructed. It will show why web*Mathematica* is both powerful and convenient.

4.1 Computation Services

The computation services provided by a web*Mathematica* server are those of *Mathematica* itself and cover many different areas. Numerical computation is well supported with functions covering topics such as differential equations, special functions, linear algebra, optimization, statistics and data analysis. *Mathematica* also provides symbolic computation, for example, algebra (solving equations), integration, differentiation and inequality processing. One of *Mathematica*'s strengths is the seamless integration of numerical and symbolic computation. *Mathematica* also contains a high-level, interactive, functional programming language, good for rapid prototyping but which scales to large computations. This allows for new algorithms and techniques to be added to the system.

The client side extension mechanisms that were discussed in the previous section can be used to call on the computation from a web*Mathematica* server. One example is the use of an applet built from JavaView, to render the integral curve of an ODE. JavaView is a 3D geometry viewer and a geometric software library written in Java, more information is found at `http://www.javaview.de`. In the example shown below the web*Mathematica* server solves the ODE and returns a sample of the solution to the applet.

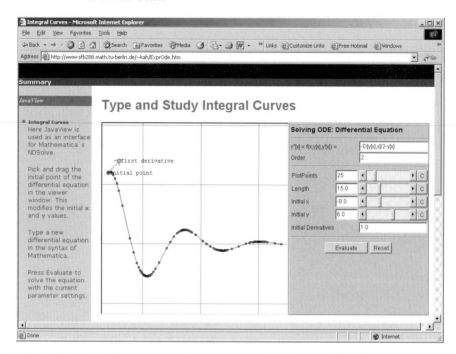

Fig. 1. Example of a computation service provided by a web*Mathematica* server. The web*Mathematica* server solves an ODE and returns a sample of the solution to a JavaView applet for rendering. (see Color Plate 44 on page 449)

There are several ways that a client such as the applet shown above can call to a web*Mathematica* server. One way would involve the use of the Java URL API. The applet would make a direct call to the server passing parameters controlling the computation on the URL following a '?'. A code fragment is shown below.

```
URL url = new
    URL("http://server/webMathematica/MSP/Examples/Compute
        ?arg1=x&arg2=y");

InputStream in = url.openStream();

while( (len = in.read( b, 0, 1024)) != -1) {

}
```

The result would be returned in a Java stream as binary or text data. The client would read and extract the necessary information from the stream. If the server was to format the results into XML, this might be a convenient way to validate and then extract results. Both Java and *Mathematica* provide considerable support for working with XML.

4.2 Visualization Services

Mathematica has long contained a variety of powerful visualization features. These are based around a graphics language that can be used for constructing components in two- and three-dimensions. There are many functions in the system that make use of the graphics language to support various visualization, charting and plotting features.

web*Mathematica* can make use of all of these visualization functions. In fact it offers a quick and simple way to do plotting on a web site. web*Mathematica* supports image based graphics as well as providing a Java applet to support three-dimensional plotting. A fragment of an MSP that demonstrates image based graphics is shown below. In this example, the *Mathematica* graphics function `Plot` is invoked, this returns a graphics object built using the *Mathematica* graphics language. The web*Mathematica* command, `MSPShow` takes this graphics object and converts it into an image representation.

```
<form action="Plot" method="post">
  Enter a function (eg Sin[x]):
  <input type="text" name="fun" size="10">
  Enter an upper limit (eg 10):
  <input type="text" name="num" size="3">
  <br/>

  <%Mathlet
        MSPBlock[{$$fun,$$num},
              MSPShow[ Plot[ $$fun, {x,0,$$num}]]] %>

  <br/>
  <input type="submit" name="button" value="Evaluate">
</form>
```

An example of graphics generated with web*Mathematica* is shown below. The user enters some parameters, to specify the rates of reactions and clicks on the visualize button. The request is sent to the web*Mathematica* server which calls *Mathematica* to solve the problem and generate a visualization of the result. The server returns a web page that displays the image.

This use of visualization technology is based on the use of images. These are a simple way to render pictures that are widely supported by many web browsers. Disadvantages to the use of images include the fact that they do not support any form of interaction, they may not match styles set locally in a browser, and they do not always transfer to other applications, such as

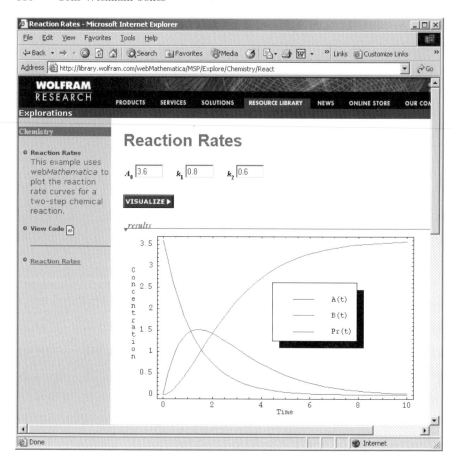

Fig. 2. An example of graphics rendered with web*Mathematica*. The user enters some parameters, to specify the rates of reaction and clicks on the visualize button. (see Color Plate 45 on page 450)

print media, very satisfactorily. In general, it is better to render results in the client rather than on the server. Some of these problems can be solved by the use of applets as shown above, at the expense of greater development cost.

SVG offers an alternative. SVG is a language for describing two-dimensional graphics in XML. A web*Mathematica* site could convert its graphics object into SVG, instead of an image. The SVG would be sent to a browser where it could be rendered by one of the plug-in technologies that are available. This whole process is quite analogous to the way that MathML can be used, as is discussed in the next section.

4.3 Visualization of Mathematical Notation

There are various different notational systems for representing mathematical concepts that have evolved along with the development of the abstract ideas of mathematics itself. It is interesting to note that these notational systems can be inconsistent from one area of mathematics to another and can be inconsistent within a given area. With the development of modern computer technology, computer support for mathematical notation has evolved. The typesetting system TEX, was one important step that allowed the production of high-quality printed material using various important elements of mathematical notation such as two-dimensional layout and extended character sets.

Another important step was the development of the typesetting system in *Mathematica*. This supports interactive typeset notation based on a language that can be processed to extract semantic information. Consequently it is a feasible input and interchange medium in addition to displaying mathematics. Being fully integrated with the *Mathematica* programming language it can also be extended to support new formats. One issue in the use of *Mathematica* typesetting is that it requires the use of *Mathematica* specific technology. However, many of the advantages of *Mathematica* typesetting are also available in MathML. In general, *Mathematica* objects are closely related to XML and this makes *Mathematica* a good system for supporting MathML. It also means that many of the concepts for working with *Mathematica* typesetting also apply to MathML.

MathML MathML, a W3C recommendation, is intended to assist the use and re-use of mathematical and scientific content on the Web. It can also be used as an interchange format for mathematical computation systems. MathML is designed to encode either the presentation or the semantic content of mathematics. Technically MathML is an application of XML. This allows it to make ready use of the many technologies that are available for working with XML. Some of these are very relevant for web usage, especially as more web browsers support features such as XSLT. In fact the support of XSLT, combined with the latest plug-in architectures, makes the issue of rendering MathML by general web technology quite feasible.

One convenient way to see a MathML representation is to use the *Mathematica* printing code, which can format the typesetting language into a MathML representation, as shown below.

$$\frac{\sqrt{x+y}}{5} //\texttt{MathMLForm}$$

```
<math>
<mfrac>
  <msqrt>
    <mrow>
      <mi>x</mi>
```

```
            <mo>+</mo>
            <mi>y</mi>
          </mrow>
        </msqrt>
        <mn>5</mn>
      </mfrac>
    </math>
```

Rendering MathML For MathML to be useful on the web it needs to work conveniently in browsers. This means that the technology needs to be readily available, and that it is easy to integrate MathML with HTML. In addition MathML documents need to avoid referencing technology specific solutions.

For rendering MathML in browsers there are now a number of solutions. These include, Active-X controls, plug-ins and a number of browsers that support MathML natively. The MathML web site, http://www.w3.org/Math/, lists a number of solutions that provide a good coverage of major browsers and computer systems.

To integrate MathML in HTML an author should not be required to write their document in a way that is specific for one particular rendering technology. This can be achieved by the use of XLST stylesheets. The author writes the document as an XML document, mixing MathML with XHTML (the XML compliant successor to HTML), and references an XSLT stylesheet. When the browser receives the document, it processes the stylesheet and at this time all decisions concerning the actual rendering technology can be made. Such an XSLT stylesheet has been developed and works well with many of the MathML rendering mechanisms; it is in the process of being made publicly available on the W3 web site.

Using MathML with web*Mathematica* There are several ways that web*Mathematica* can work with MathML. One involves using web*Mathematica* to generate documents that contain MathML, while another would use MathML as input to web*Mathematica*. Generation of MathML is supported by the web*Mathematica* formatting functions, drawing on the MathML support that is built into *Mathematica*. In addition web*Mathematica* can also accept MathML as input, all of the functions provided for working with input to *Mathematica* will accept input formatted as MathML. It is more difficult to arrange a source of MathML, since it is not very useful for a user to enter raw MathML directly. It is possible to use some of the web integrated tools that can work as MathML editors, and to integrate these with the web browser. This all requires a technology specific solution, the problem is not solved as neatly as it is for rendering MathML.

The example below shows a web *Mathematica* server formatting the result of a computation into MathML. The MathML is then sent to the browser where it is rendered with MathPlayer, a MathML display engine.

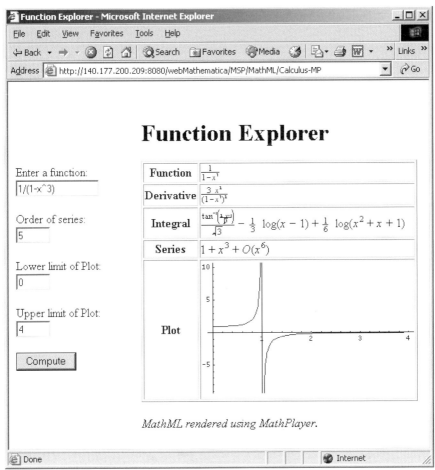

This example makes use of some of the web*Mathematica* utility functions for working with MathML. An example of a Mathlet that makes use of these commands formatting an expression into MathML is shown below.

```
<%Mathlet
    MSPBlock[{$$fun},
        MSPFormat[ Integrate[$$fun, x],
                    TraditionalForm, RawMathML]] %>
```

5 Summary

web*Mathematica* is a technology that can be used to support web based computation and visualization services. It is driven by a technology called *Mathematica Server Pages*, which makes it very easy to mix *Mathematica* and HTML. web*Mathematica* can be readily integrated with many client and server web technologies and can make use of interchange languages such as SVG for graphics, and MathML for mathematical notation. It is somewhat unusual in supporting the use of a mathematical package, such as *Mathematica*, for server based computation, and in providing direct web server support for MathML. Some examples of the use of web*Mathematica* for computation, visualization as well as mathematical notation were given.

6 References

1. web*Mathematica* site:
   `http://www.wolfram.com/products/webmathematica/`
2. Tom Wickham-Jones. "web*Mathematica*: Using *Mathematica* over the Web", *The Mathematica Journal*, 8, 2001, p. 178.
3. MathML: `http://www.w3.org/Math/`
4. Apache Tomcat: `http://jakarta.apache.org/tomcat/index.html`
5. *Mathematica* Graphics Gallery:
   `http://library.wolfram.com/graphics/`
6. Tom Wickham-Jones. *Mathematica* Graphics Techniques and Applications, Springer Verlag, ISBN: 0387940472.
7. JavaView: `http://www.javaview.de/`
8. SVG: `http://www.w3.org/Graphics/SVG/`
9. Florian Cajori. A History of Mathematical Notations, Dover, ISBN: 0486677664.

Part VI

Education and Communication

Films: A Communicating Tool for Mathematics

Michele Emmer

Dipartimento di Matematica, Università di Roma "La Sapienza", Piazzale A. Moro, 00185 Rome, Italy. *emmer@mat.uniroma1.it*

To Valeria
To Fred
   "*Multimedia* refers to an information system based on different communication instruments such as text, graphics, animation, sound, used specifically for didactics, information, and artistic purposes. Use of different mass communication instruments. Where mass media means all the communication instruments of the cultural industry, such as the press, cinema, television." This definition can be found in the Italian dictionary, *Lo Zingarelli*, vocabolario della lingua italiana (1996).

Keywords. mathematical films, multimedia, mathematics and art

1 Introduction

Undoubtedly one of the most important phenomena of the past few years has been the great proliferation of different types of mass media, of increasingly sophisticated technological instruments, which, thanks to the ability of the multimedia industry, are considered absolutely necessary by the *public* immediately after their appearance on the market. Take mobile phones for example: can one live without mobile phones today?

   Naturally all these technological marvels concern only a part of the human population, those who live in developed industrial countries in Europe, Asia and America. That part of the global population that uses the Internet since many years as a source of information, as an instrument to communicate and to work. In other words, users of the *New Economy*.

Among the first persons to be interested in the use of new technologies in research and in the diffusion of knowledge, were the mathematicians. Just think of one of the first films with computer-animation realised by Thomas Banchoff and Charles Strauss in 1976: the *inside-out* movement of a four-dimensional cube, a hypercube.

The *International Mathematical Union*, announcing that 2000 was the International Year of Mathematics, on the 6th May 1992 in Rio de Janeiro, set itself three principal aims. First of all "the great challenges of the Twenty-first century"; just as, at the World Conference in Paris in 1900, David Hilbert had listed a series of great problems that the mathematicians would have had to face during the course of the Twentieth century, one of the aims of the International Year of Mathematics is to focus the attention of mathematicians on the great challenges for the new century. Another aim of the Year: the key for development. Pure Mathematics and Applied Mathematics are the most important keys for development. Which means a great effort for scientific education, specially in those countries in which access to scientific knowledge is difficult.

Finally, a last aim is to relaunch "the image of mathematics". In the modern society of information, in which mathematical scientific knowledge has an increasingly important role, mathematicians and mathematics, however, do not seem to exist, they do not seem to be present. Mathematics and mathematicians are practically never mentioned in the media.

On the 11th November 1997, the general conference of UNESCO approved resolution 29C/DR126 and decided to sponsor the International Year of Mathematics.

Article 26 of the *Universal Declaration of Human Rights* states: "Everyone has the right to education. Education shall be free, at least in the elementary and fundamental stages. Technical and professional education shall be made generally available and higher education shall be equally accessible to all, on the basis of merit.

Education shall be directed to the full development of the human personality and to the strengthening of respect for human rights and fundamental freedoms. It shall promote understanding, tolerance and friendship among all nations, racial and religious groups, and shall further the activities of the United Nations for the maintenance of peace."

One of the big problems, when using modern technologies, is that often the word *novelty* is used instead of *efficiency*. In other words, in order to follow the latest technical novelties, the production of new cultural products is attempted, and their only intent, or at least their main intent, is not that of communicating knowledge better, and mathematics in particular, but rather of realising up-to-date products. As if modern progress were to pursue technical innovation, as the important value.

Obviously it is not so. It is sufficient to note that the market of mass media has to constantly produce new items to rouse the interest of the consumers, and therefore it must continuously change the standard technical characteristics in order to *force* people to continually buy different items.

The above is to say that next to the realisation and utilisation of new technologies for scientific communication, people must carefully reflect on how these new instruments must be utilised to effectively increase knowledge and not in order to flatten, and standardise knowledge itself all over the world.

It is an obvious statement, but it is worth repeating over and over again: technology is a means, an instrument, not the goal. The goal must be the diffusion of knowledge, of scientific knowledge, of culture. The diffusion of a basic culture that enables one to acquire knowledge, that is the only way to obtain multimedia products that are effectively new and not mere improvements of media that are already existent. There is no progress without knowledge.

Also in mathematics, in the past few years there has been a great diffusion of media, specially visual media. In the field of communicating for mathematics, often the images that were realised had a great visual impact, and attracted attention, due to their *complex* nature and *beauty*, but were not very useful to increase the comprehension of those using them, because training and teaching were secondary to the visual impact. The images were realised even without the explicit wish of the author, to strike the attention, not the imagination and fantasy.

Therefore it is useful not only to show the latest images that have been obtained, but also to discuss the utility of some multimedia products. So in the same way as one normally writes reviews on mathematics magazines about articles and books, likewise, all the multimedia products should be analysed and commented, specially those for students and teaching staff.

It is obvious that every multimedia instrument is useful when it utilises a specific technique in an essential manner. By this I mean that Otherwise it would not be possible to understand why we insist of giving lessons and making people read books. Lessons and books are both instruments which, in my modest opinion, will always remain the most effective multimedia means, for the obvious reason that they enable an interactivity (just think of the faces of the students, their reactions, that the person who is teaching the lesson can capture in *real time*) that no multimedia instrument, that is more or less modern, can ever obtain; and books, where the speed of comprehension is set by the person who is reading.

2 Mathematics: a Special Case?

To avoid a too general view, I would like to deal with the visual media for communicating mathematics in particular.

In order to do this, I think the best thing is to illustrate some examples from my own experience, trying to clarify the motivations that led me to these experiences. I shall not deal with the use of calculators in didactic mathematics, a topic I dealt with for 10 years in a European consortium.

In an article which is published in the Proceedings of the Meeting in Maubeuge, *Art and Mathematics* (September 2000) [1], I explained how the project *Matematica ed arte (Mathematics and art)* was born in 1976. The project started from the very beginning as a visual multimedia project, obviously with the multimedia instruments that were available at the time. Multimedia instruments have changed progressively since then, and there is a project, that is now at an advanced stage, with INFM, (Istituto Nazionale di Fisica della Materia - the Italian National Institute of Physics of Matter), for an interactive web-site called "Archimedes", in which young boys and girls aged 12 to 16 will find information regarding some of the important themes of the various sectors of science [2]. Which was unthinkable 25 years ago. However the project has remained unvaried in its general lines, and I can say that it still works due to the cultural idea on which it is based.

3 The Mathematics and Art Project

The *Mathematics and Art* project started in 1976. Or better, that year I started thinking of the project. The project was to make films/videos in which to compare the same theme from a mathematical and an artistic point of view, asking the opinion of mathematicians and artists. Not just filming long discussions between artists and scientists on the theme, that is so vague, of the connections between art and science, but a real confrontation on the visual ideas of the artists and mathematicians. *To make the invisible visible* like the artist David Brisson says in the film *Dimensions* [3] made in 1984 with Thomas Banchoff. So the general plan of the project was almost clear: to make films on the visual relationships of the forms created by artists and mathematicians. The themes of the first two films were soap bubbles and topology, in particular the Möbius band. To have more visual ideas and objects to film we finally decided to include the connections between mathematics and architecture, all the other sciences, in particular biology and physics, without excluding literature and even poetry. And, why not, also cinema. From the very beginning of the project there was the idea of focusing on the cultural aspect of mathematics, the influence and the connections of mathematics and culture, of course starting from the point of view that mathematics has always played a relevant role in culture, being an important part of it. All this was realised using the most important visual tool: filming. As these were the general lines of the project, it was quite natural to consider as a part of it the organisation of exhibitions (many were made in the next years), congresses and seminars, the publishing of books (with many illustrations!), even for students in mathematics, in history of art, in architecture. Today, 25 years later, it is easy to say that the project went far beyond the expectations. Starting from 1997, we organised an annual congress on *Mathematics and Culture* at the University of Ca' Foscari in Venice. (http://www.mat.uniroma1.it/ venezia2002; venezia2003 starting October 2002) Every year proceedings are published by Springer Verlag. Another project started in October 2000 at the University of Bologna: a two-day congress on *Mathematics, Art, Technology, Cinema* [4], an exhibition on M. C. Escher (with 60 original works), Lucio Saffaro, Oscar Reuteswaard and computer graphics, with a catalogue, a two-month programme of fiction and non-fiction films regarding mathematics; from *Will Hunting* to *Morte di un matematico napoletano*, from *Möbius* to *PI* and *Cube*. (http://www.dm.unibo.it/bologna2000)

A new Math Film Festival was organized at the Piccolo Teatro in Milan march 2002 in connections with the play on the paradox of mathematics "Infinities" by Luca Ronconi (www.piccoloteatro.org). The play was a great success and a Spanish edition was on stage in Valencia. A larger Festival was organized in a cinema in Rome, May - June 2002, partially supported by the Istituto di Alta Matematica and the Dipartimento di matematica of the University "La Sapienza" (http://www.mat. uniroma1.it/cinema2002) [5].

At first, a not so precise idea of a congress like the one I organize in Venice, was included in the project of the Seventies. In 1976, the whole project seemed very absurd for many reasons:

- to make a film was (and is) very expensive; one thing was very clear to me. I did not intend to make an amateur-film. I wanted to make a real high quality professional movie, and all the technicians involved had to be well qualified.

- I had started my professional career at the university and one of the most difficult things to do in an Italian university is to be involved in a field that connects two or more different areas. It can be the very quick end of your work at the university. This is still true today. But I was lucky because I was working on the Calculus of Variations and Minimal Surfaces, a field of great importance in the Seventies.
- trying to obtain the collaboration of Italian mathematicians (for the reasons illustrated in the previous point) was very hard. It was considered not very professional for a mathematician to be involved in such a project. During the last ten years I have been invited to many Italian universities to show and discuss my movies. But when I first showed one of my movies in Rome, in 1981, to a public audience, mathematicians of my department told me that it was not good for the reputation of our department. This is the main reason why almost all my movies have been made abroad, in Europe, in the USA, in Canada, Japan, even in India. And the same is true for the publication of books and proceedings of congresses organised abroad or with the help of non-Italian mathematicians. This is the reason why it has been possible to organise the congress Mathematics and Culture in Venice, only in the last five years, and not before. And in a few years time the congress has become an important traditional meeting for mathematicians and students.
- as it is clear from the previous remarks, to obtain funds and support for the project from the Italian institutions, was a desperate feat (and still is, in a sense).

Notwithstanding all this, we started the project. The themes of the first two films were clear enough.

In those years I had already discovered the works of the Dutch graphic artist Maurits Cornelis Escher. From the first time I saw his engravings, my purpose was to make a film only on him. My idea was to use the animation technique in order to make his works really three-dimensional. Something that Escher himself suggested; he personally was involved in a short film with several animations of his works before his death in 1972.

A few years ago a CD-ROM on Escher: *Escher Interactive*. (H. N. Abrams, Inc. Byron Press Multimedia Company) was made. In the CD-ROM there are the animations of some of Escher's works, and also some games. While the documentation part is very interesting, including a very vast *Gallery* of the works of the Dutch artist, the animations and the games are a demonstration of how it is possible to use an interesting instrument, a priori, to realise a product that is not interesting, it has no sense, one could say. In that part of the CD-ROM, Escher, Escher's works are transformed into graphic games of no interest, the graphics are quite approximated, and not only are they unable to produce the three-dimensional effect of the original works, but they completely cancel the atmosphere of mystery and fantasy, the dramatic feeling that arises from ambiguity, which is one of the essential components of Escher's works. I think Escher would have refused to let anyone use his works in this way. [6]

A clear message must be that no reproduction, however perfect it may be, can render the original work of art. The style, colour, technique, in other words all that makes the *miracle* of the unique work of art. Surely it can be used as documentation, but nothing more.

My idea was to film all over the world, where the artists and mathematicians involved in the film were working. What we needed for the project was a title; it was quite natural to choose the general title "Mathematics and Art"

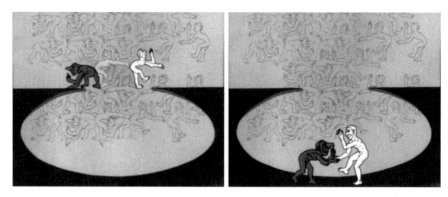

Fig. 1. Images from video "The Fantastic World of M.C. Escher" [21]

The reasons why I started thinking of it are essentially two. The first: in 1976, I was at the University of Trento, in the North of Italy. I was working in that area called the *Calculus of Variations*, in particular *minimal surface* and *Capillarity problems*. I had graduated from the University of Rome in 1970 and started my career at the University of Ferrara, where I was very lucky to start working with Mario Miranda, the favourite graduate student of Ennio De Giorgi; then I met Enrico Giusti and Enrico Bombieri. It was the period in which in the investigations of Partial Differential Equations, of the Calculus of Variations and the Perimeter theory, first introduced by Renato Caccioppoli and then developed by De Giorgi and Miranda, the Italian school of the Scuola Normale Superiore of Pisa was one of the best in the world. And in the year 1976, Enrico Bombieri received the Fields medal. By chance I was in the right place at the right time. All the mathematicians world-wide who were working in these areas of research had to be updated about what was happening in Italy.

Always in 1976, Jean Taylor proved a famous result that closed a conjecture that was raised experimentally by the Belgian physicist Joseph Plateau over a hundred years before: the types of singularities of the edges that soap films generate when they meet. Plateau had experimentally observed that the angles generated by the soap films are only of two kinds. Jean Taylor, using the Theory of Integral Currents introduced by Federer, and then by Allard and Almgren, was able to prove that the result was true.

Let's get back to Jean Taylor and to the year 1976. In 1976, the journal Scientific American asked Jean Taylor and Fred Almgren (they got married a few months before) to write a paper on the more recent results on the topic of Minimal Surfaces and Soap Bubbles. A professional photographer was asked to take the pictures for the paper. The same year, Jean Taylor and Fred Almgren were invited to the University of Trento as visiting professors and during the summer, they gave a summer course in Cortona, near Arezzo. I already knew both of them, maybe the one I knew better was Fred, who in his Swedish manner, was always very kind with me.

When Almgren and Taylor came to Trento in 1976 the issue of the Scientific American had just been published. The pictures of the article and the cover were really

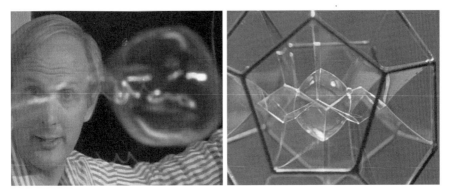

Fig. 2. Images from video "Soap Bubble" [22] showing Fred Almgren and a soap film in a dodecahedron.

beautiful and interesting. I do not remember why, but looking at the pictures I had the idea of making a film on soap film and to show its shapes and geometry in the greatest possible detail, more closely and using the rallenti technique. Both Almgren and Jean Taylor were very interested in my project. In any case my idea was not to make a *small* scientific film, a sort of scientific commercial just to show some small experiments with soap bubbles and soap films. I have never been able to stand these short films on mathematics (which have fortunately disappeared with the diffusion of computers), made to illustrate theorems or results of plane geometry or similar topics. These films are very boring and not very useful, not even for teaching mathematics at all the levels. I was attracted by the phenomena of soap films because they were visually interesting and I thought that the technique of filming them would have increased the general interest and fascination about them. I was not at all interested in just filming a lesson by Almgren and Taylor, with them explaining their results, inserting a few images of soap bubbles and soap films here and there. Almgren and Taylor shared my opinion. The project was not making any progress, because the motivation for making a film like this was not clear to me. Which was the purpose, if any; just the fascination of soap films? For which audience. And what did the length of the film have to be?

Now the second reason. I was working at the University of Trento while my family, Valeria and sons, lived in Rome. Every Friday I left Trento to go to Rome (seven hours by train) and then on Monday, I travelled back to Trento. I have been always a lover of art, of any kind, of any culture and period. Of course there are some artists that I prefer. When I was in Trento, I read in a newspaper of an exhibition in Parma, dedicated to one of the most important artists of this century: Max Bill. I already knew some of the sculptures of the Swiss artist but I had not visited a large exhibition like the one in Parma before. As the town of Parma was more or less on my way from Trento to Rome, I decided to stop on my way back to Rome to see the exhibition. Bill's topological sculptures were a real discovery for me. Years before, I had seen a large exhibition of the works of Henry Moore in Florence and of many other artists, but Bill's works almost immediately gave me the impression of *Visual Mathematics*. The *Endless Ribbon*, that Möbius Band, enormous and made of stone, granite, was a real revelation. Its shape, its physical nature, three-dimensionally real, making it live in space. A mathematical form, alive. This was the idea that was missing: mathematics, mathematicians in all the historical periods and in all the civilisations have created shapes, forms, relationships. Some of these shapes and relationships are really visual, they can be made visible. The idea for the great success of the use of computer graphics in some sectors of mathematics. In these same years the mathematician Thomas Banchoff was making his first short films in animation of mathematical surfaces but at that time I was not aware of his work.

Fig. 3. Images from video "Moebius Band" [23] showing Max Bill in his Zurich studio and a Möbius band.

At the end of 1979, I was able to show a first and preliminary version of *Soap Bubbles* at a Scientific Film festival at the CNRS in Paris. Also the second movie on the Möbius band was almost ready. In the film on soap bubbles I asked the collaboration of the Italian artist Arnaldo Pomodoro, who has always been fascinated by the theme of Spheres, while in the film on the Möbius band, apart from Max Bill, I filmed the works of Corrado Cagli, of the French designer Möbius.

I contacted Max Bill, writing him a letter. He was very kind; he invited me with my troupe to his house in Zurich and he gave me permission to film everything I was interested in, including his fabulous collection of contemporary art. With one exception: it was strictly forbidden to film a little window in which there was his collection of forms, topological forms, made of paper. Very small objects, his Data Base for future works. He was afraid that someone could see his projects and copy them. We then became friends, we made two exhibitions together, and another film on *Ars Combinatoria*. We both were in the board of editors of the journal *Leonardo*, that was published by Pergamon press at that time, and subsequently by MIT Press. For my book, *The Visual Mind: Art and Mathematics*, MIT Press, [7] Bill rewrote the title and made some changes to his famous paper originally written in 1949, *A mathematical approach to art*. Two of Bill's works are reproduced on the front and back cover of the book. A new volume *The Visual Mind 2* will be published, always by MIT Press, in 2003. The volume will be dedicated to Valeria and Max Bill.

Of course it is very hard to describe a film using words, it is almost impossible, and even not correct. If it is almost impossible to describe a film using words, it is good, because it means that the film has been made really using a visual technique, mixing, images, sounds, music in an essential and possibly unique way. If a film can be narrated it means that something is not working well from the visual point of view. One thing was really clear to me: in making the films, all words, all explanations had to be reduced to the minimum, or even be absent if possible. Whenever possible, images must speak for themselves. If, for its nature, art does not need explanations, mathematics too has to be presented almost without words. A film is not the best tool to explain and to learn. A film can, in a short amount of time, give ideas, suggestions, stimuli, emotions. A film can generate interest, even enthusiasm. Looking at an interesting, pleasant film can stimulate the audience to learn more, both in the artistic and the mathematics fields. In this sense I consider my films educational, but only with this meaning.

This, on the contrary, is the secret of their success, as in the case of the movie Soap Bubbles, even 20 years after the film was made. In fact the most beautiful sequences I have ever made, the soap films dancing to Weber's Rosenkavalier waltz were included in the Video Math festival selection for the World Mathematical Congress in Berlin [8] in 1998, and in the European Congress in Barcelona in 2000 [9].

4 Mathematics and Fiction Films

It is no surprise that in recent years mathematicians have figured large in the world of show business - in movies and theatre, as well as in books. Books dealing with mathematicians have had enormous success all over the world - books such as Simon Singh's "Fermat's Last Theorem" [10, 11], "The Number Devil: a Mathematical Adventure" by Hans Magnus Enzensberger [12], "Uncle Petros and the Goldbach Conjecture" by Apostolos Doxiadis [13, 14].

Not to mention successful plays about mathematicians in the theatre. Starting with "Arcadia" [15] by Tom Stoppard, Oscar-winner for the screenplay for "Shakespeare in love". Stoppard, who has a passion for physics and mathematics, made a videocassette with the mathematician Robert Osserman on behalf of MSRI (Mathematical Science Research Institute), Berkeley, California. [16, 17] In the video, Osserman and Stoppard discuss the mathematical aspects of "Arcadia" while actors perform the scenes on another part of the stage. Obviously Stoppard said he didn't know anything about the mathematics referred to in his play, and that he was only interested in whether the plot worked for theatre or movie performance. However, there's no doubt that he took the trouble to find out something about the subject. In "Arcadia", Stoppard imagines the story of a self-taught mathematical prodigy. The 13-year-old heroine, Thomasina Coverly, discovers the set which subsequently came to be known as the *Mandelbrot set* - as well as intuiting the first inklings of fractals. All this in 1809, many years ahead of time. Obviously, instead of the *Mandelbrot set*, it is called the *Coverly set*. Thomasina's mathematical intuitions are rediscovered by a 20th Century mathematician, Valentine, one of her descendants. The plot centers on Lord Byron whose wife Annabella was interested in mathematics. But there was no doubt about the mathematical talent of

Byron's daughter, Ada, who worked with Charles Babbage in the first attempts to use machines to carry out calculations. The figure of Thomasina is inspired by Ada and her tragic end. Ada signed her mathematical works with the initials A.L.L.. Only thirty years after her death was it discovered who was concealed behind that pseudonym. Thomasina, as Valentine discovers, had begun to penetrate the secrets of what today is called the chaos theory, of dynamic systems and non-Euclidean geometries. Her tragic death at the age of sixteen prevented her from achieving the mathematical reputation she was destined for. Clearly, this play by Tom Stoppard does not only deal with mathematics. The author is such an ingenious and inventive writer, with flights of fancy that are really exceptional. The play deals not only with mathematics, but also with gardens, aristocrats, duels, and Byron who is evoked but never present, and misunderstandings, such as when the young mathematician Valentine and a historian try to understand from documents what happened at the time of Thomasina, reconstructing past events with absurd results. The only person who fully appreciates Thomasina's precocious talent is Valentine, the mathematician from our own day.

But the real boom in plays dealing with mathematicians took place in 2000 and 2001. In 2000 (perhaps because it was World Mathematics Year?) there were several plays being performed at the same time in New York, on Broadway or at off-Broadway theatres, dealing with mathematicians. The New York Times on June 2, 2000, had a two-page article in the theatre supplement entitled "Science Finding a Home On-stage". The writer of the article Bruce Webern made the forecast that one of the off-Broadway shows, called *Proof*, would be very successful. And that is what has happened. *Proof* by David Auburn [18] opened in late May 2000 at the Manhattan Theatre Club. Theme: the world of mathematics.

The "proof" referred to in the title concerns a problem of number theory, but the author never gets to the bottom of the mathematical problem (why should one?). Auburn says that his play doesn't attempt to "prove theorems". But the encounter with mathematicians, rather than furnishing specific mathematical information, helped the author and the actors to realise that mathematics is not an arid subject; mathematicians enjoy themselves, they discuss, argue, they get excited. "It was a surprise for all of us." Auburn also confessed that he didn't do very well in math at school. He says that today we live in a technological age, in which technology itself produces a host of "dramas". Maybe the "two cultures" division is breaking down. The play was so successful that, from October 2000 it moved to a large Broadway theatre, and dates for the USA tour have been fixed right through 2002. *Proof* has also been officially recognised. It won three Tony Awards for best play, best actress, Mary-Lousie Parker, and best director, Daniel Sullivan. In addition, the play won the Pulitzer Prize for theatre in 2001.

In December 2001, Ron Howard's film "A Beautiful Mind" [19] was released in the USA. It is the story of the mathematician John Nash who won the Nobel Prize for economics. He is the mathematician responsible for the well-known theorem of regularity by De Giorgi-Nash, a famous example of a theorem proved by two mathematicians using different techniques, working separately and unbeknown to one another, but in the same period. This example is referred to by the Italian mathematician De Giorgi, who has a page in the book on Nash, and in the video interview I made with him in 1996. In March 24, the film received four Oscar including best film, best director, best non original script. Crowe playing the part of a mathematician is a very significant sign. As the film was very successful, producers will begin searching for more stories about mathematicians, even though Nash's life was rather special. [20]

5 Final Comments

In this paper I was interested in putting together some of the ideas that formed the basis for the beginning and the making of the project *Mathematics and Art*: for the exhibitions, for the books, for the congresses and in particular for the film-series. Many are the multimedia instruments that were used: from cinema to videos, from computer graphics to three-dimensional animation. Without giving any instrument greater privilege, but trying to use them all in the best possible manner. The fact of having greatly focused on the cultural aspect of mathematics, besides the visual impact, has enabled the different realisations to be seen by a vast public, from children to university students, from the teaching staff to the art museums, from mathematics congresses to film festivals. This was one of the aims I had set myself clearly from the very beginning, to succeed in realising multimedia instruments that could be useful for the diffusion of mathematical knowledge, that could arouse an interest in mathematics, which could be used at the same time, due to their visual, narrative, expressive and aesthetical nature. And also as documentation. Undoubtedly, the instrument I gave most privilege to was the cinema (and videos) with which it is possible to show a film in a large hall with many persons in the dark, obtaining an attention that is difficult to obtain with any other multimedia instrument. The cinema is very similar to a book, for its capacity to involve a person, obviously if the book and the film are interesting. Or even a beautiful conference or lesson. A full cinema-theatre in which a film on mathematics is being shown, is a great emotion for the audience and also for the persons who realised the film.

By this, I do not want to say that all the multimedia instruments are not interesting and cannot be useful. However, great care must be taken in choosing the persons in order to use the most suited technique and instruments. A very important observation is that those who have an idea, who realise beautiful scientific images, are not always also able to realise a multimedia product that is useful. So very often the so-called multimedia experts do not understand anything about the topic on which they are realising a multimedia product, and realise products that are totally useless if not harmful.

The solution was never given *a priori*, one must try to combine all the requirements, the visual aspect and the informative one, the spectacular aspect and accuracy in the information.

In any case, if a good multimedia project is lacking, the realisation of the same in an effective and involving manner, any message regarding mathematics or art or any other topic, will never attract the attention of those who must use it.

References

1. M. Emmer, Mathematics and Art: the Project, in C. Bruter, ed. Mathematics and Art, Proceedings of the Maubeuge conference, Springer, Berlin (2002).
2. S. Di Sieno, M. Emmer, G. M. Todesco Matematica, Progetto Archimedes, INFN (2000-2002)
3. M. Emmer, Dimensions, video, 27 minutes (1982)
4. M. Emmer & M. Manaresi, eds. Matematica, arte, tecnologia, conema, Springer Italia, Milano (2002; see in particular the 150 pages edited by M. Emmer on Mathematics and Cinema. English edition, to appear.
5. M. Emmer, A Beautiful Mind: recensione, B.U.M.I. (agosto 2001) sez. A, vol. IV-A,pp. 331-339. _____, (italian) presentazione, cura rassegnae schede La perfezione visibile: matematica e cinema", Piccolo Teatro e Politecnico di Milano (marzo 2002). _____, (italian) Matematica al cinema, Circuito cinema (marzo 2002), Venezia, presentazione, cura della rassegna e schede.
6. H.S.M. Coxeter, M. Emmer, R. Penrose, M. Teuber, eds. M.C.Escher: Art and Science, North-Holland, Amsterdam (1986). M. Emmer, D. Schattschneider, eds. M.C. Escher's Legacy, Springer, Berlin (2002) with a CD Rom.
7. M. Emmer, ed. The Visual Mind, The MIT Press, Cambridge, Mass (1993).
8. M.Emmer Soap Bubbles: Homage to Fred Almgren, in H-C. Hege & K. Polthier, eds. VideoMath Festival at ICM98, Springer, Berlino, 1998, 3'.
9. M. Emmer Soap Bubbles , in Xambo-Descamps & S. Zarzuela Santiago Video and Multimedia at 3ecm, Springer, 2000, video & DVD, 6'
10. S. Singh, Fermat's Last Theorem, It. ed. Rizzoli (1997).
11. S. Singh, director, Fermat's Last Theorem, produced by John Lynch for BBC Horizons, UK (1996) _____, L'ultimo teorema di Fermat. Il racconto di scienza del decennio, in M. Emmer, ed., Matematica e cultura 2, Springer Italia, Milan (1998) p. 40-43.
12. H. M. Enzensberger, Der Zahlenteufel, Carl Hanser Verlag, München (1997).
13. A. Dioxadis, Uncle Petros and the Goldbach Conjecture, Faber & Faber, London (2000).
14. A. Dioxadis, La poetica di Euclide: le analogie tra narrativa e dimostrazione matematica, in M. Emmer, ed., Matematica e cultura 2002, Springer Italia, Milan, (2002), p179-186.
15. T. Stoppard, Arcadia, Faber & Faber, London (1993).
16. Mathematics in Arcadia: Tom Stoppard in conversation with Robert Osserman, The Mathematical Science Research Institute, Berkeley, February 19, 1999, video.
17. R. Osserman, La matematica al centro della scena in M. Emmer, ed., Matematica e cultura 2002, Springer Italia, Milan, (2002), p. 85- 93.
18. D. Auburn, Proof : a Play, Faber & Faber, London (2000).
19. R. Howard, director, A Beautiful Mind, screenplay by Akiva Goldsman, played by Russel Crowe, Ed Harris, Jennifer Connely, produced by Brian Grazer and Ron Howard for DreamWorks Pictures, Universal Pictures, Imagine Entertainment, USA (2001).
20. A. Goldsman, A Beautiful Mind: the Shooting Script, Newmarket Press, New York (2002). S. Nasar, A Beautiful Mind, Simon & Schuster, New York (1998) H. W. Kuhn & S. Nasar, eds., The Essential John Nash, Princeton University Press, Princeton (2002). H. W. Kuhn, Math in the Movies: a Case study, in M. Emmer, ed. Matematica e cultura 2003, Springer, to appear.

21. M. Emmer, "The Fantastic World of M.C. Escher", video (1994). Distribution by Springer Verlag (2000) and in the USA by Acorn Media Publ. Inc. © M. Emmer and M.C. Escher Foundation for all Escher reproductions.
22. M. Emmer, "Soap Bubbles", video (1979). See also [8] and [9].
23. M. Emmer, "Moebius Band", video (1979).

The Potentials of Math Visualization and their Impact on the Curriculum

Beau Janzen

Zipheron Design Labs, 503 West 122nd St #9, New York, NY 10027, USA.
beau@zipheron.com

Summary. Complex abstract concepts have traditionally been conveyed with formulae or equations, but visualization tools offer new potentials to encapsulate abstract ideas in visual form. Visualizations can aid in learning with their ability to build a mental model through intuitive and harmonious representation, imbuing an idea with palpability and context. The potentials offered in this new communication medium not only have the ability to reshape the nature of the information that is being taught, but stand to redefine the content of the curriculum itself.

Keywords. mathematical videos, education, Newtonian mechanics

1 Introduction

Computer visualization offers us an unprecedented potential for communication of complex ideas. The only way in which we can address how we should harness these new potentials is to start from the core nature of education. Since we are no longer limited by the shortcomings of blackboards and filmstrips, we have to take a fresh look at what we hope to communicate. Why do we learn math, and what does it take to understand a mathematical concept?

Nearly three years ago, I began production on a computer-animated video describing the phenomenon of Foucault's pendulum. I'd like to use my experiences in production of this video as a starting point for discussion since I believe it offers an excellent framework for demonstrating current perceptions of the role of mathematical visualization, uncovering potentials for what it can be, and addressing the ramifications that new communication possibilities bring to the curriculum.

I believed Foucault's pendulum was an ideal topic for a video for several reasons. First of all, every explanation that I've seen on the subject consists of one of two things. The forces that cause the direction of the swinging pendulum to twist over time are sometimes explained with a set of equations that keep the physics hidden from the learner in disembodied formulas. Other times, an explanation is given that says something to the extent of "the Earth is turning below the swinging pendulum". This explanation is at best confusing, if not completely misleading. I realized that the rotational inertial frames involved in producing the phenomenon were virtually impossible to illustrate effectively with flat drawings. This was a content point that computer animation alone could convey, and would be an excellent example to illustrate the potential I see in the medium.

In addition, the Coriolis force affecting the pendulum was an elegant application of basic Newtonian mechanics. With a foundation of these core building blocks of physics, I reasoned that a video on Foucault's pendulum would be a welcome tool for all high school physics teachers. Examples such as clacking billiard balls might provide a good initial framework for delineating Newton's laws, but alone, would only give a learner a dim impression of what the laws truly mean. Something such as Foucault's pendulum could give a remarkable example of their power and far-reaching applications. I believe that there's a real sense of awe in knowing that, armed with only basic Newtonian mechanics, Foucault was able to create concrete proof that the Earth was spinning, something that no previous astronomer looking up to the heavens had been able to do.

In preparation for the video on Foucault's pendulum, I worked out an initial method for breaking down and visualizing the phenomena. I wanted to run these initial plans by an expert to get input and insure I was on the right track. A reputable observatory close to where I lived had a working Foucault pendulum, and upon my request, a member of their research staff agreed to help me.

I showed him some of my initial plans, and the whole idea of what I was trying to do seemed to elude him. I believe he felt I was missing the obvious. He wrote out for me the basic trigonometric function that would quantify how much the pendulum would rotate at a given latitude and said I just needed to follow this equation. I tried to explain that I didn't want to merely show the equation, but demonstrate how it worked and to plainly illustrate the geometric relationship that the equation quantifies.

I realized that this researcher and I were approaching the problem from two different directions. He obviously had a mental grasp of the phenomenon, but the only way he could think about it and describe it was strictly from the perspective of the equation itself and the results the equation could predict.

While equations are essential tools for calculation, they are often very poor vehicles for introducing concepts. Calculation and communication are two very different things.

There are a series of cognitive steps through which a learner must advance before computation of a problem holds any meaning. Before a student can be asked to find the solution to an equation, they must first understand the answer to other questions. What is the problem they face? What are the variables involved? Is there an observable pattern in the relationship of these variables? Is there a summative statement than can be made to describe this pattern? What are implications of the statement's ability to describe or predict? Once the learner sees these steps as the reason why and how an equation exists, then and only then can they find use in calculation. However, it is fully possible to train someone to follow the recipe of an equation and crank through numbers without the student having any idea as to what they are doing.

To effectively teach a mathematical concept, two distinct aspects of the concept must be conveyed: palpability and context.

2 Palpability

A mathematical concept is cognitively palpable to a learner when they have created an accurate mental model of it and can thoroughly describe the relationship without merely reciting the equation.

Quite often, the concept contained in a mathematical formula isn't intuitively palpable even if the learner knows what the formula is quantifying. If a student were presented with the inverse square law describing how light dims as it travels through space, they face a complex mental task. While there are only the two variables of distance and brightness and just two mathematical operations being conducted, there is an elaborate numerical relationship for the learner to cognitively digest. By itself, the equation provides only a mysterious recipe, and says nothing immediately appreciable about light and space.

3 Context

The context of a mathematical concept is the reason why it exists. We must never loose sight that equations are tools that we created; they are a technology. They were not spontaneously generated, but were derived as a means of encapsulating an observed relationship. It is impossible to see the value of a tool when taken away from the context of its use. Yet, we expect students to grasp the value of a mathematical concept when its purpose is typically self-referential and there is little if no insight as to why or how that technology was developed in the first place. The raison d'etre for math must come from outside of itself. An equation is the end result of quantifying an observed phenomenon. Without the foundation of this process of quantification, an equation holds no meaning.

Therefore, I believe that a learner should be introduced to mathematical concepts without numerical formulac as the primary communication vehicle since formulae on their own offer little if no palpability or context. In creating formulae and equations, we have created a numerical language to help encapsulate it in an abstract form. It is this leap into purely numerical abstraction that has the potential to break down communication to students.

What is needed is a communication tool that reencapsulates the relationship of a mathematical equation into a concrete form and is structured by the same train of thought that developed the idea in the first place. While there is nothing intrinsic in visual models that gives them palpability or context,

Fig. 1. Light shining in a one-dimensional universe.

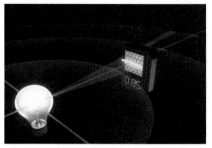

Fig. 2. Light spreading out in concentric circles in a two-dimensional universe.

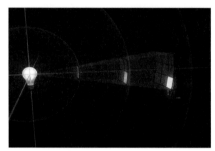

Fig. 3. Light spreading into spheres in a three dimensional universe.

they do offer a more flexible medium for communication of these principles. The structure of a mathematical visualization should be driven by the need to communicate palpability, and the building blocks of that structure should come from the need to communicate the context.

To design and orchestrate the visual elements, the designer should draw upon the skills of as many communication disciplines as possible: engineering psychology, color theory, gestalt psychology, and the artistic sensibility employed by graphic designers, cinematographers, and computer interface designers.

The use of these visualization tools can build a new visual language for quantitative communication. While equations gain their syntax by the need for computation, visualization tools should gain their syntax by the need to understand.

The syntax and vocabulary of imagery may be unique for each idea conveyed, and it may be necessary throughout the course of a lesson to build a visual vocabulary just as one would build a verbal one. For example, for an animation I created describing the inverse square law, I didn't feel I should immediately start by showing light falling off in three-dimensional space. To build the audience into the concept, I start with illustrating light in a one

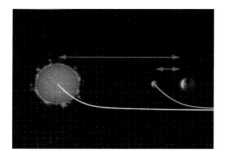

Fig. 4. The gravities of the sun and moon weakening with the inverse square relationship as they travel the distance to the earth.

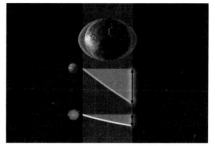

Fig. 5. Illustration of the change in strength of gravity from the near side to the far side of the earth.

and then two-dimensional space. (see Figs. 1,2) This helps build the idea that the way in which light spreads as it travels through space is dependent on the shape of the space. At that point, I felt the learner could progress to seeing how light spreads out in spheres in three-dimensional space with an inverse square relationship. (see Fig. 3 and Color Plate 10(c) on page 431) Furthermore, I was able to harness the visual vocabulary already established to help illustrate an inverse cube relationship describing how gravity affects the tides. With gravity falling off in the same manner as light, we see the familiar inverse square graphed lines extending from the sun and moon representing the strength of their gravity as it travels to the earth. (see Fig. 4 and Color Plate 10(d) on page 431) We can then look at these lines and measure the slope of their tangent when they reach the earth. (see Fig. 5) The greater the slope, the more stretching they create in the tides. The concept of the inverse square law is therefore not just passively observed, but actively reinforced by looking into the way that it works to explain a phenomenon.

I'd like to include an anecdote on the importance of effectively communicating one's ideas. Newton was the first to derive the concept of tidal forces being driven by an inverse cube relationship, which Halley claimed as one of the finest achievements of the Principa. However, Newton did little to demonstrate his point, so scholars were at a loss to understand it. Eighteenth-century astronomy textbooks therefore omitted this idea since the authors had no means of following it.[1]

The physicists that read the Principa were presumably capable of understanding the inverse cube relationship, but they did not have insight on the process of its derivation and could not create a mental model of how it worked. The idea held no palpability for them. Therefore, although they might be able to accurately plug numbers into the equation, the idea was meaningless. If scholars are unwilling to accept mathematics blindly, then why should we expect students to do so?

4 Implementation

With these new possibilities for conveying information, we need to consider their role in the teaching process. Mathematical visualizations are communication tools, and do not necessarily dictate any pedagogical style. It is not my intention to prescribe a particular method in which these tools should be used, but instead, start to uncover the underlying potentials that mathematical visualization can bring to learning. As a starting point, we can consider how the unique strength of these visualization tools can best augment the needs of existing teaching approaches.

For instance, discovery-based learning has gained widespread popularity in the past decade. The goal of this approach is not just that the student become able to retain a specific content point, but also that they build their ability to experiment and problem solve.

One of the potential pitfalls of this approach is the possibility for misunderstanding and confusion. If a student is given a light meter and a light bulb, they might be able to derive the inverse square law through experiment, but this is no guarantee that they also will be able to construct an accurate mental model of what is happening. They could even uncover the correct formula and through observation be convinced of its veracity, but still not have an appreciation of what makes it true. A mathematical visualization could augment this teaching strategy by helping illustrate and quantify the invisible elements at work and insuring that the learner has an accurate mental model of the abstract relationships.

In addition, if one of the primary goals of discovery-based learning is to build experimentation and problem solving skills, the students must have a foundation on which to build this kind of thinking. When students see mathematical ideas presented in their context, they have a window into the underlying thought process that constructed them in the first place. Seeing the problem solving involved in the creation of a mathematical formula not only gives context and support to the formula, but also similarly gives context and support for the idea of problem solving itself.

With this in mind, visualization tools are not well suited to model every scenario. We must consider when they should not be used and in what context they are ineffectual.

When I was working with the researcher on potential ways to illustrate the Coriolis force as manifested in the Foucault pendulum, he suggested what he thought was an obvious solution. He said I should create a model globe and show a pendulum swinging from a position on the North Pole. I should show this pendulum turning clockwise 360 degrees in one day. Then, I should show another pendulum suspended at the equator that demonstrated no rotation.

As powerful as mathematical visualization is, it's a poor medium to use for merely reproducing phenomena. The purpose of visualizations should be to illustrate the invisible, not to merely mimic observable reality.

Fig. 6. We start with simple example of Newton and his apple. A visual vocabulary is established that shows how the apple will move from the combination of motion given to it.

Fig. 7. Coriolis on a merry-go-round. Three factors effect the ball's motion, and the ball's resulting path is a straight line when seen from above.

Fig. 8. Coriolis' perspective on the merry-go-round, seeing the ball apparently fly away in an arc.

Fig. 9. If an artilleryman is at the North Pole, the earth's axis of rotation is perpendicular to his feet, and is represented by a perfectly circular red arrow.

For instance, the idea that in a vacuum, a large rock will fall at the same speed as a feather might seem unintuitive. Illustrating this concept to a student by means of an animated representation would do nothing to help convey the idea. Since this contrived environment can be made to do anything, the rock and feather could just as easily fall upwards, tumble sideways, or be choreographed to dance a Viennese waltz. In this case, an abstracted representation is a poor tool for conveying the concept; a student would have to believe the validity of the illustration on faith alone. An actual demonstration of a rock falling next to a feather would be infinitely more impacting on a learner.

The same is true for the Foucault pendulum. There is no reason to reproduce physical results when the graphics could be put to more practical use in quantifying the impalpable, such as the relative rotational inertial frames. Showing time-lapse film footage of a working Foucault pendulum might prove the veracity of the phenomenon to a student, but simply observing it rotate helps explain the mechanics involved as much as observing the turning hands of a clock explains the function of the hidden clockwork. Mere reproduction is not enough since observation of the actual phenomenon itself is often not enough.

What is necessary is to make the motion driving the pendulum a concrete visual element. The vocabulary of these elements is first built in my video with Newton dropping and throwing his apple. (see Fig. 6 and Color Plate 11(a) on page 432) The components that combine to create the overall motion of the apple are delineated with white arrows. He repeats the same exercises on a moving train, and we see the results from his perspective and from that of an outside observer. The idea is then applied to a rotational inertial frame of reference with Coriolis on a merry-go-round (see Fig. 7 and Color Plate 11(b) on page 432) We see that from his perspective on the merry go round, the ball he throws appears to fly away in an arc. (see Fig. 8) This same scenario is now repeated with Poisson's calculations of how the Coriolis force might effect the aiming of cannonballs. At the North Pole, the artilleryman experiences perfectly circular rotation, observed by a circular arrow seen below his feet. (see Fig. 9 and Color Plate 11(c) on page 432) He therefore sees his cannonballs apparen! tly veer to the right. (see Fig. 10 and Color Plate 11(d) on page 432) At the equator, he now experiences linear movement from the earth, observed by the round arrow, which now appears flat when viewed on its side. (see Fig. 11 and Color Plate 11(e) on page 432) If we trace how this circular arrow appears to us at various latitudes, take the thickness of the resulting ellipses and line them up as a graph, we can see how the effect of the Coriolis force is shaped like a sine curve. (see Fig. 12 and Color Plate 11(f) on page 432)

One thing to note, at no point in the narration of my animation do I bring in vocabulary words like vector, centrifugal force, or inertial frame. These concepts are dealt with visually, and there is no need to add in potentially confusing verbal vocabulary that the learner may not already know. At this time, the focus is on building a visual vocabulary to communicate the ideas. Once the learner grasps the concept, afterwards they can be introduced to the words that encompass the ideas they already understand.

By the end of the animation, the student has seen the generation of a sine curve. They have a mental model of the phenomenon, and are ready to see how this might be summarized into a formula. They can see how to describe particular latitudes on the earth with triangles, and how ratios in those triangles would quantify the effect. Finally, they can learn how we have given those ratios in the triangles special names - which we call trigonometric functions.

As additional reinforcement, the simple harmonic motion of the pendulum itself could provide another, entirely different means of producing a sine curve. Through one experiment, there are now two ways to show the context of quantifying a trigonometric function. The leaner can now begin to absorb

Fig. 10. From his point of view on the earth, the cannon appears to bank to the right.

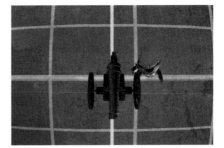

Fig. 11. At the equator, the artillery-man sees the circular red arrow on its side, and therefore feels the earth's rotation is straight, linear motion.

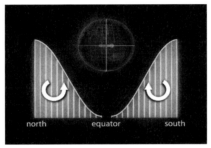

Fig. 12. If we trace how this circular arrow appears to us at various latitudes, take the thickness of the resulting ellipses and line them up as a graph, we can see how the effect of the Coriolis force is shaped like a sine curve.

the trig functions since they have been given the foundation of palpability and context.

One application of visualization tools that has been put into widespread use in schools is the interactive simulator. For example, a student studying projectile motion could be presented with an interactive computer "game" in which they would be able to control the angle and firing force of a cannon. Modifying these variables, the student could shoot virtual cannonballs, observe the simulated results, and hopefully develop a mental model from this experience.

Results have shown that while this type of interactive simulation can be helpful, clusters of students may be able to master the semantics of the simulator but not take the extra step in absorbing the actual concepts and theories.[2]

I would propose that this shortcoming in the design of some interactive graphics comes from not striving to do more than mimic observable reality or serve as graphing calculators. The visuals remain more "self-contained" and don't help build mental models transferable to outside experience. While interactivity is a powerful new variable for learning, we need to insure that the models which are being simulated bring a compelling veracity and applicability to the learner.

We need to insure that we never let ourselves become infatuated with the technology and think that the newer the software, the clearer the communication it will provide. The example of teaching projectile motion provides us with an interesting case history. Baron Franz von Uchatius communicated the concept that a projectile will travel in a parabolic arc to his artillery class at the Vienna Academy of Science in 1883. He cranked glass slides on which he had painted cannonballs through a primitive zoopraxiscope. This technology was truly cutting-edge at the time and did offer an excellent new communication tool. If the presentation had incorporated Eadweard Muybridge's 1881 high-speed sequential photography to make his slides, he would bring to this teaching tool the added benefit of veracity. With photographs of actual projectiles in motion, the students wouldn't have to rely on faith that the painted slides were accurate. Therefore, technology at this rather primitive le! vel was able to model this concept with a paplabilty and veracity. Sadly, this means that many educational computer animations today fail to do what could be done with the technology of 1881.

5 Impact on the Nature of the Curriculum

Even if one doesn't fully believe Marshall McLuhan's insight that "the medium is the message"[3], it is important to consider how using such mathematical visualizations to communicate concepts would impact the content itself. To provide a frame of reference, we should consider how our traditional communication tools shape the nature of our curriculum.

Traditionally, when a student is introduced to a new mathematical relationship, they are presented with an equation and possibly are lead through various proofs to illustrate its validity. Although the proofs are thorough and accurate, a student is only going to bring a novice understanding to the lesson and might have little ability to see any potential shortcomings in these proofs. Therefore, uncertainties are used to define uncertainties. Ultimately, the student is asked to rely on cyclical logic as proof for an equations' validity.

By itself, there is often nothing in the nature of an equation that makes it appear "correct". Even with a relationship as basic and harmonious as the Pythagorean theorem, a student could look at $a^2 - b^2 = c^2$ and not necessarily see anything inherently wrong with it. The statement is several steps removed from its harmony.

Equations and numerical notations are implicit in their meaning, but not necessarily explicit. While numerical notation provides a smooth framework for computation, it often takes ideas into a more and more abstracted framework. The abstracted syntax given to equations is solely for the purpose of computation, and not for explanation to novices. As students progress through the math curriculum, their material can seem further and further removed from a concrete, cognitive foundation and becomes embodied fully in the abstract.

When given only a foundation of abstracted numerical proofs, the student often must accept the material they are taught on faith alone. Their teacher says it is true, and it was printed in their textbook, therefore it must be true.

This basic approach of accepting facts blindly runs completely contrary to the core nature of scientific inquiry – one cannot start with the assumption that a fact is correct and work myopically towards proving that it is true. While faith in numbers is important, faith must be earned for its power to be real.

The ramifications of having faith as a basis for math understanding go beyond a learner's ability to later conduct a proper scientific study. If the basic animus of mathematical and scientific thinking is hidden from a learner, then the school is not preparing them to assimilate logical, quantitative thinking into the way they view the world. A public lead to believe that science and math can be accepted on blind faith is susceptible to scientific abuse. With our increasingly technocratic society, we cannot afford to leave learners unprepared to deal with such scientifically rich issues as our medical, energy, and economic policies.

Beyond the issue of asking students to accept the math curriculum on face value is the issue of how the learner sees math integrated into their environment. Traditionally, students are taught a mathematical concept and then, occasionally, shown a " real world" application of it. This basic approach in which math is a study unto itself that is later imposed onto observable reality gives a misleading view. The study of mathematics is a constantly evolving language created to describe and understand our environment. It therefore seems pointless to attempt to teach math apart from the needs it is designed to fill. To try and do so is similar to teaching a person a foreign language without communicating the meaning of the words they are learning. It is possible for a language student to conjugate a verb or construct a grammatically correct sentence without understanding the meaning of the words and the message they are communicating. We are asking students do the same meaning! less task when we ask them to solve equations while not understanding how they could be relevant to their environment.

Even if applications are imposed on a concept after it has been studied, the process is still misleading and confusing. Mathematics is not generated and then imposed on reality. To expand on the linguistic metaphor, people did not invent the word " dog" and then go looking for something to apply that name to. The mathematics that a student learns is a language of inquiry and exploration, and the question must be posed before a student can begin to answer it with mathematics.

At the core of a student's math and science curriculum should be an understanding that careful, critical quantitative thinking can produce accurate models of our world. The concepts a student is asked to learn may be difficult and perhaps seem illogical at first, but ultimately, they contain a harmony. Equations and theorems work because they have a harmony. A graphic model has the potential to explicitly illustrate the harmony in abstract material. The harmony in the relationship can take on a real aesthetic beauty and can immediately look "right".

This type of communication extends beyond the idea of being able to appeal to students who are predisposed to visual learning. By communicating content through a graphic medium, the content has the potential to become infused with its own harmony and provide the learner with a useful mental model. When an idea is communicated with a harmonious mental model as opposed to memorization of seemingly arbitrary recipes, it has a greater potential for retention, recall, and ultimately, utilization.

Studies have shown that a learning approach that focuses on semantic and visual relationships and creating a useful mental model promotes long-term memory representation much more effectively than an approach based on rote recitation.[4]

Some visualization tools have been created that have tapped into the harmony in math, but they fall short of creating this semantic mental model. For instance, the Project Mathematics! series created at CalTech has been able to encapsulate the harmony of mathematical relationships very well. However, the approach used is still one in which ideas are developed in a disembodied, abstract form before, occasionally, being imposed on reality. For instance, when an idea such as the Pythagorean theorem is conveyed, proofs are illustrated that clearly communicate the harmony in the theorem. But, the videos fail to take the extra step and provide a context for the concept. The Pythagorean theorem is a landmark idea in mathematics with far-reaching ramifications in measurement and quantification. Without this context, the theorem is merely a novel magic trick.

These kinds of animations that don't pursue communication of context are similar in nature to an abstract expressionist painting. The work can show beauty and harmony and be crafted with great skill as to have a strong impact on the observer. However, it's in its nature to have little or no relevance to the outside world. It's created as a self-referential work and does not render a clear view of an outside concrete idea.

With the aid of computer graphics, some media has been successful in doing the task that chalkboards and book diagrams have traditionally attempted to do, but the nature of the information has remained unchanged.

While some of the parts of ideas may be better conveyed to learners, the ideas are still removed from their purpose.

The way we present information will effect how a learner remembers it. Ultimately, our goal is to create an emancipated learner who can recall the math they have learned to solve the myriad of problems they encounter. Therefore, mathematical relationships should be stored not as declarative memory (knowledge of facts) but as procedural memory (knowledge of how to do something). Procedural memory is much more helpful in solving novel tasks since it deals with the underlying process as opposed to facts which might have little transient value.[5]

When a student is able to see beyond the formulas to the harmony in mathematical relationships, they can start to use numbers as a creative tool and a means for discovery. Seeing the harmony in math along with its context can give the material significance. It becomes a real, useful concept and not a quagmire of figures on a page. With significance comes the potential for personal relevance, a core concept in Academic Rationalism and humanistic Curriculum Philosophy[6]. The curriculum is no longer a bitter vitamin pill a student must swallow, but a tool for self-empowerment.

Showing a student that their curriculum is significant and relevant is a basic way to promote motivation. If the curriculum doesn't show enough respect to the student to demonstrate why it is worth learning, then how can we ask a student to respect the curriculum enough to learn it?

6 Impact on the Content of the Curriculum

After several months of designing and animating, I assembled a rough-cut version of my Foucault's pendulum video. The concept flow and the visual communication was at a level that it was ready for formative evaluation. The video was shown to several high school physics teachers to get their comments on how effective a tool this video would be to them and what suggestions they might have for its completion.

The teachers all said they were familiar with the Foucault pendulum and were interested in seeing how I described it. After seeing the video, the teachers admitted that they realized they hadn't fully understood the phenomenon before. As was the case with the researcher at the observatory, their understanding existed only within the confines of the equation. At a certain level, there was a logical impasse where the cause and effect were not questioned and it "just was that way".

The goal of the video was to show the phenomenon outside the confines of the equations and illustrate the mechanics of what caused the effect. ¿From the response of the teachers, it had done its job. However, I would learn that this would not necessarily earn it acceptance.

Even though the teachers liked the video and said it illustrated its points well, they weren't sure if they would want to use it. They believed that the video would be effective in conveying the topic to their students, but that the video featured applications of material that were typically too advanced for their curriculum. Their curriculum did encompass Newton's first law, centrifugal force, and relative inertial frames, but these ideas were not combined in the application of Foucault's pendulum. Granted, Foucault's pendulum is one application out of hundreds that could be used to illustrate these basic science content points. The reason I was given for its exclusion, however, was not that it was irrelevant or too arcane. I was told that Foucault's pendulum is "too advanced a topic" for students. If the component parts are already covered, and the idea was clearly communicated, then what was the basis for this judgment?

Visualization tools now not only stand to recreate the nature of the curriculum, but the content of the curriculum itself. I believe that our curriculum, in its present state, has been formed not as much by what we felt students should know, as it has by what we felt we could communicate to them. Just because we have been unable to effectively communicate an idea via our older teaching tools, does that make the idea itself any less important?

The potential in visualization tools demands that we reassess our entire curriculum. We must ignore tradition and dogma and look at our information through fresh eyes. We need to judge ideas not our previous capacity to convey them, but by their ability to demonstrate relevant tools for self-empowerment, provide mental connections that draw the curriculum together, and invoke a sense of marvel and awe in learning.

7 Conclusion

Within the past fifty years, we have seen a shift in the way we create our tools. Through the middle of the last century, the design of our input and feedback mechanisms on our tools, whether they be the design of typewriter keys or the layout of a pilot cockpit, were given little consideration as to how they could be most easily used by the human operator. The philosophy of training users was that of "designing the human to fit the machine". After World War II, the field of engineering psychology was born and shifted emphasis towards designing machines to operate within the framework of human performance.[7]

Engineering psychology has formalized the field of cognitive ergonomics, and we now develop ways that are more efficient for our machines to display information to us. It is now time for us to take the next step and use our machines to redesign the information itself. With the possibilities offered by mathematical visualization, we can strive to not only make information easier to communicate to learners, but also imbue that information with a new significance.

References

1. Nicholas Kollerstron. *Newton's Forgotten Lunar Theory.* Green Lion Press, 2000.
2. Joseph C. Principe, Neil Eulino and Curt Lefebvre. *An Interactive Learning Environment for Adaptive Systems Instruction.* University of Florida, Gainsville.
3. Marshall McLuhan. *Understanding Media.* MIT Press, p. 7, 1994.
4. F. G. Halasz and T. P. Moran. *Mental Models and Problem Solving in Using a Calculator.* In A. Jada (ed.), Human Factors in computing systems: Proceedings of CHI 1983 Conference, pp. 212–216. New York: Association for Computing Machinery, 1983.
5. F. I. M. Craik and R. S. Lockhart. *Levels of Processing: A framework for memory research.* Journal of Verbal Learning and Verbal Behavior, 11, 671-684, 1972.
6. Eliot W. Eisner. *The Educational Imagination: On the Design and Evaluation of School Programs.* (3rd Ed.) New York, MacMillan, 1994.
7. Christopher W. Wickens. *Engineering Psychology and Human Performance* Second Ed. (p. 4) Harper Collins Publishers Inc, 1992.

Appendix: Color Plates

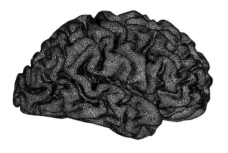 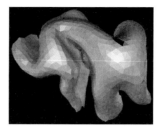

(a) Right hemisphere and subsurface.

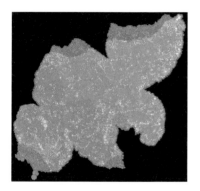 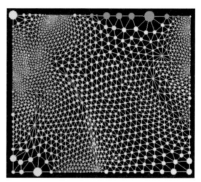

(b) Radii packing of hemisphere and inversive distance packing of a subsurface.

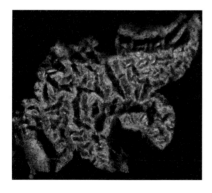 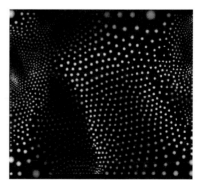

(c) Packing with bump map texture.

Fig. 1. Quasi-conformally mapping of the human brain to a planar domain. *(Bowers, Hurdal, p. 28)*

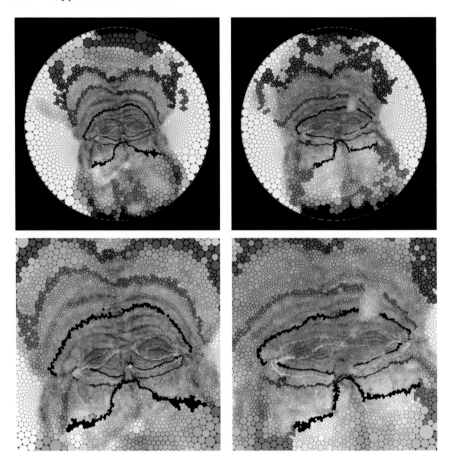

Fig. 2. Mapping two different cerebellum of the human brain. *(Bowers, Hurdal, p. 29)*

Fig. 3. Comparison of different surface evolution models. From left to right the initial surface, the result of the mean curvature motion, the result employing an isotropic nonlinear diffusion coefficient and the resulting surface under the new edge preserving anisotropic evolution using the diffusion coefficient *(8, p. 250)* are depicted. The different results are evaluated for the same time $t = 0.0032$ and the parameters where chosen as $\sigma = 0.02$ and $\lambda = 5$. The diameter of initial surface is chosen to be 1. *(Clarenz, Diewald, Rumpf, p. 248)*

Fig. 4. Example for the evolution of a surface with texture information under the combined diffusion *(9, p. 252)* and *(11, p. 253)*. Because of the dependency of the diffusion coefficient a_G on the texture during the evolution geometry edges develop in areas of high texture gradients. The parameters are chosen as $\tau = 0.00001$, $\sigma = 0.0045$, $\lambda = 20$, $\mu = 2$, and the diameter of the surface is scaled to 1. On the left the initial surface is shown with and without texture information, and on the right two timesteps for $t = 0.000015$ and $t = 0.00003$ are depicted. *(Clarenz, Diewald, Rumpf, p. 253)*

Fig. 5. From left to right the initial surface and three timesteps of the anisotropic geometric evolution using the diffusion coefficient *(8, p. 250)* are shown for a venus head consisting of 268714 triangles. The evolution times are 0.00005, 0.0001, and 0.0002 and the parameters are $\lambda = 10$, $\sigma = 0.02$. *(Clarenz, Diewald, Rumpf, p. 250)*

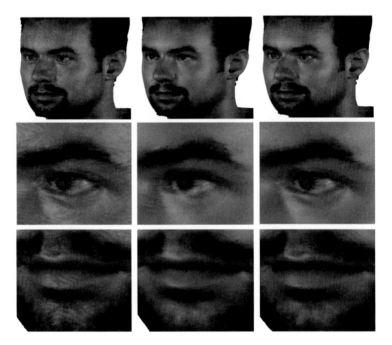

Fig. 6. On the left a surface obtained from a laser scan with an onscribed photographic texture and moderate superimposed isotropic noise is depicted (on top the whole surface, below magnified parts of it). This surface is considered as initial surface for the pure geometric evolution and for the combined geometry and texture evolution. In the middle the result of the pure geometry evolution and on the right the result of the combined model are shown. In both cases the same evolution timestep $t = 0.00012$ is considered. *(Clarenz, Diewald, Rumpf, p. 254)*

Fig. 7. The view of a sample hyperbolic manifold, with geometry $\mathbb{H}^3$, as seen on one wall of ALICE. This hyperbolic space is tiled by right-angled dodecahedra, meeting 8-to-a-vertex, just as Euclidean space can be tiled with cubes. *(Francis, Goudeseune, Kaczmarski, Schaeffer, Sullivan, p. 312)*

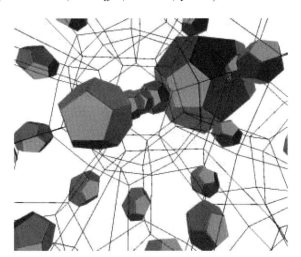

Fig. 8. The view of a sample spherical manifold, with geometry $\mathbb{S}^3$, as seen on one wall of ALICE. This space, formed by identifying opposite faces of a dodecahedron, is almost a counter-example to the Poincaré conjecture. An inhabitant of this space would see 120 repeating images of each object in the space, as in this regular tiling of the sphere by 120 dodecahedra. *(Francis, Goudeseune, Kaczmarski, Schaeffer, Sullivan, p. 313)*

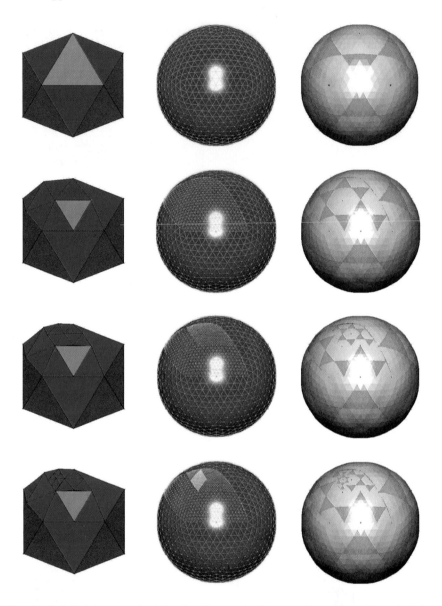

Fig. 9. Subdivision invariant local refinement of a smooth surface. From left to right are shown the input triangulation, the smooth surface together with its Bézier control mesh and the color shaded Bézier control mesh: to each input triangle correspond 4 Bézier patches, the central one is shaded in red. From top to bottom the successive refinement of the icosahedron input mesh is shown. *(Hahmann, Bonneau, Yvart, p. 197)*

(a) Light shining in a one-dimensional universe.

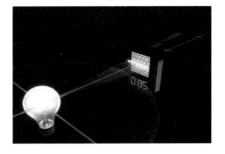

(b) Light spreading out in concentric circles in a two-dimensional universe.

(c) Light spreading into spheres in a three dimensional universe.

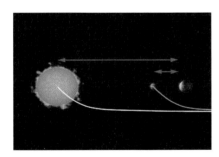

(d) The gravities of the sun and moon weakening with the inverse square relationship as they travel the distance to the earth.

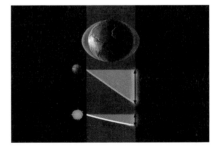

(e) Illustration of the change in strength of gravity from the near side to the far side of the earth.

Fig. 10. Still frames from video "The Inverse Square Law" *(Janzen, p. 411)*

(a) Simple example of Newton and his apple. A visual vocabulary is established that shows the combination of motion.

(b) Coriolis on a merry-go-round. Three factors effect the ball's motion. The resulting path is a straight line when seen from above.

(c) Artilleryman at the North Pole. The earth's rotation is shown by a circular red arrow.

(d) From his point of view on the earth, the cannon appears to bank to the right.

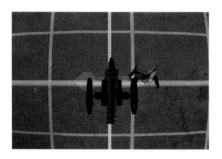

(e) At the equator, the artilleryman sees the circular red arrow on its side.

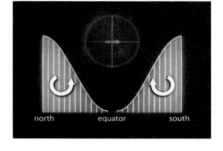

(f) Ellipticity of the arrow at various latitudes. The thickness of the ellipses shows the Coriolis force.

Fig. 11. From video "Foucault's Pendulum: A Turning Point" *(Janzen, p. 414)*

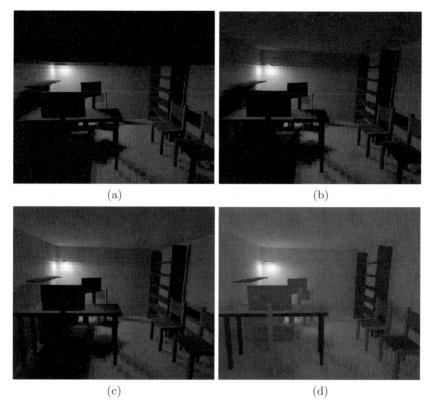

Fig. 12. First iterations (9 iterations (a), 15 iterations (b), 100 iterations (c)) and final result (d) of progressive radiosity for an office discretized in 11825 quadrangular patches. *(Leblond, Rousselle, Renaud, p. 263)*

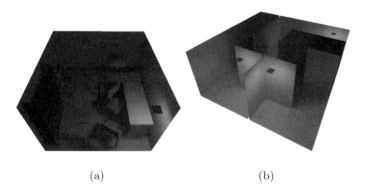

Fig. 13. The two test scenes show an office (a) and a maze (b). *(Leblond, Rousselle, Renaud, p. 275)*

Fig. 14. The gradient vector field on a figure eight knot model. *(Lewiner, Lopes, Tavares, p. 95)*

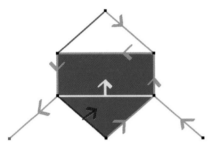

Fig. 15a. A cell complex with its discrete gradient vector field. *(p. 97)*

Fig. 15b. The Hasse diagram with the pairing. *(p. 97)*

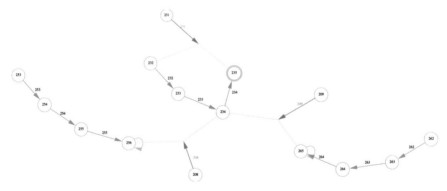

Fig. 16. A part of the dual hypertree resulting while processing a solid 2-sphere. *(Lewiner, Lopes, Tavares, p. 100)*

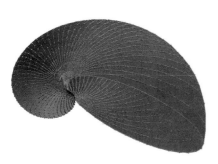

Fig. 17. A discrete gradient vector field on a shelf model. *(p. 106)*

Fig. 18. A discrete gradient vector field on a simple 2-sphere. *(p. 106)*

(a) The gradient vector field.

(b) First step: opening the torus along its meridian

(c) First step: opening the torus along its meridian

(d) Further steps continue on edges and vertices

Fig. 19. A decomposition of a torus. *(Lewiner, Lopes, Tavares, p. 108)*

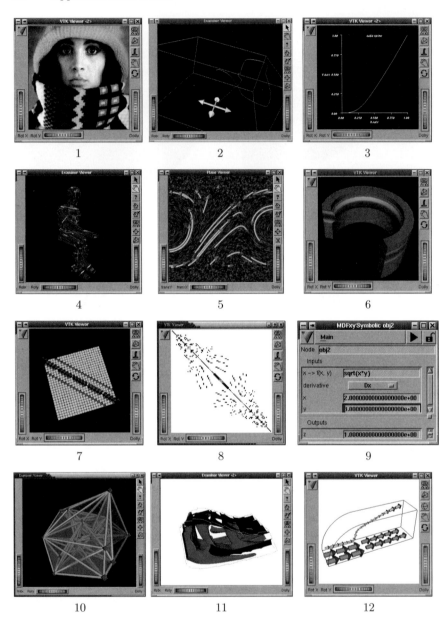

Fig. 20. NumLab visualization of various computational results. *(Maubach, Telea, p. 319)*

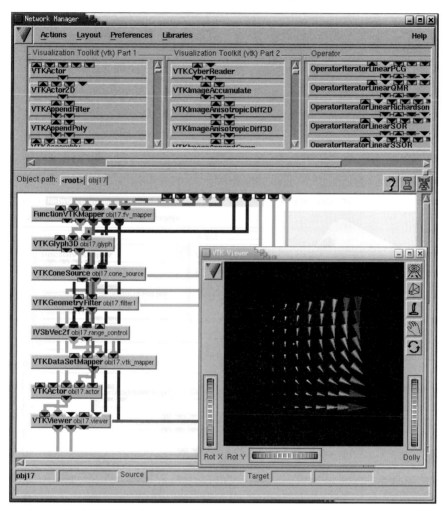

Fig. 21. The modules hidden inside group module `FunctionVTKViewerField` of NUMLAB. The viewer at the end of the pipe-line visualizes the result. *(Maubach, Telea, p. 333)*

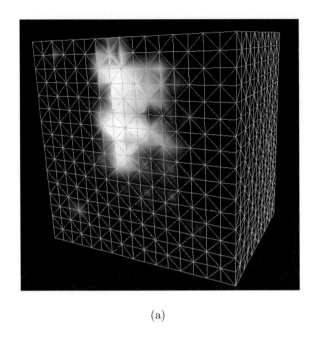

(a)

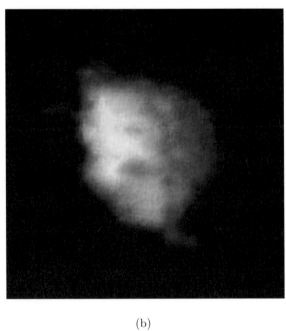

(b)

Fig. 22. Volume visualization of tumor dataset. *(Mello, Velho, Cavalcanti, Silva, p. 354)*

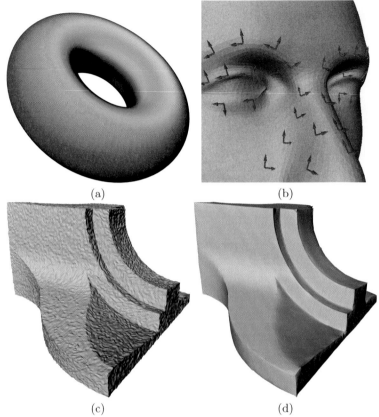

Fig. 23. Some applications of our discrete operators: (a) mean curvature plot for a discrete surface, (b) principal curvature directions on a triangle mesh, (c-d) automatic feature-preserving denoising of a noisy mesh using anisotropic smoothing. *(Meyer, Desbrun, Schröder, Barr, p. 48)*

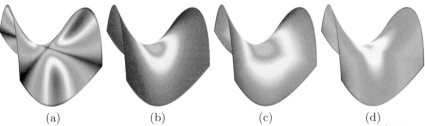

Fig. 24. Curvature plots of a triangulated saddle using pseudo-colors: (a) Mean, (b) Gaussian, (c) Minimum, (d) Maximum. *(Meyer, Desbrun, Schröder, Barr, p. 50)*

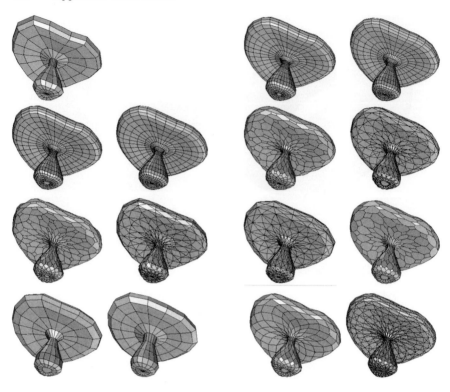

Fig. 25. First row, left: The mesh of a mushroom which is used for illustration of the different subdivision schemes. First row, right pair: The Doo/Sabin and the Catmull/Clark scheme. Second row, left pair: The VEF1/F-scheme and the VFE1/F/F-scheme. Second row, right pair: The EF2/F-scheme and the EF2/F/F-scheme. Third row, left pair: The VE2/F-scheme and the VE2/F/F-scheme. Third row, right pair: The F/VF2-scheme and the VF2/F-scheme. Fourth row, left pair: The F-scheme and the F/F-scheme. Fourth row, right: The E-scheme and the E/EF2-scheme. *(Müller, Rips, p. 216)*

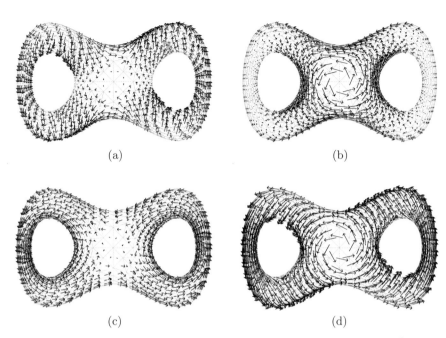

Fig. 26. Decomposition of a tangential vector field (d) on a pretzel in $\mathbb{R}^3$ in a rotation-free component (a) and a divergence-free component (b). The original vector field (d) was obtained by projection of a flow around the z-axis onto the tangential space of the curved, 2-dimensional surface. The harmonic component (c) belongs to an incompressible, rotation-free flow around the handles of the pretzel. *(Polthier, Preuß, p. 130)*

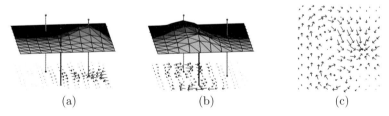

Fig. 27. Test vector field (c) decomposed in rotation-free (a) and divergence-free (b). The three vertical lines in (a) and (b) indicate the centers of the original potentials for comparision with the extrema of the calculated potential functions. *(Polthier, Preuß, p. 129)*

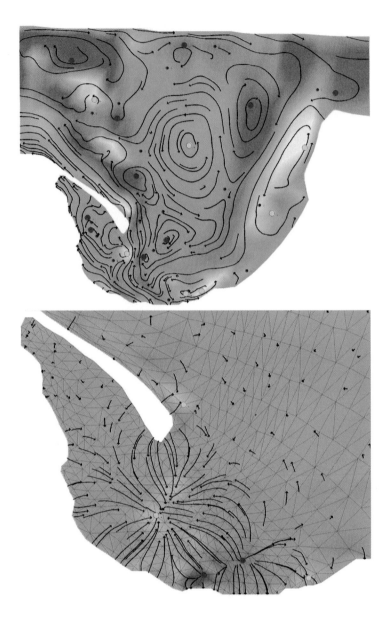

Fig. 28. Automatic identification of vector field singularities using a Hodge decomposition of a horizontal section of a flow in Bay of Gdansk. Rotation-free component (bottom) with sinks and sources, which come from vertical flows, and divergence-free component (top). The big dots indicate the location of sinks and sources (top) respectively vortices (bottom). The small dark dots mark saddle points (top and bottom). The bay is colorshaded by its discrete rotation (top) and divergence (bottom). *(Polthier, Preuß, p. 130)*

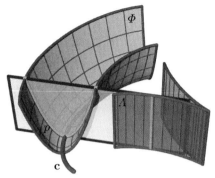

Fig. 29. When the normal plane of the planar curve **c** rolls on the evolute cylinder Λ of **c**, the parabola p generates the moulding surface Φ. *(Pottmann, Hofer, p. 223)*

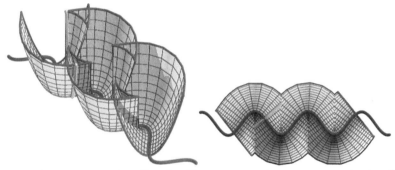

Fig. 30. A moulding surface Φ with self-intersections in an axonometric view (left) and viewed from below, where the visible points correspond to the graph Γ of the squared distance function to a sine curve (right). *(Pottmann, Hofer, p. 224)*

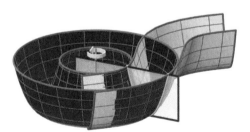

Fig. 31. Surface of revolution Ψ which has second order contact with the moulding surface Φ at all points of $p(t_0)$. *(Pottmann, Hofer, p. 226)*

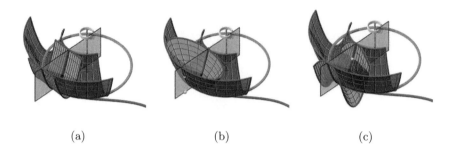

(a)   (b)   (c)

Fig. 32. (a) The graph surface Γ_0 of F_0 is a parabolic cylinder with rulings parallel to the curve tangent. (b) The graph surface Γ_d of F_d is an elliptic paraboloid for $s > 0$. (c) The graph surface Γ_d of F_d is a hyperbolic paraboloid for $s < 0$. *(Pottmann, Hofer, p. 227)*

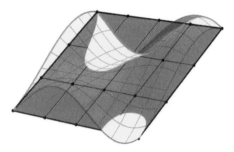

Fig. 33. Approximation of a surface patch (light colored) by a B-spline surface (dark colored): Initial position of B-spline surface. *(Pottmann, Hofer, p. 239)*

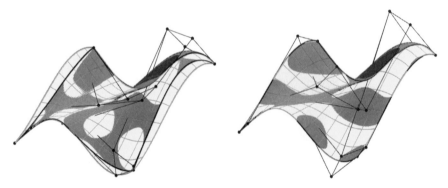

Fig. 34. Approximation of a surface patch by a B-spline surface; final position after five iterations without boundary approximation (left), and with boundary approximation (right). *(Pottmann, Hofer, p. 240)*

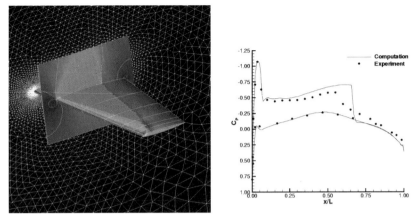

Fig. 35. On the left, the Mach number contours for an ONERA M6 wing. On the right, comparison of the between numerical (on continuous line) and experimental (dotted) pressure distributions on a section of the wing. *(Quarteroni, Sala, Sawley, Parolini, Cowles, p. 370)*

Fig. 36. Mach number contours on the surface of a Falcon aircraft and on selected cutting planes. The asymptotic Mach number is 0.45, with one degree of angle of attack. *(Quarteroni, Sala, Sawley, Parolini, Cowles, p. 371)*

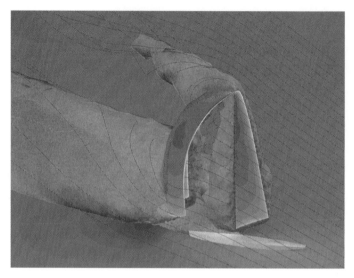

Fig. 37. Streamlines and total pressure iso-surfaces around the downwind sails of a racing yacht, illustrating the flow separation behind the sails and the generation of trailing vortices. The boat speed is 10 knots (1 knot = 0.514 m/s), and the true wind speed is 15 knots at an angle of 160 degrees. *(Quarteroni, Sala, Sawley, Parolini, Cowles, p. 372)*

Fig. 38. Contours of surface height indicating the waves generated around a Wigley hull at the water-air interface. *(Quarteroni, Sala, Sawley, Parolini, Cowles, p. 373)*

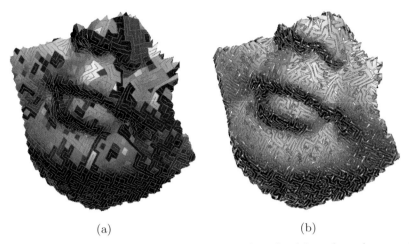

Fig. 39. Every connected manifold quadrilateral mesh without boundary can be represented as a single Hamiltonian generalized triangle strip cycle by splitting each face along one of its diagonals, and connecting the resulting triangles along the original mesh edges. (a) An arbitrary choice of face diagonals produces several cycles. (b) Cycles are then joined to form a single cycle by flipping diagonals. *(Taubin, p. 85)*

Fig. 40. Not every Eulerian circuit corresponds to a Hamiltonian triangulation. Edges that are opposite to each other on a face cannot be contiguous in the Eulerian circuit. We used two different colors to visualize the circuit intersections (or lack of), but both red and green dual edges belong to the same Eulerian circuit. The Eulerian circuit on the left does not correspond to any Hamiltonian triangulation. *(Taubin, p. 78)*

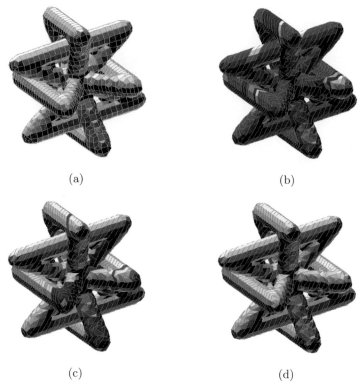

(a)   (b)

(c)   (d)

Fig. 41. The algorithm of figure (4, p. 74) applied to a more complex mesh (a). A random choice of diagonals produces a triangulation with 73 cycles (b). After 30 diagonal flips the resulting triangulation still has 43 cycles (c). After 42 additional flips we obtain a triangulation with a single Hamiltonian cycle (d). *(Taubin, p. 78)*

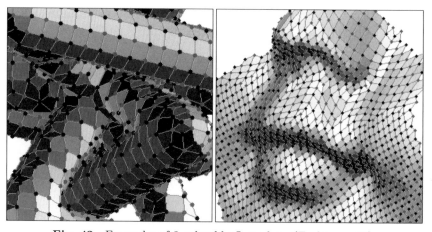

Fig. 42. Examples of 2-colorable Q-meshes. *(Taubin, p. 80)*

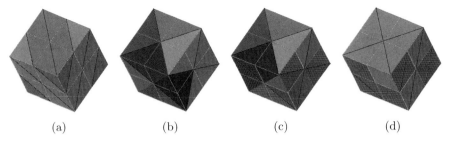

(a)   (b)   (c)   (d)

Fig. 43. Construction of Hamiltonian T-strip cycles on subdivision meshes. (a) Choosing all the diagonals of the coarse quadrilateral faces parallel to the diagonal chosen for the corresponding coarse face produces two parallel cycles with little vertex locality. (b) Flipping all the marching edges of these two cycles produces a large number of small cycles. (c) Some of these diagonals must be flipped back to link all these cycles into a single one. (d) The resulting Hamiltonian T-strip. *(Taubin, p. 88)*

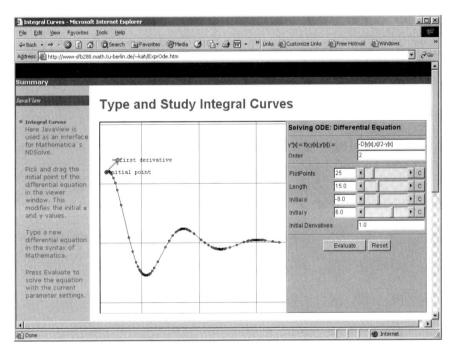

Fig. 44. Example of a computation service provided by a web*Mathematica* server. The web*Mathematica* server solves an ODE and returns a sample of the solution to a JavaView applet for rendering. *(Wickham-Jones, p. 384)*

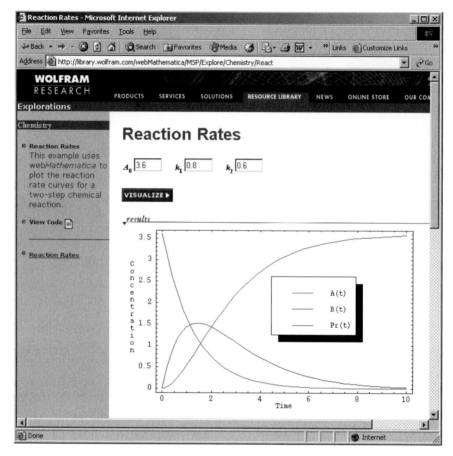

Fig. 45. An example of graphics rendered with web*Mathematica*. The user enters some parameters, to specify the rates of reaction and clicks on the visualize button. (*Wickham-Jones, p. 386*)

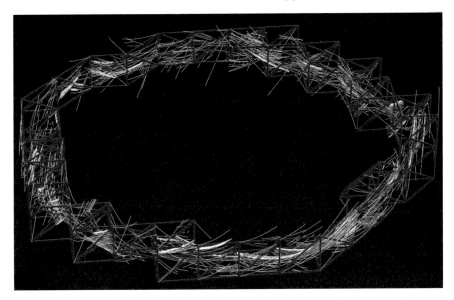

Fig. 46. Loop including cell cycle and backward integrations. *(Wischgoll, Scheuermann, p. 157)*

Fig. 47. Loop in a 3D vector field with streamsurfaces. *(Wischgoll, Scheuermann, p. 158)*

Index

Printing: Saladruck Berlin
Binding: Stürtz AG, Würzburg